Computer Algebra Recipes

An Advanced Guide to
Scientific Modeling

Richard H. Enns
George C. McGuire

Computer Algebra Recipes

An Advanced Guide to
Scientific Modeling

Springer

Richard H. Enns
Simon Fraser University
Department of Physics
Burnaby, B.C. V5A 1S6
Canada
renns@sfu.ca

George C. McGuire
University College of Fraser Valley
Department of Physics
Abbotsford, BC V2S 7M9
Canada
george.mcguire@ucfv.ca

Cover design by Mary Burgess.

Library of Congress Control Number: 2006936017

Additional material to this book can be downloaded from http://extras.springer.com.

ISBN-10: 0-387-25768-3   e-ISBN-10: 0-387-49333-6
ISBN-13: 978-387-25768-6   e-ISBN-13: 978-0-387-49333-6

Printed on acid-free paper.

9 8 7 6 5 4 3 2 1

springer.com   (EB)

PREFACE

A computer algebra system (CAS) not only has the number "crunching" and plotting capability of traditional computing languages such as Fortran and C, but also allows one to perform the symbolic manipulations and derivations required in most mathematically based science and engineering courses. To introduce students in these disciplines to CAS-based mathematical modeling and computation, the authors have previously developed and classroom tested the text *Computer Algebra Recipes: A Gourmet's Guide to the Mathematical Models of Science* [EM01] based on the Maple CAS. Judging by course evaluations and reader feedback, the response to this book and the computer algebra approach to modeling has been very favorable. With the release of several new versions of Maple since this text was published and the authors' accumulation of many insightful comments and helpful suggestions, a second up-dated edition seemed expedient. However, incorporating all the changes would make an already lengthy book even longer. So the topics of the *Gourmet's Guide* have been reorganized into two new stand-alone volumes, an already-published *Introductory Guide* [EM06] and this *Advanced Guide*.

In this book, we explore mathematical models involving linear and nonlinear ordinary and partial differential equations (ODEs and PDEs). This volume, which may be used either as a course text or for self-study, features an eclectic collection of Maple computer algebra worksheets, or "recipes," that are systematically organized to illustrate graphical, analytical, and numerical techniques applied to ODE/PDE-based scientific modeling. No prior knowledge of Maple is assumed, the early recipes introducing the reader to the basic Maple syntax, the subsequent recipes introducing further Maple commands and structure on a need-to-know basis.

The recipes are fully annotated in the text and in most cases presented as "stories" or in a historical context. Each recipe typically takes the reader from the analytic formulation of an interesting mathematical model to its analytic or numerical solution and finally to either a static or animated graphical visualization of the answer. Every recipe is followed by a set of problems that can be used to check one's understanding or develop the topic further. For your convenience, the recipes are included on a CD located in the inside back cover.

Contents

INTRODUCTION

A. Computer Algebra Systems

Man is still the most extraordinary computer of all.
John F. Kennedy, former American president. Speech, 21 May 1963.

Unlike traditional programming languages such as Fortran and C, a computer algebra language such as Maple allows one to compute not only with numbers, but also with symbols, formulas, equations, and so on. Using a computer algebra system (CAS), symbolic computation can be done on the computer, replacing the traditional pen-and-paper approach with the keyboard/mouse and computer display. By entering short, simple, transparent commands on the computer keyboard (which will be referred to as the "classic" approach), or by selecting mathematical symbols from a palette with the mouse, the CAS user can quickly and accurately generate symbolic input and output on the computer screen. Mathematical operations such as differentiation, integration, and series expansion of functions can be done analytically on the computer.

Because it also has numerical capability, a CAS allows the student or the researcher to tackle all aspects of mathematical modeling, from analytic derivation and manipulation of the model equations to the analytic or numerical solution of those equations, to the plotting or animation of the results. One of the most powerful computer algebra systems currently available is Maple 10, which will be used in this text. Useful reference books to this CAS are the Maple user manual [Map05] and the introductory and advanced programming guides [MGH+05].

In the two volumes of *Computer Algebra Recipes*, we present classic Maple worksheets, or "recipes," that demonstrate how a CAS can serve as a valuable adjunct tool in easily deriving, solving, plotting, and exploring interesting, modern scientific models chosen from a wide variety of disciplines ranging from the physical and biological sciences to the social sciences and engineering. The present book is the second volume and concentrates on mathematically more advanced models involving linear and nonlinear ordinary and partial differential equations (ODEs and PDEs). The classic Maple worksheet interface, which requires less computer memory than the standard interface, is used to generate all the mathematical and graphical output shown in this text.

1

B. Computer Algebra Recipes

The mathematics is not there till we put it there.
Sir Arthur Eddington, *The Philosophy of Physical Science*, 1939

The ODE and PDE recipes in our computer algebra "menu" have been organized into three sections of increasing mathematical sophistication. The **Appetizers** illustrate phase-plane portraits and analysis, the **Entrees** deal with linear and nonlinear ODEs and linear PDEs, and the **Desserts** feature the hunt for solitons and some nonlinear diagnostic tools. The recipes are intended to fulfill not only a useful and serious pedagogical purpose but also to titillate and stimulate the reader's intellect and imagination. Associated with each recipe is an important scientific model or method and usually some historical background or an interesting story featuring an engineering or science student who will guide you through the steps of the recipe. These storybook characters are fictitious composites of some of the more likeable, industrious, and brighter students that the authors have had the privilege of teaching over the years.

Every topic or story in the text contains the Maple code or recipe to explore that particular topic. To make life easier for you, all recipes have been placed on the CD-ROM enclosed within the back cover of this text. The recipes are ordered according to the chapter, section, and subsection (story) number. For example, the recipe **01-2-3**, entitled *Rössler's Strange Attractor*, is associated with Chapter 1, Section 2, Subsection 3. Although the recipes can be directly accessed on the CD by clicking on the appropriate worksheet number, it is recommended that you access them through the menu index file, **00menu**. All recipes may be conveniently accessed from this menu using the hyperlinks. Complete instructions on how to do this may be found in the menu file.

The computer code on the CD is unannotated, so you will have to read the text in order to understand what the code is trying to accomplish. The code has been imported into the text and here is accompanied by detailed explanations of the underlying modeling concepts and computational methods.

The recommended procedure for using this text is first to read a given topic/story for overall comprehension and enjoyment. If you are having any difficulty in understanding a piece of the text code, then you should execute the corresponding Maple worksheet and try variations on the code. Keep in mind that the same objective may often be achieved by a different combination of Maple commands from those used by the authors. After reading the topic, you should execute the worksheet (if you have not already done so) to make sure the code works as expected. At this point feel free to explore the topic. Try rotating any three-dimensional graphs or running any animations in the file. See what happens when changes in the model or Maple code are made and then try to interpret any new results. This book is intended to be open-ended and merely serve as a guide to what is possible in mathematical modeling using a CAS, the possibilities being limited only by your own background and desires.

Each topic or story is self-contained and generally done completely, from the derivation to the solution to the plot and accompanied by a thorough discussion of the steps and results. Since arriving at the answer is more important in our opinion than the method used, one will encounter recipes in which analytic derivation of the model equations occurs, followed by a numerical solution because an analytic solution doesn't exist. Although brief introductions, which generally include some definitions of terminology and short explanations of underlying concepts, are given for each main topic area, this text is not intended to teach you everything that you want to know, for example, about methods of solving ODEs or PDEs. Neither is it intended to teach you about the myriad subject areas of science or engineering. Instead, it is meant to serve as a guide to how these topics and areas can be handled using a CAS. However, this book is not just any ordinary guide. It is a *gourmet's guide*! It presupposes that the reader has learned or is about to learn about various scientific models and/or methods, and we are providing the computer algebra tools to enable you to solve complex scientific problems more easily, to attain greater understanding, and to explore the frontiers of science that interest you.

At the end of most recipe subsections, there are related problems where you, the reader, can check your mastery of the scientific computation and computer algebra techniques presented in the recipes. The problems also allow you to explore new frontiers and challenge you to invent and solve "What happens if ...?" problems. The purpose of this text is not only to teach computer-assisted computational techniques useful to engineering and science students, but to whet the student's curiosity and put some fun back into the pursuit of a science education. For maximum satisfaction and learning, it demands an interactive approach by the reader. Although the stories were designed to be interesting or amusing to read, the Maple recipes must be run, the models explored, and the problems solved. Some things never change in the learning process!

C. Introductory Recipe: Boys Will Be Boys

Giving money and power to government is like giving
whiskey and car keys to teenage boys.
P. J. O'Rourke, American journalist (1947–)

To give you some idea of what a typical computer algebra recipe looks like and to introduce some basic Maple syntax, consider the problem that follows. The recipe that solves this problem is not on the CD, so after reading this section you should open up classic Maple 10 on your computer, type in the recipe, and execute it.

Richard's grandson Daniel throws a small ball with an initial speed of 15 m/s towards a 3.5-meter-high fence located 20 meters from the ball's initial position. The ball leaves Daniel's hand at a height of 2 m above the level ground and just clears the top of the fence. The gravitational acceleration is $g = 9.8$ m/s^2. The ball may be regarded as a point particle and air resistance neglected.

(a) At what angle ϕ with the horizontal is the ball thrown?

(b) How long does it take the ball to reach the fence?

(c) Plot the entire trajectory, then animate the motion of the ball, including the fence in the animation.

Let's choose the origin to be on the ground below the initial position of the ball and take the x-coordinate to be horizontal and the y-coordinate vertical. To begin the recipe, we first clear Maple's internal memory of any previously assigned values (other worksheets may be open with numerical values given to some of the same symbols being used in the present recipe). This is done by typing in the **restart** command after the opening prompt ($>$) symbol, ending the command with a colon (:), and pressing Enter (which generates a new prompt symbol) on the computer keyboard.

```
>   restart:
```

All Maple command lines must be ended with either a colon, which suppresses any output, or a semicolon (;), which allows the output to be viewed.

Next, the given parameter values are specified. For example, the initial x-coordinate of the ball is entered, the symbolic name **xb** being placed to the left of the Maple assignment operator (:=). The numerical value (0) of the coordinate is placed on the right-hand side of the operator and the output suppressed here with a command-ending colon. Assigned quantities can be mathematically manipulated. In a similar manner, the numerical values of the ball's initial y-coordinate (**yb**), the horizontal location (**xf**) of the fence, the fence's height (**yf**), the initial speed (**V**) of the ball, and the magnitude of the gravitational acceleration (**g**) are entered. Because the command entries are short, we have chosen to place them all on the same prompt line, separating the entries by a space for reading clarity.

```
>   xb:=0: yb:=2: xf:=20: yf:=3.5: V:=15: g:=9.8:
```

Using the symbol * for multiplication, we express the horizontal (**vx**) and vertical (**vy**) components of the ball's initial velocity in terms of the unknown angle ϕ. The Maple input syntax **phi** is used to generate the Greek letter ϕ in the output. Note that the assigned value (15) of **V** is automatically substituted.

```
>   vx:=V*cos(phi); vy:=V*sin(phi);
```

$$vx := 15\cos(\phi) \qquad vy := 15\sin(\phi)$$

Using the standard kinematic relations [Oha85], we calculate the ball's x and y coordinates at arbitrary time t. The symbols +, -, /, and ^ are used for addition, subtraction, division, and exponentiation. Note that the decimal coefficient of t^2 in the output is given to 10 digits, Maple's usual default accuracy.

```
>   x:=xb+vx*t; y:=yb+vy*t-(1/2)*g*t^2;
```

$$x := 15\cos(\phi)\,t \qquad y := 2 + 15\sin(\phi)\,t - 4.900000000\,t^2$$

Setting **x=xf** in the **solve** command, the time $t = tf$ for the ball to reach the fence is determined in terms of ϕ.

```
>   tf:=solve(x=xf,t); #time to reach fence
```

$$tf := \frac{4}{3}\,\frac{1}{\cos(\phi)}$$

A comment, prefixed by the pound sign #, has been added to the command line. Short comments are useful for later reference or for others to read and understand the purpose of a Maple command.[1]

A transcendental equation *eq* for ϕ results on evaluating y at $t = tf$ and equating the result to *yf*.

```
>  eq:=eval(y,t=tf)=yf;
```

$$eq := 2 + \frac{20\sin(\phi)}{\cos(\phi)} - \frac{8.711111111}{\cos(\phi)^2} = 3.5$$

The transcendental equation is solved for ϕ, the result being labeled Φ.

```
>  Phi:=solve(eq,phi); #angles in radians
```

$$\Phi := 0.9917653601,\ 0.6538908144,\ -2.149827293,\ -2.487701839$$

Four angles, expressed in radians, are generated in the above output Φ, the default accuracy being 10 digits. Since the initial angle must be above the horizontal, only the positive answers are acceptable as solutions for this problem. Since there are two positive results, this means that Daniel could throw the ball at two different angles to just clear the fence. To proceed, we shall select one of the positive answers, say the second one in Φ. This is done by entering Phi[2]. You can look at the first positive angle by changing this entry to Phi[1]. If desired, the angle $\phi \approx 0.65$ radians can be converted to degrees using the convert command with units as the second argument.

```
>  phi:=Phi[2]; theta:=convert(phi,units,radian,degree);
```

$$\phi := 0.6538908144 \qquad \theta := 37.46518392$$

In this case, Daniel throws the ball at an angle $\theta \approx 37\frac{1}{2}$ degrees to the horizontal. It should be mentioned that the convert command is very useful for converting an expression from one form to another, the form of conversion being dictated by the choice of second argument. To see the types of conversions possible with Maple, click on the convert command in the worksheet, then on **Help** in the tool bar, and finally on **Help on convert**.

Using the floating-point evaluation command, evalf, we numerically evaluate the time to reach the fence, which is found to be about 1.68 seconds.

```
>  tf:=evalf(tf); #time to reach fence
```

$$tf := 1.679846933$$

To plot the entire trajectory, the time T for the ball to hit the ground must be determined. This is accomplished by setting $y = 0$ and solving for t, which produces two answers.

```
>  T:=solve(y=0,t);
```

$$T := -0.1981184874,\ 2.060197767$$

---

[1]Longer or more detailed comments may be inserted into a worksheet by clicking on **Insert** in the tool bar at the top of the computer screen, then on **Execution Group**, on either **Before Cursor** or **After Cursor**, on **Text**, and finally typing in the comments.

The negative answer is the time at which the ball would have had to been thrown if it had started from ground level. The second answer is the relevant one here and is now selected. The ball hits the ground after 2.06 seconds.

> `T2:=T[2]; #time to hit ground`

$$T2 := 2.060197767$$

The entire trajectory of the ball is now plotted using the `plot` command. The ball's horizontal and vertical coordinates must be entered as part of a Maple "list," the list entries being separated by commas and the entire list enclosed in square brackets. Maple preserves the order of items in a list. In the present case, the list entries are the x-coordinate, the y-coordinate, and the time range.[2]

If no optional arguments are specified in the `plot` command, a figure will be generated in which Maple chooses its own horizontal and vertical scaling. Constrained scaling can be achieved by including the option `scaling=constrained` as an additional argument.[3]

On executing the `plot` command, we generate a computer plot similar to that reproduced in Figure 1, with y plotted vertically and x horizontally.

> `plot([x,y,t=0..T2],scaling=constrained);`

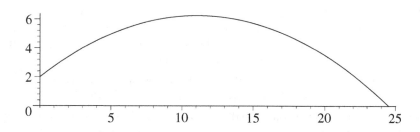

Figure 1: Trajectory of the ball.

Note that the horizontal and vertical axes are not labeled in the figure. Axis labels can be added by including the option `labels=["x(in m)","y(in m)"]` in the `plot` command. Try it and see.[4]

If you wish to learn more about the `plot` command and its optional arguments, click the left mouse button on the `plot` command and then on **Help** at the top of the computer screen. Clicking on the entry **Help on plot** opens up a help page with information about this command structure. The various

---

[2] This is a *parametric* plot, because x and y depend parametrically on the time t.

[3] If this optional argument is omitted, constrained scaling can be achieved in another way. Click your left mouse button on the computer plot. This opens a tool bar at the top of the computer screen with various options. Clicking on **Projection** and then on **Constrained** produces a plot with constrained scaling.

[4] The double quotes in the `labels` option indicate a Maple "string," a sequence of characters that has no value other than itself. A string cannot be assigned to, and will always evaluate to, itself. Omit the double quotes and note what happens when the `plot` command is executed.

plotting options available can be found by clicking on the underlined hyperlink **plot/details** that appears on the help page, and then on **plot[options]**. Try including some other options in the `plot` command. For example, change the color of the ball's trajectory from the default red to, say, blue.

The final part of the problem is to animate the trajectory of the ball with the vertical fence included. The fence is now plotted, but not displayed, being assigned the name `fence`. The coordinates of the bottom and the top of the fence are entered as Maple lists and then formed into another list, i.e., one has a "list of lists." The fence will be plotted as a vertical (default red) line.

```
> fence:=plot([[xf,0],[xf,yf]]):
```

To produce specialized plots, such as an animated one, we must access the plots library package. This is done by entering the command `with(plots)`. The preface `with` always indicates a Maple library package is being "loaded" into the worksheet. Library packages are extremely important, since they contain approximately 90% of Maple's mathematical knowledge. Normally, we would end the command with a colon, but here a semicolon is used to see what plot commands are contained in the plots library.[5]

```
> with(plots);
```

Warning, the name changecoords has been redefined

[*Interactive, animate, animate3d, animatecurve, arrow, changecoords,...*]

The animate command, appearing in the above list, will be used to animate the motion of the ball. The syntax is `animate(plot command,[plot arguments],` `time range,options)`. The `pointplot` command will plot the ball as a point. The plot arguments `symbol=circle` and `symbolsize=14` instruct Maple to plot the point as a size-14 circle.[6] Included in the plot arguments list are the ball's x- and y-coordinates given as a list of lists. The time range is $t = 0$ to $T2$. In the options, the number[7] of frames is taken to be 200, the frames being equally spaced in time. The option `background=fence` causes the fence to appear as background in each time frame. Finally, the scaling is constrained.

```
> animate(pointplot,[[[x,y]],symbol=circle,symbolsize=14],
     t=0..T2,frames=200,background=fence,scaling=constrained);
```

When the `animate` command line is executed on the computer, the initial frame of the animation will appear on the screen. Clicking on the picture with the left mouse button places the picture in a viewing box and opens up an animation bar at the top of the screen. The animation is started by clicking on the arrowhead (▷) and stopped by clicking on the square (□).

---

[5]Only a partial list of plot commands is shown here in the text, as indicated by the dots. Note that a warning message is also produced that informs us that the name `changecoords` has been redefined in the current release of Maple. If desired, warnings can be removed by inserting the command `interface(warnlevel=0)` prior to loading the library package. From now on, all such warnings will generally be artificially removed in the text.

[6]The default symbol is a diamond and the default size is 10.

[7]The default number is 25 frames. 200 frames produces a smoother animation.

D. Maple Help

We teachers can only help the work going on,
as servants wait upon a master.
Maria Montessori, Italian educator, *The Absorbent Mind*, 1949.

We have already seen in the introductory recipe how Maple's Help can be used to learn more about the Maple commands that appear in the text recipes. If you wish to find out what other help is available, click on **Help** at the top of the computer screen and then on the entry **Using Help**. A help page opens with a number of hyperlinks that you should explore.

Two of the more important hyperlinks are entitled **Perform a Topic Search** and **Perform a Full Text Search**. Here we shall give two simple examples of using these searches, leaving the full descriptions of the search types for you to read. It should be noted that neither type of search is case-sensitive.

Our first example illustrates a topic search, which locates help based on a keyword that you specify. For example, suppose that you wanted the correct form of the command for taking a square root. Click on **Help**, then on **Topic Search**, making sure that the Auto-search box is selected. Depending on the programming language used, the square root command could be `sqr`, `sqrt`,...,... On typing the first couple of letters, `sq`, in the Topic box, Maple will display all the commands starting with `sq`. Double click on `sqrt` or, alternatively, single click on `sqrt` and then on OK. A description of the square root command will appear on the screen. You can then close the Help window and proceed with programming your recipe.

The second example illustrates a full-text search. Suppose, for example, that you wish to find the command for analytically or numerically solving an ODE. In the Help window, click on **Full Text Search**. Then type ode in the **Word(s)** box and click on **Search**. Double clicking on **dsolve** produces a description of the `dsolve` command for solving ODEs along with several examples. If, for example, you want to know how to find a numerical solution, click on the hyperlink **dsolve,numeric**.

The approach employed in the introductory recipe may also be used to find information about unfamiliar mathematical functions that appear in the Maple output. If, for example, the output contained the word "EllipticF," you may find out what this function is by clicking on the word to highlight it, then on **Help**, and finally on **Help on EllipticF**. You will find that EllipticF refers to the incomplete elliptic integral of the first kind, which is defined in the Help page. The same Help window may also be opened by typing in a question mark followed by the word and a semicolon, e.g., `?EllipticF;`.

Maple's Help is certainly not perfect, and on occasion you might feel frustrated, but generally it is helpful and should be consulted whenever you get stuck with Maple syntax or are seeking just the right command to accomplish a certain mathematical task.

E. How to Use This Text

Begin at the beginning ... and go on till you come to the end.
Lewis Carroll, *Alice's Adventures in Wonderland* (1865)

The recommended procedure for most readers, particularly for someone who is new to CASs in general and to Maple in particular, is to follow the advice given by the king to the white rabbit in *Alice's Adventures in Wonderland*. Start with the **Appetizers**, then go on to the **Entrees**, and finish off with the **Desserts**. In the early recipes of the **Appetizers** you will be introduced to more of the basic features of the Maple system and see further examples of Maple's Help. Keep in mind that although we have made every effort to keep this book self-contained, it is impossible for us to teach you everything you would wish to know about programming with the Maple CAS. After all, that it is not the primary purpose of this text. Further, the choice of Maple library packages and associated command structures is to some extent dictated by the choice of subject matter. Since the topics of *Computer Algebra Recipes* have now been split between two volumes, you might wish to consult the *Introductory Guide* for use of library packages not emphasized in this *Advanced Guide*. And, at the risk of sounding immodest, the former volume also contains a wonderful selection of scientific models.

Of course, if you are already a Maple expert, feel free to pick and choose those topics and recipes in this volume that interest you or are relevant to your own scientific tastes or goals.

No matter what approach to using this text is taken, we hope that you will enjoy the wide range of interdisciplinary topics and stories that we have presented. Before beginning your intellectual journey through this book, let us paraphrase a well-known saying from the world of sports with these words of advice:

You can't learn the great game of scientific modeling by being a spectator. You must play the game!

We trust that as you sample and explore the various recipes on which our menu is based, you will enjoy the intellectual "feast" that we have prepared and presented in this *Advanced Guide to the Mathematical Models of Science*.

Bon Appetit!
Richard and George,
Your CAS chefs

Part I

THE APPETIZERS

A man ceases to be a beginner in any given science and becomes a master in that science when he has learned that... he is going to be a beginner all his life.
R. G. Collingwood, British philosopher, *The New Leviathan*, 1942

In a time of drastic change it is the learners who inherit the future. The learned usually find themselves equipped to live in a world that no longer exists.
Eric Hoffer, American philosopher, *Reflections on the Human Condition*, 1973

To know yet to think that one does not know is best; Not to know yet to think that one knows will lead to difficulty.
Lao-Tzu, Chinese philosopher, 6th century BC

Chapter 1

Phase-Plane Portraits

Every portrait that is painted with feeling
is a portrait of the artist, not of the sitter.
Oscar Wilde, Anglo-Irish playwright, novelist, and poet (1854–1900)

Consider a system of two first-order coupled ODEs of the general structure

$$\dot{X} \equiv \frac{dX}{dt} = P(X,Y), \quad \dot{Y} \equiv \frac{dY}{dt} = Q(X,Y), \tag{1.1}$$

where P and Q are known functions of the dependent variables X and Y, and the independent variable has been taken to be the time t. In other model equations, the independent variable could be a spatial coordinate, e.g., the Cartesian coordinate x. For compactness, the dot notation of (1.1) will often be used in our text discussion for time derivatives, one dot denoting d/dt, two dots standing for d^2/dt^2, and so on. Superscripted primes on the dependent variable indicate a spatial derivative, e.g., $Y' \equiv dY/dx$, $Y'' \equiv d^2Y/dx^2$, etc.

The 2-dimensional ODE system (1.1) is said to be *autonomous*, meaning that P and Q do not depend explicitly on t. If there is an explicit dependence on the independent variable, the equations are said to be *nonautonomous*. Our goal in the following section is to illustrate a simple graphical procedure for exploring *all possible solutions* of equations (1.1) for specific forms of P and Q. In the second section, we shall show that a 2-dimensional nonautonomous system of ODEs can be cast into a 3-dimensional autonomous system and present some interesting examples of the latter.

1.1   Phase-Plane Portraits

Some biological models of competing species are naturally of the *standard form* (1.1). For example, a simple ODE model of the temporal evolution of interacting rabbit and fox populations might be given by the system

$$\dot{r} = 2r - 0.04\,rf, \quad \dot{f} = -f + 0.01\,rf, \tag{1.2}$$

where $r(t)$ and $f(t)$ are the numbers of rabbits and foxes per unit area at time t. If the "interaction" terms involving rf are omitted in (1.2), the remaining

ODEs are linear (first order) in r and f, respectively, and are said to be *linear* ODEs. Inclusion of the higher-order interaction terms changes the equations into *nonlinear* ODEs. That is to say, $P(r, f) \equiv 2r - 0.04\, rf$ and $Q(r, f) \equiv -f + 0.01\, rf$ are nonlinear functions of the dependent variables r and f. Nonlinear models, such as this one, are usually difficult or impossible to solve analytically, so that one must resort to graphical or numerical means to obtain a solution.

Other models, involving second-order ODEs, can often be recast into the standard form, e.g., models arising from Newton's second law of the structure

$$\ddot{x} = F(x, \dot{x}), \tag{1.3}$$

where x is the displacement and F is the force per unit mass. Setting $\dot{x} = v$, which is the velocity, (1.3) can be expressed as the coupled first-order system

$$\dot{x} = v, \quad \dot{v} = F(x, v). \tag{1.4}$$

For example, consider the damped, simple harmonic oscillator (SHO) equation

$$\ddot{x} + 2\,\gamma\,\dot{x} + \omega_0^2\, x = 0, \tag{1.5}$$

describing the oscillations of a mass m attached to a light spring (spring constant k) that obeys Hooke's law and experiences a frictional force (damping coefficient γ) linear in the velocity. The characteristic frequency ω_0 is equal to $\sqrt{k/m}$. On setting $\dot{x} = v$, this second-order linear ODE may be rewritten in standard form with $P(x, v) \equiv v$ and $Q(x, v) \equiv F(x, v) = -2\,\gamma\,v - \omega_0^2\, x$.

Whether linear or nonlinear, a graphical approach can be used to view *all possible solutions* of those ODE systems that can be put into the standard form. This graphical procedure has proved especially important in the investigation of nonlinear systems, less so for linear systems, since the latter can usually be solved analytically. Since equations (1.1) do not depend explicitly on t, the independent variable may be eliminated by dividing one equation by the other to form the ratio

$$\frac{dY}{dX} = \frac{Q(X,Y)}{P(X,Y)}. \tag{1.6}$$

If the Y-versus-X plane is considered, this ratio represents the slope of the *trajectory* of the ODE system at the point (X,Y) in the plane. The X-Y plane is referred to as the *phase plane* and the trajectory as a *phase-plane trajectory*. For the rabbits–foxes system the phase plane shows the fox number (f) versus the rabbit number (r), while for the SHO the phase plane is a plot of velocity (v) against displacement (x).

The time evolution of any possible motion of the ODE system may be pictured by systematically filling the phase plane with a grid of uniformly spaced arrows indicating the direction of increasing time and the slope at each grid point. For a given set of initial conditions, the subsequent temporal evolution of the system may be traced out by moving from one arrow to the next and drawing an appropriate line for the trajectory. Pictures created in this manner are called *phase-plane portraits*. Since an arrow at any point in the phase plane is tangent to the trajectory at that point, the grid of arrows is referred to as the *tangent field*.

The phase-plane analysis of ODE systems, particularly those that are non-linear, can be aided by first locating the *stationary* or *fixed points* (X_0, Y_0) of the system and then identifying their *topological nature*, i.e., the characteristic shape of the trajectories in the vicinity of the fixed points. At these points all the derivatives are zero. Thus, from equation (1.1) they are found by solving

$$P(X_0, Y_0) = 0, \quad Q(X_0, Y_0) = 0. \tag{1.7}$$

At a stationary point, it follows from equation (1.6) that the slope of the trajectory is of the form $0/0$ and thus indeterminate. At any other point (referred to as an *ordinary point*) of the phase plane, a trajectory has a definite slope.

For linear systems that can be expressed in standard form there will be only one stationary point. For example, setting $P = Q = 0$ for the SHO yields a stationary point at the origin of the v versus x phase plane.

For nonlinear systems, there can be more than one fixed point. As an example, let's use Maple to determine these points for the rabbits–foxes system.

>   restart:

The forms of P and Q for equations (1.2) are entered,

>   P:=2*r-0.04*r*f;

$$P := 2r - 0.04\,r\,f$$

>   Q:=-f+0.01*r*f;

$$Q := -f + 0.01\,r\,f$$

and the `solve` command applied to the set of equations[1] $P = 0$, $Q = 0$. A Maple "set" is enclosed in "curly" ({ }) brackets. Unlike a Maple list, the order of the items is not preserved in a Maple set. The unknowns r and f are also entered as a set.

>   solve({P=0,Q=0},{r,f});

$$\{r = 0., f = 0.\}, \{r = 100., f = 50.\}$$

From the output, we see that there are two stationary points, one at the origin of the f-r phase plane and the other at $r = 100$ rabbits, $f = 50$ foxes.

Every stationary point has a certain topology in its neighborhood that dictates the nature of the phase-plane trajectories near that point. Thus, identifying the nature of each stationary point was an important task historically, since it allowed investigators in the precomputer era to sketch all the possible phase–plane trajectories from a knowledge of the location of the stationary points and their types. In the modern computer era we can let software packages like Maple do the graphing and analysis for us.

For ODE systems that can be put into standard form, what types of stationary points are possible? It turns out that there are only four types of so-called *simple*[2] stationary points, which are schematically illustrated in Figure 1.1. For the first three types (the vortex, focal, and nodal points) we shall

---

[1]Note that it isn't necessary to explicitly set P and Q equal to zero, since Maple will automatically assume that this is so unless you specify otherwise.

[2]The precise definition of the phrase "simple" will be given in Chapter 2.

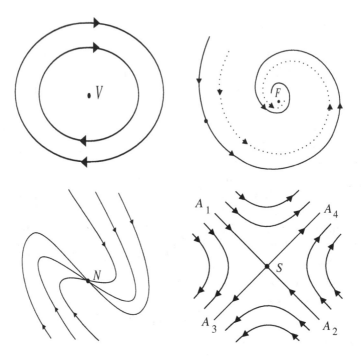

Figure 1.1: Curves near a vortex (V), focal (F), nodal (N), saddle (S) point.

use the SHO equation expressed in standard form as an explanatory tool, assuming that the reader already has some idea of the qualitative nature of the solutions as the damping coefficient γ is varied.

For $\gamma = 0$, a SHO released from rest outside the origin will oscillate indefinitely about the origin with no change in amplitude. If the velocity is plotted versus the displacement, the phase-plane trajectories near the origin will qualitatively look like those shown in the top left of Figure 1.1, the arrows indicating the direction of increasing time. As the SHO passes through the origin, the velocity will be a maximum, going to zero as the system reaches its turning points. For each choice of initial displacement, a different closed loop will be traced out in the phase plane. In this case, the origin (labeled V) is said to be a *vortex point* or *center*. Since a continuum of initial conditions is possible, a vortex point is surrounded by a continuum of closed loops.

For small γ below some critical threshold, the SHO will oscillate about the origin with an ever-decreasing amplitude, asymptotically approaching the origin as $t \to \infty$. The origin is said to be a *stable equilibrium point*. Two representative trajectories are shown in the top right of Figure 1.1. As time progresses, each trajectory approaches the stationary point F at the origin along a (different) spiral path. The point F is an example of a stable *spiral point* or *focal point*. An *unstable focal point* would be one for which the trajectory winds off the stationary point as $t \to \infty$.

As γ is increased, a critical damping is reached, beyond which the behavior of the SHO system changes. The SHO will no longer oscillate indefinitely about the origin but eventually approach it from a definite direction as $t \to \infty$. This is schematically illustrated in the bottom left of Figure 1.1 for six different initial conditions. The origin, labeled N, is in this case an example of a stable *nodal point*. For an unstable nodal point, the time arrows would be reversed.

The fourth type of possible fixed point, which will be illustrated in the following **Romeo and Juliet** recipe, is called a *saddle point*.

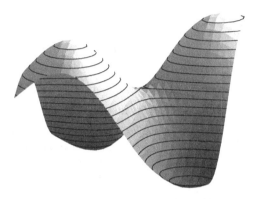

Figure 1.2: Saddle-point topography.

If you have done any alpine hiking, you might know that a "saddle" is a ridge between two mountain peaks or summits as illustrated in Figure 1.2, acquiring its name because the local topography resembles that of a horse saddle. The saddle point is the low point in the saddle, where in two opposite directions you would go downhill toward the two valleys and in the two transverse directions uphill toward the two peaks. If one characterized these four directions by arrows pointing in the direction of decreasing elevation, the arrows would point away from the saddle point in the first case and toward it in the second. If given an infinitesimal nudge, a particle would tend to move away from the saddle point in the first case and toward it in the second.

A saddle point (labeled S) in the phase plane is a two-dimensional analogue of this topology, as schematically indicated in the bottom right of Figure 1.1, the arrows pointing in the direction of increasing time. The four trajectories labeled A_1S, A_2S, A_3S, and A_4S are examples of so-called *separatrixes*, dividing the area about the saddle point into four distinct regions with the trajectories in each region evolving with time in the directions indicated. A representative point approaches S along A_1S and A_2S as $t \to +\infty$, while it departs from S along A_3S and A_4S at $t = -\infty$.

1.1.1   Romeo and Juliet

I am convinced we do not only love ourselves in others
but hate ourselves in others too.
G. C. Lichtenberg, German physicist, philosopher (1742–1799)

The mathematician Steven Strogatz [Str88] [Str94] has suggested a simple dynamic model to create different scenarios for the love affair between Romeo and
Juliet. In his model, $R(t)$ and $J(t)$ represents Romeo's love/hate for Juliet and
Juliet's love/hate for Romeo, respectively, at time t. Positive values of R and
J indicate love, while negative values indicate hate. The love affair equations
take the form

$$\dot{R}(t) = a\,R + b\,J, \qquad \dot{J}(t) = c\,R + d\,J,$$

where a, b, c, and d are real coefficients that may have either sign. For the sake
of definiteness, let's take $a = 2$, $b = 1$, $c = -1$, and $d = -2$ in the following
problem, leaving other coefficient values for you to explore.

(a) Is the ODE system linear or nonlinear? Locate the fixed point(s).

(b) Create a tangent field plot and identify the nature of the fixed point(s).

(c) Create a phase-plane portrait that contains the four trajectories corresponding to the following initial conditions: **(i)** $R(0) = -0.25$, $J(0) = 1$,
(ii) $R(0) = -0.27$, $J(0) = 1$, **(iii)** $R(0) = 0.27$, $J(0) = -1$, and finally
(iv) $R(0) = 0.25$, $J(0) = -1$. Consider $t = 0$ to 4 time units.

(d) Plot R versus t over the interval $t = 0$ to 2 for initial condition **(i)**.

(e) Derive analytic solutions for $R(t)$ and $J(t)$ for initial condition **(i)**.

To plot the tangent field and create the phase-plane portrait, the `dfieldplot`
and `phaseportrait` commands, respectively, will be used. These specialized
differential equation plotting tools are contained in the DEtools library package, which is now loaded. The colon may be replaced with a semicolon to
display the complete list of available commands in this package.

```
>   restart: with(DEtools):
```
The general love affair differential equations are entered in `de1` and `de2`, each
first-order derivative with respect to t being entered with the `diff` command.

```
>   de1:=diff(R(t),t)=a*R(t)+b*J(t);
```

$$de1 := \frac{d}{dt}\,R(t) = a\,R(t) + b\,J(t)$$

```
>   de2:=diff(J(t),t)=c*R(t)+d*J(t);
```

$$de2 := \frac{d}{dt}\,J(t) = c\,R(t) + d\,J(t)$$

Before proceeding further with solving our problem, a few words should
be said about Maple's derivative command. Using `exp`, `ln`, and `sin` to enter
the exponential, natural logarithm, and sine functions, suppose that Romeo's
love/hate for Juliet depended on time in the following way.

```
> Romeo:=exp(-t)*ln(1+t)*sin(t)/sqrt(1+t^2);
```

$$Romeo := \frac{e^{(-t)} \ln(1+t) \sin(t)}{\sqrt{1+t^2}}$$

The first derivative of *Romeo* with respect to t is now taken, the "inert" form, Diff, being used on the left-hand side to display the derivative, the "active" form, diff, employed on the right to explicitly perform the differentiation.

```
> derivative:=Diff(Romeo,t)=diff(Romeo,t);
```

$$derivative := \frac{d}{dt} \left(\frac{e^{(-t)} \ln(1+t) \sin(t)}{\sqrt{1+t^2}} \right) = -\frac{e^{(-t)} \ln(1+t) \sin(t)}{\sqrt{1+t^2}}$$

$$+ \frac{e^{(-t)} \sin(t)}{(1+t)\sqrt{1+t^2}} + \frac{e^{(-t)} \ln(1+t) \cos(t)}{\sqrt{1+t^2}} - \frac{e^{(-t)} \ln(1+t) \sin(t) t}{(1+t^2)^{(3/2)}}$$

The extension of Maple's syntax to higher-order derivatives is straight-forward. The active form of the second derivative of *Romeo* is diff(Romeo,t,t) or, alternatively, diff(Romeo,t$2). As an exercise, you should calculate, say, the seventh time derivative of *Romeo* and see how easy it is to generate the new answer. A calculation by hand would be very tedious and prone to error.

Returning to our problem, since the right-hand side (rhs) of both *de1* and *de2* depend linearly on R and J, the ODEs are linear. By visual inspection, there is one stationary point at $R = J = 0$. This is confirmed by applying the solve command to the rhs of the two ODEs.

```
> statpoint:=solve({rhs(de1),rhs(de2)},{R(t),J(t)});
```

$$statpoint := \{J(t) = 0,\ R(t) = 0\}$$

The given coefficient values are now specified,

```
> a:=2: b:=1: c:=-1: d:=-2:
```

which are automatically substituted into the two ODEs.

```
> de1; de2;
```

$$\frac{d}{dt} R(t) = 2 R(t) + J(t)$$

$$\frac{d}{dt} J(t) = -R(t) - 2 J(t)$$

The dfieldplot command is used to plot the tangent field.

```
> dfieldplot([de1,de2],[R(t),J(t)],t=0..1,R=-1..1,J=-1..1,
    dirgrid=[30,30],arrows=MEDIUM);
```

The first and second arguments are Maple lists of the ODEs and dependent variables. The time range has been taken to be from $t = 0$ to 1 and the plotting range for both R and J is from -1 to $+1$. The dirgrid option specifies the number of horizontal and vertical mesh points (30×30 here) to use for the tangent arrows. The mimimum is 2×2 and the default is 20×20. The option arrows=MEDIUM produces "full" arrowheads, the default being "half" arrowheads. To learn more about dfieldplot, highlight this command with your mouse, then click on **Help**, and on **Help on dfieldplot**.

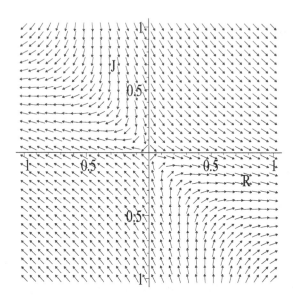

Figure 1.3: Tangent field for Romeo and Juliet's love affair.

The tangent field for Romeo and Juliet's love affair is shown in Figure 1.3. Inspecting the graph and comparing with Figure 1.1, it is seen that the stationary point at the origin is a saddle point. By changing the coefficient values, the nature of the stationary point and therefore the nature of the love affair can be altered. This is left as a problem at the end of the section for you to explore.

The four initial conditions are now entered,

```
>   ic1:=(R(0)=-0.25,J(0)=1): ic2:=(R(0)=-0.27,J(0)=1):
    ic3:=(R(0)=0.27,J(0)=-1): ic4:=(R(0)=0.25,J(0)=-1):
```

and the **phaseportrait** command is used to create the phase-plane portrait for the four initial conditions. These conditions are entered as a list of lists. The time step size is taken to be 0.05. The default is to divide the time range into 20 equal steps. So the default step size here would be $4/20 = 0.2$. Again MEDIUM arrows are chosen, which are colored red. The trajectories are colored blue using the `linecolor` option. A complete list of options may be found under DEplot, whose Help page may be accessed though the topic search.

```
>   phaseportrait([de1,de2],[R(t),J(t)],t=0..4,[[ic1],[ic2],
    [ic3],[ic4]],stepsize=0.05,dirgrid=[30,30],R=-1..1,J=-1..1,
    arrows=MEDIUM,color=red,linecolor=blue);
```

The phase-plane portrait for Romeo and Juliet's love affair is reproduced in Figure 1.4. Asymptotically, the trajectories approach the separatrixes of the saddle point at the origin, the separatrixes dividing the phase plane into four different "flow" regions for the tangent arrows. Thus, for example, one can see that for any initial condition in the lower right region, R will remain positive and J negative. Romeo's love for Juliet is unrequited!

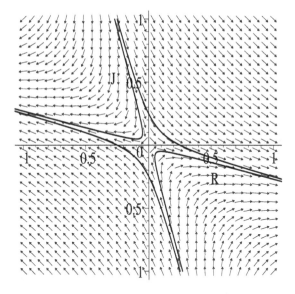

Figure 1.4: Phase-plane portrait for the love affair.

By introducing the option `scene=[t,R]` into `phaseportrait`, a plot of R vs. t can be produced. The resulting picture for the first initial condition is shown in Figure 1.5, Romeo's initial hate turning to love. Changing R to J in the `scene` option will generate J versus t. Try it and see how Juliet responds.

```
> phaseportrait([de1,de2],[R(t),J(t)],t=0..2,[[ic1]],
    stepsize=0.2,scene=[t,R],color=red,linecolor=blue);
```

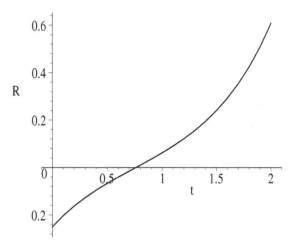

Figure 1.5: R versus t for the love affair.

Now, the ODEs in Strogatz's model are linear with constant coefficients, so they can be solved analytically. If proceeding by hand, one can solve *de1* for $J(t)$ and substitute the result into *de2*, yielding a second-order ODE in $R(t)$ alone. Assuming a solution for $R(t)$ of the form $e^{\alpha t}$ yields a quadratic equation in α, with two roots α_1 and α_2. Then R would involve a linear combination of $e^{\alpha_1 t}$ and $e^{\alpha_2 t}$, the two arbitrary coefficients being evaluated with the initial condition. With $R(t)$ completely determined, $J(t)$ is then easily evaluated.

The Maple differential equation solve command, `dsolve`, has such *standard* analytic methods of attack built into its solution algorithm. Applying this command to the ODE set, subject to the first initial condition, generates the following analytic solutions for $R(t)$ and $J(t)$.

> `solution:=dsolve({de1,de2,ic1},{R(t),J(t)});`

$$solution := \left\{ R(t) = -\left(\frac{1}{2} - \frac{7\sqrt{3}}{24}\right)\sqrt{3}\, e^{(\sqrt{3}\, t)} + \left(\frac{1}{2} + \frac{7\sqrt{3}}{24}\right)\sqrt{3}\, e^{(-\sqrt{3}\, t)} \right.$$

$$- 2\left(\frac{1}{2} - \frac{7\sqrt{3}}{24}\right) e^{(\sqrt{3}\, t)} - 2\left(\frac{1}{2} + \frac{7\sqrt{3}}{24}\right) e^{(-\sqrt{3}\, t)},$$

$$\left. J(t) = \left(\frac{1}{2} - \frac{7\sqrt{3}}{24}\right) e^{(\sqrt{3}\, t)} + \left(\frac{1}{2} + \frac{7\sqrt{3}}{24}\right) e^{(-\sqrt{3}\, t)} \right\}$$

In this case, one can see by inspecting the output that $\alpha = \pm\sqrt{3}$. If the analytic forms of either $R(t)$ and $J(t)$ are to be further manipulated, the *solution* must be assigned. Otherwise, entering `R(t)` or `J(t)` will not produce the analytic solutions, but just the symbols $R(t)$ and $J(t)$ in the output.

> `assign(solution):`

Having assigned the *solution*, it can now be checked by, for example, subtracting the right-hand side (`rhs`) of *de1* from the left-hand side (`lhs`) of *de1* and simplifying with the `simplify` command. The result is 0 as expected.

> `check:=simplify(lhs(de1)-rhs(de1));`

$$check := 0$$

You can check that a zero result occurs if *de1* is replaced with *de2*.

PROBLEMS:
Problem 1-1: Recasting into first-order ODEs
Recast each of the following second-order ODEs into a first-order ODE system and identify P and Q. Which systems are nonlinear? For each autonomous system, locate all the stationary points. All parameters are real and positive.

(a) Airy equation: $y'' - x\,y = 0$;

(b) "soft" spring equation: $\ddot{x} + (1 - x^2)\,x = 0$;

(c) Hermite equation: $y'' - x\,y' + n\,y = 0$;

(d) Rayleigh equation: $\ddot{x} - \epsilon\,(1 - \dot{x}^2)\,\dot{x} + x = 0$;

(e) confluent hypergeometric equation: $x\,y'' + (\gamma - x)\,y' - \alpha\,y = 0$;

(f) plane pendulum equation: $\ddot{\theta} + \sin\theta = 0$.

Problem 1-2: Symbolic differentiation
Display and evaluate the following derivatives:

(a) $\dfrac{d^5}{dx^5}\left(x^6\,\ln(x)\cos(x)\,e^{-x^2}\right)$, (b) $\dfrac{d^9}{dx^9}\left(\dfrac{x^{11}\,\tanh(2\,x)}{\sqrt{1+x^4}}\right)$

Problem 1-3: Alternative love affairs
For each of the following possible love affairs involving Romeo and Juliet:

- produce a tangent field plot using the `dfieldplot` command and identify each stationary point and its stability;
- produce a phase-plane portrait with the specified initial condition;
- use the `phaseportrait` command and scene options to plot $J(t)$ and $R(t)$;
- use the `dsolve` command to derive the analytic solution;
- discuss how the love affair evolves with time.

(a) $\dot{R} = -2\,R + J, \quad \dot{J} = -R - 2\,J, \quad R(0) = -1, J(0) = 1$.
(b) $\dot{R} = J, \quad \dot{J} = -R, \quad R(0) = 5, J(0) = -2$.
(c) $\dot{R} = J, \quad \dot{J} = -R + J, \quad R(0) = 0.1, J(0) = 0$.
(d) $\dot{R} = 2\,R + J, \quad \dot{J} = R + 2\,J, \quad R(0) = 0.2, J(0) = -0.1$.

1.1.2 There's No Damping Vectoria's Romantic Heart

Don't waste time trying to break a man's heart; be satisfied if you can just manage to chip it in a brand new place.
Helen Rowland, American journalist, *A Guide to Men*, "Syncopations," 1922

Vectoria, a physics major at the Metropolis Institute of Technology (MIT), is working on her computer algebra assignment while waiting for a phone call from her boyfriend Mike, who is returning from a summer job with an archaeological dig in a remote area of Asia. In particular, she is asked to use the `DEplot` command to illustrate the change in the phase-plane trajectory of the damped simple harmonic oscillator (SHO) as the damping coefficient is varied.

Entering **DEplot** in Maple's Topic Search and clicking **OK**, Vectoria finds that the `DEplot` command is contained in the DEtools library package, which she now loads.

```
>   restart: with(DEtools):
```
The SHO equation results on applying Newton's second law of motion to a unit mass acted on by a Hooke's-law restoring force, $F_{\text{hooke}} = -\omega^2\,x(t)$, and a velocity-dependent *Stokes's drag force*, $F_{\text{drag}} = -\beta\,(dx/dt)$. Here $x(t)$ is the displacement of the mass from equilibrium, ω is the frequency, and β is the

damping or drag coefficient. These two forces are now entered, square brackets being used for each assigned force to produce the relevant subscript.

>   F[hooke]:=-omega^2*x(t); F[drag]:=-beta*diff(x(t),t);

$$F_{\text{hooke}} := -\omega^2\, x(t) \qquad F_{\text{drag}} := -\beta \left(\frac{d}{dt}\, x(t) \right)$$

Equating the second time derivative of $x(t)$ to the sum of the forces generates the SHO equation, de.

>   de:=diff(x(t),t,t)=F[hooke]+F[drag];

$$de := \frac{d^2}{dt^2} x(t) = -\omega^2\, x(t) - \beta \left(\frac{d}{dt} x(t) \right)$$

Vectoria now relates the velocity $v(t)$ to the displacement in $de2$,

>   de2:=diff(x(t),t)=v(t);

$$de2 := \frac{d}{dt} x(t) = v(t)$$

and substitutes this result into de to create a coupled system ($de2$ and $de3$) of two first-order ODEs, linear in x and v.

>   de3:=subs(de2,de);

$$de3 := \frac{d}{dt} v(t) = -\omega^2\, x(t) - \beta\, v(t)$$

To plot the phase-plane trajectory, the parameter values must be specified. Vectoria takes $\omega = 1$ and three different β values, namely, $b_1 = 0$, $b_2 = 0.2$, and $b_3 = 3$.

>   omega:=1: b[1]:=0; b[2]:=0.2; b[3]:=3;

$$b_1 := 0 \qquad b_2 := 0.2 \qquad b_3 := 3$$

A "do loop" is used to create a phase-plane portrait using the DEplot command for each of the three β values. The general syntax for a do loop is

for <name> from <expression> by <expression> to <expression>
while <expression> do <statement sequence> end do

where the *<statement sequence>* is the main body (the DEplot command here) of the do loop. In the following do loop, *<name>* is the index i, the first *<expression>* is 1, the second *<expression>* is missing so i automatically increments by 1, the third *<expression>* is 3, and there is no conditional *while <expression>* present. The syntax for the DEplot command is the same as for the phaseportrait command used in the **Romeo and Juliet** recipe, the time range here being from $t = 0$ to 50.

>   for i from 1 to 3 do

>   DEplot([de2,subs(beta=b[i],de3)],[x(t),v(t)],t=0..50,
>   [[x(0)=1,v(0)=0]],stepsize=0.05,x=-1.1..1.1,v=-1.1..1.1,
>   dirgrid=[30,30],arrows=MEDIUM,color=red,linecolor=blue):

>   end do;

On execution of the do loop, the phase-plane portraits shown in Figures 1.6 to 1.8, are produced. For zero damping $(\beta = b_1 = 0)$, the phase-plane trajectory

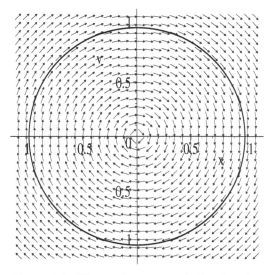

Figure 1.6: Phase-plane portrait for $\beta = 0$.

displayed in Figure 1.6 is a closed loop circling a vortex fixed point at the origin. The tangent field is also shown, the arrows pointing in the direction of increasing time.

For $\beta = 0.2$, the phase-plane trajectory shown in Figure 1.7 is a spiral that

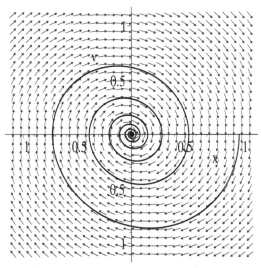

Figure 1.7: Phase-plane portrait for $\beta = 0.2$.

asymptotically approaches a stable focal point at the origin. This behavior is characteristic of *underdamping.*

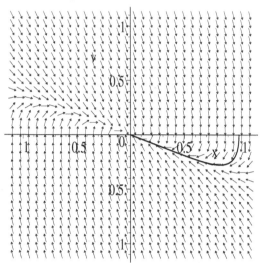

Figure 1.8: Phase-plane portrait for $\beta = 3$.

Finally, for $\beta = 3$, the phase-plane trajectory shown in Figure 1.8 approaches a stable nodal point at the origin, a behavior characteristic of *overdamping.* It is left as an exercise for you to determine the *critical damping* threshold between under- and overdamping.

Unlike the **phaseportrait** command, **DEplot** may also be used to produce solution curves, but not tangent fields, for single higher-order ODEs.

Taking, say, $\beta = b_2 = 0.2$, Vectoria now uses **DEplot** to generate the $x(t)$ solution curve shown in Figure 1.9 for the second-order ODE *de*, given the initial condition[3] $x(0) = 1$, $\dot{x}(0) = 0$. The time range is from $t = 0$ to 50, and the time step size in the underlying numerical scheme is taken to be 0.05.

```
>  beta:=b[2]:
>  DEplot(de,x(t),t=0..50,[[x(0)=1,D(x)(0)=0]],stepsize=0.05);
```

The numerically derived solution curve in Figure 1.9 decreases in amplitude in an oscillatory manner, again characteristic of the underdamped SHO.

Finally, since the SHO equation *de* is linear with constant coefficients, an analytic solution can be easily obtained for $x(t)$ using the **dsolve** command.

```
>  dsolve({de,x(0)=1,D(x)(0)=0},x(t));
```

$$x(t) = \frac{1}{33}\sqrt{11}\,e^{\left(-\frac{t}{10}\right)}\sin\left(\frac{3\sqrt{11}\,t}{10}\right) + e^{\left(-\frac{t}{10}\right)}\cos\left(\frac{3\sqrt{11}\,t}{10}\right)$$

---

[3]Note that the derivative condition $\dot{x}(0) = 0$ is entered as D(x)(0)=0, where D is the differential operator. The differential operator D is more general than diff. It can represent derivatives evaluated at a point and can differentiate *procedures.*

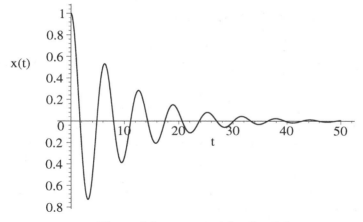

Figure 1.9: x versus t for $\beta = 0.2$.

Vectoria leaves it as a problem for you to obtain the analytic solution for the critically damped case.

Unfortunately, Vectoria must leave us for now, since her cell phone has just rung. Mike's plane has just landed and she's off to the Metropolis International Airport to pick him up after he clears immigration and customs.

PROBLEMS:

Problem 1-4: Critical damping

Given $\omega = 1$, what β value corresponds to critical damping of the SHO? Make a phase-plane portrait for this case using the DEplot command and the initial condition $x(0) = 1$, $\dot{x}(0) = 0$. Use this command to plot $x(t)$. Then obtain the analytic solution.

Problem 1-5: Competition for the same food supply

Two biological species competing for the same food supply are described by the following nonlinear population number equations:

$$\dot{N}_1 = (4 - 0.0002\,N_1 - 0.0004\,N_2)\,N_1, \quad \dot{N}_2 = (2 - 0.00015\,N_1 - 0.00005\,N_2)\,N_2.$$

(a) Locate all the stationary points.

(b) Create a tangent field plot that includes all the stationary points. Identify the nature of these points.

(c) Create a phase-plane portrait that includes several representative trajectories that support your identification of the stationary points.

(d) Use the scene option to plot $N_1(t)$ and $N_2(t)$.

(e) Attempt to obtain an analytic solution of the ODE system.

1.1.3   Van der Pol's Limit Cycle

In order to be able to set a limit to thought, we should have to find both sides of the limit thinkable (i.e., we should have to be able to think what cannot be thought).
Ludwig Wittgenstein, Austrian philosopher (1889–1951)

In the first two recipes, the ODEs were linear and, because they had constant coefficients, were easily solved analytically. In this recipe, we look at the historically important nonlinear *Van der Pol* ODE, for which no analytic solution exists. Balthasar Van der Pol was a Dutch electrical engineer who pioneered the development of experimental nonlinear dynamics in the 1920s and 1930s using electrical circuits and discovered several important nonlinear phenomena.

For example, he found that certain nonlinear circuits containing vacuum tubes could begin to spontaneously oscillate even though the energy source was constant, the oscillations evolving into a stable cycle, now called a *limit cycle*. When these circuits were driven with a signal whose frequency was near that of the limit cycle, the resulting periodic response shifted its frequency to that of the driving signal. The circuit became *entrained* to the driving signal. Entrainment is the basis of the modern pacemaker, which is used to stabilize irregular heart beats, or *arrhythmias*.

In the September 1927 issue of the journal *Nature*, Van der Pol and van der Mark reported that an "irregular noise" was heard at certain driving frequencies, probably one of the first experimental reports of *deterministic chaos*.[4]

Here, we shall look at a modern electrical circuit [Cho64] that is governed by the Van der Pol equation and can produce his limit cycle. The circuit, involving a battery (voltage V_B), inductor L, resistor R, capacitor C, and a tunnel diode D, is shown on the left of Figure 1.10. The tunnel diode has a *nonlinear* current (i_D)-voltage (V_D) curve similar to that shown on the right.

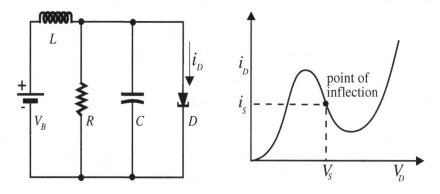

Figure 1.10: Left: tunnel diode circuit. Right: Current-voltage curve for diode.

---

[4]This historical information is taken from the IEEE History Center (www.ieee.org).

The battery voltage V_B is adjusted to coincide with the inflection point V_S of the i_D vs. V_D curve, i.e., $V_B = V_S$. Near this operating point, one may write $i = -a v + b v^3$, where $i = i_D - i_S$ and $v = V_D - V_S$, and a and b are positive.

The governing Van der Pol (VdP) ODE, which will presently be derived, is

$$\ddot{x} - \epsilon (1 - x^2) \dot{x} + x = 0, \quad \epsilon > 0, \tag{1.8}$$

with x proportional to v. Equation (1.8) is just the simple harmonic oscillator equation for unit frequency and mass with an amplitude-dependent damping term. For $x < 1$, the damping contribution is negative, so that oscillations tend to grow, while for $x > 1$ the damping is positive, tending to reduce the oscillations. The negative damping is responsible for the growth of any small spontaneous circuit "noise" into stable oscillations, i.e., into a stable limit cycle.

Let's now derive the VdP equation and demonstrate the growth of a small input signal into a limit cycle for a typical tunnel diode, 1N3719, for which $a = 0.05$ and $b = 1.0$ in SI units.

The DEtools and PDEtools packages are loaded. The former contains the DEplot3d command, which is a three-dimensional generalization of the DEplot command. The PDEtools package contains the dchange command, which will allow us to easily make a somewhat complicated variable transformation.

> restart: with(DEtools): with(PDEtools):

The time-dependent tunnel diode current and voltage expressions are entered.

> i[D]:=i[S]-a*v(t)+b*v(t)^3; V[D]:=V[S]+v(t);

$$i_D := i_S - a v(t) + b v(t)^3 \qquad V_D := V_S + v(t)$$

The voltage drop across both the resistor R and the capacitor C is the same as across the diode D, i.e., $V_R = V_D$ and $V_C = V_D$. The voltage drop across the inductor L is $V_L = V_B - V_D = V_S - V_D$, the latter form being entered.

> V[R]:=V[D]: V[C]:=V[D]: V[L]:=V[S]-V[D];

$$V_L := -v(t)$$

By *Ohm's law*, the current through the resistor is $i_R = V_R/R$. The current through the capacitor is $i_C = C\,(dV_C/dt)$.

> i[R]:=V[R]/R; i[C]:=C*diff(V[C],t);

$$i_R := \frac{V_S + v(t)}{R} \qquad i_C := C\left(\frac{d}{dt} v(t)\right)$$

Using *Kirchhoff's current rule*, eq1 states that the current leaving L must be equal to the sum of the currents entering R, C, and D.

> eq1:=-i[L](t)+i[R]+i[C]+i[D]=0; #Kirchhoff's current rule

$$eq1 := -i_L(t) + \frac{V_S + v(t)}{R} + C\left(\frac{d}{dt} v(t)\right) + i_S - a v(t) + b v(t)^3 = 0$$

Differentiating *eq1* with respect to t eliminates the inflection point current i_S.

> eq2:=expand(diff(eq1,t)/C);

$$eq2 := \frac{\frac{d}{dt} v(t)}{C R} - \frac{\frac{d}{dt} i_L(t)}{C} + \left(\frac{d^2}{dt^2} v(t)\right) - \frac{a\left(\frac{d}{dt} v(t)\right)}{C} + \frac{3 b v(t)^2 \left(\frac{d}{dt} v(t)\right)}{C} = 0$$

From the definition of inductance, one has $di_L/dt = V_L/L$, which is substituted into *eq2*. This yields a second-order ODE entirely in terms of the potential $v(t)$.

```
>   de:=subs(diff(i[L](t),t)=V[L]/L,eq2);
```

$$de := \frac{v(t)}{C\,L} + \frac{\frac{d}{dt}v(t)}{C\,R} + \left(\frac{d^2}{dt^2}v(t)\right) - \frac{a\left(\frac{d}{dt}v(t)\right)}{C} + \frac{3\,b\,v(t)^2\left(\frac{d}{dt}v(t)\right)}{C} = 0$$

Then *de* is put into more compact form by collecting the first derivatives. The resulting ODE is the *unnormalized* form of the VdP equation.

```
>   de1:=collect(de,diff(v(t),t)); #unnormalized VdP equation
```

$$de1 := \left(\frac{1}{C\,R} - \frac{a}{C} + \frac{3\,b\,v(t)^2}{C}\right)\left(\frac{d}{dt}v(t)\right) + \frac{v(t)}{C\,L} + \left(\frac{d^2}{dt^2}v(t)\right) = 0$$

To obtain the *normalized* (dimensionless) VdP equation, a transformation will be made to new variables. First a characteristic frequency $\omega = 1/\sqrt{LC}$ is introduced by making the following substitution into *de1*.

```
>   de2:=subs(v(t)/(C*L)=omega^2*v(t),de1);
```

$$de2 := \left(\frac{1}{C\,R} - \frac{a}{C} + \frac{3\,b\,v(t)^2}{C}\right)\left(\frac{d}{dt}v(t)\right) + \omega^2\,v(t) + \left(\frac{d^2}{dt^2}v(t)\right) = 0$$

Inspecting the structure of *de2*, we are led to introduce a dimensionless time $\tau = \omega\,t$, and voltage $x(\tau) = \sqrt{(3\,b)}\,v(t)/\sqrt{(a - 1/R)}$. The transformation from the "old" $(t, v(t))$ to the "new" $(\tau, x(\tau))$ variables is entered.

```
>   tr:={t=tau/omega,v(t)=x(tau)*sqrt(a-1/R)/sqrt(3*b)}:
```

The **dchange** command allows us to apply the transformation to *de2*. The result is then multiplied by the factor $\sqrt{(3\,b)}/(\omega^2\,\sqrt{(a - 1/R)})$.

```
> sqrt(3*b)*dchange(tr,de2,[x(tau),tau])/(omega^2*sqrt(a-1/R)):
```

Using the ditto operator, %, to refer[5] to the last computed result, we collect $dx(\tau)/d\tau$ terms and factor the result.

```
>   de3:=collect(%,diff(x(tau),tau),factor);
```

$$de3 : \frac{(x(\tau) - 1)\,(x(\tau) + 1)\,(a\,R - 1)\left(\frac{d}{d\tau}x(\tau)\right)}{\omega\,C\,R} + x(\tau) + \left(\frac{d^2}{d\tau^2}x(\tau)\right) = 0$$

Introducing the dimensionless parameter $\epsilon = (a\,R - 1)/(\omega\,C\,R)$, the normalized Van der Pol equation results.

```
>   vdp:=subs((a*R-1)=epsilon*(omega*C*R),de3); #VdP equation
```

$$vdp := (x(\tau) - 1)\,(x(\tau) + 1)\,\epsilon\left(\frac{d}{d\tau}x(\tau)\right) + x(\tau) + \left(\frac{d^2}{d\tau^2}x(\tau)\right) = 0$$

To make a phase-plane picture, the second-order VdP equation is now rewritten as two first-order ODEs in *de4* and *de5*, by setting $y(\tau) \equiv dx(\tau)/d\tau$.

```
>   de4:=diff(x(tau),tau)=y(tau); de5:=subs(de4,vdp);
```

---

[5] To refer to the second-to-last expression, use %%, and so on.

$$de4 := \frac{d}{d\tau}x(\tau) = y(\tau)$$

$$de5 := (x(\tau) - 1)(x(\tau) + 1)\,\epsilon\, y(\tau) + x(\tau) + \left(\frac{d}{d\tau}y(\tau)\right) = 0$$

With $a = 0.05$ entered for the tunnel diode 1N3719, a necessary condition for a limit cycle to occur is that $\epsilon > 0$ or $R > 1/a = 1/0.05 = 20$ ohms. We take $R = 55$ ohms, $L = 25.0 \times 10^{-3}$ henries, and $C = 10^{-6}$ farads, and calculate ω and ϵ.

```
> a:=0.05: R:=55: L:=25.0*10^(-3): C:=10^(-6):
> omega:=1/sqrt(L*C); epsilon:=(a*R-1)/(omega*C*R);
```

$$\omega := 6324.555320 \qquad \epsilon := 5.030896278$$

The VdP equation has a fixed point at the origin of the y vs. x phase plane. Let's choose an initial condition close to this point, viz., $x(0) = 0.1$, $y(0) = 0$.

```
> ic:=x(0)=0.1,y(0)=0:
```

Instead of plotting the trajectory in two dimensions using either the DEplot or phaseportrait commands, the solution curve corresponding to the initial condition can be drawn in the three-dimensional τ vs. x vs. y space using DEplot3d with the option scene=[tau,x,y]. The line color of the trajectory is allowed to vary with τ. The resulting trajectory appears in a 3-dimensional viewing box similar to that shown in Figure 1.11. The viewing box can be rotated on the computer screen, by clicking on the box and dragging with the

```
> DEplot3d([de4,de5],[x(tau),y(tau)],tau=0..60,
    scene=[tau,x,y],[[ic]],stepsize=0.01,linecolor=tau);
```

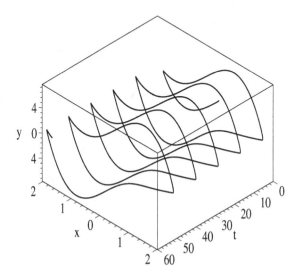

Figure 1.11: Evolution of the VdP trajectory onto a limit cycle.

mouse. The angular coordinates, θ and ϕ, of the viewing box appear in a small window near the top left of the computer screen, the default angles being $45°$, $45°$. The option orientation=[angle,angle], with the values of the two angles specified, can be inserted into DEplot3d if some other default orientation is desired. For example, choose the angles to be $\theta = 0$, $\phi = 90$ to see the phase plane.

The trajectory evolves away from the vicinity of the origin onto a closed loop, the limit cycle. You can check that the limit cycle will be obtained no matter what the choice of initial condition.[6] Can you identify the stationary point at the origin?

The nonlinear Van der Pol equation does not have an analytic solution. Applying the dsolve command to the ODE system, *de4* and *de5*, subject to the initial condition,

>   dsolve({de4,de5,ic},{x(tau),y(tau)});

produces no output for $x(t)$ and $y(t)$. Only a few nonlinear ODEs of physical interest have analytic solutions. We shall see a few examples in Chapter 4.

PROBLEMS:

Problem 1-6: Different initial conditions
With all other parameters as in the text recipe, show for a number of different initial conditions that all trajectories wind onto the limit cycle. Take the orientation that shows the y versus x phase plane.

Problem 1-7: Varying the resistance
With all other parameters as in the text recipe, investigate the behavior of the Van der Pol equation as the resistance R is varied. Choose an orientation that shows x versus τ. Discuss the results.

Problem 1-8: Tangent field
In the text recipe, use the phaseportrait command instead of DEplot3d to make a phase-plane portrait with the tangent field included.

1.2   Three-Dimensional Autonomous Systems

Although a three-dimensional plot was produced in the last example, we were still dealing with a two-dimensional autonomous system. Let's now consider a general three-dimensional autonomous ODE system of the structure

$$\dot{X} = P(X,Y,Z), \quad \dot{Y} = Q(X,Y,Z), \quad \dot{Z} = R(X,Y,Z), \tag{1.9}$$

with P, Q, and R known functions of the three dependent variables X, Y, and Z. Some systems of physical interest are naturally of this structure, while the two-dimensional nonautonomous ODE system

$$\dot{X} = P(X,Y,t), \quad \dot{Y} = Q(X,Y,t), \tag{1.10}$$

can be recast into the form (1.9) by setting $R = 1$ and imposing $Z(0) = 0$.

---

[6]If you start too far off the limit cycle, the time range may have to be increased.

1.2.1 The Period-Doubling Route to Chaos

Chaos often breeds life, when order breeds habit.
Henry Adams, American historian (1838–1918)

Nonautonomous nonlinear ODEs, such as *Duffing's equation*,

$$\ddot{x} + 2\,\gamma\,\dot{x} + \alpha\,x + \beta\,x^3 = F\cos(\omega\,t), \tag{1.11}$$

have played a very important role in the development of nonlinear dynamics. Duffing's equation is a model for the motion of a viscously damped (damping coefficient γ) spring that is subject to a nonlinear restoring force $f = -\alpha\,x - \beta\,x^3$ and is being driven by a periodic force of amplitude F and frequency ω. Depending on the signs and magnitudes of α and β, various descriptive names are usually applied to Duffing's equation:

- hard-spring Duffing equation: $\alpha > 0$, $\beta > 0$;

- soft-spring Duffing equation: $\alpha > 0$, $\beta < 0$;

- inverted Duffing equation: $\alpha < 0$, $\beta > 0$;

- nonharmonic Duffing equation: $\alpha = 0$, $\beta > 0$.

Setting $\dot{x} = y$, Duffing's equation can be written in the 2-dimensional nonautonomous form (1.10) with $P \equiv y$ and $Q \equiv -2\gamma y - \alpha x - \beta x^3 + F\cos(\omega t)$. It can be made autonomous by introducing a third dependent variable, z, and expressing Duffing's equation as the three-dimensional system

$$\dot{x} = y, \quad \dot{y} = -2\,\gamma\,y - \alpha\,x - \beta\,x^3 + F\cos(z), \quad \dot{z} = \omega, \ \text{with}\ z(0) = 0. \tag{1.12}$$

After a transient time interval, the Duffing system can, not unexpectedly, display a periodic oscillation in response to the periodic driving term. A more surprising result is that it can exhibit highly irregular, or *chaotic*, oscillatory motion that is essentially unpredictable, even though the Duffing equation is deterministic. In contrast to the periodic regime, there is an extreme sensitivity to initial conditions in the chaotic domain.

The Duffing ODE is not the only dynamical system to exhibit chaotic behavior. In general, for chaos to occur in a dynamical system, two ingredients are necessary, namely that some nonlinearity be present and that the system have at least three dynamical dependent variables (i.e., be at least three-dimensional). The study of chaotic behavior is a nonlinearly growing field, and it is not our intention to explore it in any depth in this text, although some useful diagnostic tools are briefly presented in the Desserts.

In the following recipe, Jennifer, a mathematician at MIT, will illustrate the so-called *period-doubling route to chaos* for the Duffing system. This refers to a sequence of period doublings (halving of the frequency response) that are observed when a "control" parameter is increased, ultimately ending in a chaotic regime. This period-doubling scenario is not the only route to chaos, but it is a very common one in the study of driven nonlinear ODE systems as well as other nonlinear systems that are naturally three- (or more) dimensional in nature.

For a given Duffing spring system with specified initial conditions, there are two control parameters that could be varied in the driving force, namely the frequency ω and the amplitude F. Jennifer decides to hold ω fixed, and study how the Duffing system responds as F is increased.

After loading the plots and DEtools packages, Jennifer unprotects γ,

```
>   restart: with(plots): with(DEtools): unprotect(gamma):
```
which she would like to use for the damping coefficient. Otherwise, Maple treats the entry of gamma as a request for Euler's constant. If this constant is unfamiliar to you, consult Maple's Help.

For the sake of definiteness, Jennifer considers the inverted spring system with the parameter values $\alpha = -1$, $\beta = 1$, $\gamma = 0.25$, and $\omega = 1$,

```
>   alpha:=-1: beta:=1: gamma:=0.25: omega:=1:
```
and the four force amplitudes $F_1 = 0.325$, $F_2 = 0.35$, $F_3 = 0.356$, and $F_4 = 0.42$.

```
>   F[1]:=0.325: F[2]:=0.35: F[3]:=0.356: F[4]:=0.42:
```
The F values were selected to illustrate distinctly different responses of the spring system to the driving force. To gain a preliminary understanding of what motions are possible, Jennifer decides to derive the potential energy function $V(x)$ and plot it. This may be accomplished by entering the anharmonic (nonlinear) restoring force $f = -\alpha x - \beta x^3$,

```
>   f:=-alpha*x-beta*x^3; #anharmonic restoring force
```

$$f := x - x^3$$

and performing the indefinite integral $V = -\int f \, dx$ using the int command.

```
>   V:=-int(f,x);
```

$$V := -\frac{1}{2}x^2 + \frac{1}{4}x^4$$

The potential energy V is plotted over the range $x = -1.5$ to 1.5,

```
>   plot(V,x=-1.5..1.5,tickmarks=[3,2],labels=["x","V"]);
```

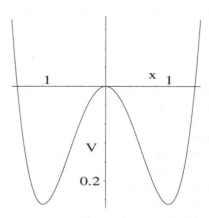

Figure 1.12: Double-well potential for an inverted-spring Duffing equation.

the resulting picture being shown in Figure 1.12. The potential is commonly referred to as the double-well potential. There are two minima, at $x = -1$ and $x = +1$, at which points $V = -\frac{1}{4}$, separated by a maximum at $x = 0$, where $V = 0$. In the absence of any driving force (set $F = 0$) or damping ($\gamma = 0$), the two minima correspond to vortex points, and the maximum is a saddle point. In this case, the spring system will oscillate in one of the two potential wells provided that the total energy is less than zero. For a total energy greater than zero, the oscillations will be back and forth between the two potential wells. These possible motions can be confirmed by making a phase-plane portrait for $\gamma = 0$ and $F = 0$ in the Duffing equation.

To make this portrait, Jennifer inserts the time-dependence of the displacement by changing the variable x to $x(t)$ in the restoring force,

```
>   f:=subs(x=x(t),f):
```
and introduces the velocity $dx/dt = y(t)$ in *eq0*.

```
>   eq0:=diff(x(t),t)=y(t):
```
The following three initial conditions are considered.

```
>   ic1:=x(0)=0.09,y(0)=0: ic2:=x(0)=-0.09,y(0)=0:
    ic3:=x(0)=-1.5,y(0)=0:
```
These conditions should produce undamped oscillatory motion in the right, left, and both potential wells, respectively. To confirm this, Jennifer applies the `phaseportrait` command to the coupled system *eq0* and $dy/dt = f$.

```
>   phaseportrait([eq0,diff(y(t),t)=f],[x(t),y(t)],t=0..100,
    [[ic1],[ic2],[ic3]],stepsize=0.1,,x=-1.5..1.5,color=red,
    linecolor=blue,arrows=MEDIUM):
```

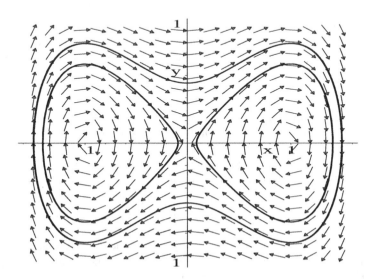

Figure 1.13: Phase portrait for inverted Duffing equation for $\gamma = 0$ and $F = 0$.

The resulting Figure 1.13 exhibits closed-loop trajectories in the phase plane characteristic of the predicted possible oscillatory motions.

The inclusion of damping and a nonzero force amplitude, on the other hand, will make the physical behavior much harder to predict. For ($\gamma \neq 0$), the two vortices will change to stable focal or nodal points, and in the absence of any energy source ($F=0$) the spring system would asymptotically ($t \to \infty$) approach one of the two minima. With the inclusion of the driving term ($F \neq 0$), which periodically pumps energy into the nonlinear system, periodic solutions are again possible but their nature much more difficult to forecast, since the periodicity of the response depends on the amplitude F chosen (for fixed ω) and the initial conditions.

To determine what happens, Jennifer now uses a do loop to construct a phase-plane graph for each of the four different F values.

```
>   for i from 1 to 4 do
```

Duffing's equation is entered for the ith force value, the result being labeled as the ith equation.

```
>   eq[i]:=diff(y(t),t)+2*gamma*y(t)-f=F[i]*cos(omega*t);
```

In the phaseportrait command, Jennifer starts the time range at $t = 100$ to eliminate any transient response of the system. The first initial condition is chosen, which corresponds to starting the system from rest in the right potential well of Figure 1.12, slightly displaced to the right of the central peak that separates the two wells. To create a phase-plane portrait for the nonautonomous case, the scene=[x,y] option is selected.

```
>   gr[i]:=phaseportrait([eq0,eq[i]],[x(t),y(t)],t=100..250,
        [[ic1]],scene=[x,y],stepsize=0.1,color=red,linecolor=blue);
>   end do:
```

The four graphs produced by this do loop are grouped together in a 2×2 array

```
>   Graphs:=array(1..2,1..2,[[gr[1],gr[2]],[gr[3],gr[4]]]):
```

and then displayed, the resulting picture being shown in Figure 1.14.

```
>   display(Graphs,tickmarks=[2,2]):
```

For the first three phase-plane portraits, corresponding to $F_1 = 0.325$, $F_2 = 0.35$, and $F_3 = 0.356$, the Duffing system executes qualitatively different periodic motions in the right potential well of Figure 1.12. Because of the mathematical uniqueness of the solutions for given initial conditions, "true" phase-plane trajectories do not cross at ordinary points of the phase plane. But here there are apparent "crossings" of the trajectories, the number increasing with increasing F. The crossings are an artifact of using a phase plane to represent the motion of the driven system. In fact, as mentioned earlier, the inverted spring system is actually a three-dimensional autonomous system. The trajectories do not cross when plotted in the three-dimensional x-y-z space.

For $F_4=0.42$, the driving-force amplitude is sufficiently large that the system clearly oscillates between both of the potential wells, its motion in the phase plane appearing to be quite chaotic.

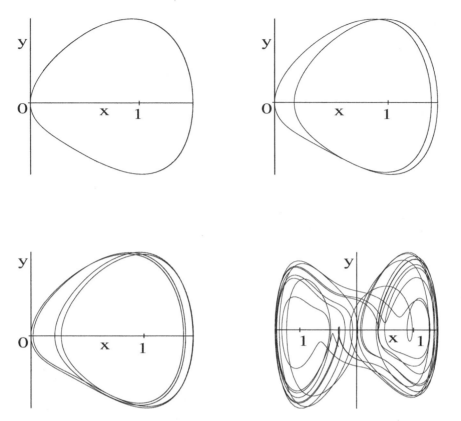

Figure 1.14: Phase-plane portraits for $F_1 = 0.325$ (top left), $F_2 = 0.35$ (top right), $F_3 = 0.356$ (bottom left), and $F_4 = 0.42$ (bottom right).

To produce a deeper understanding, Jennifer reruns the file with the scene option in the ith graph, `gr[i]`, replaced with `scene=[t,x]` and the time range shortened to $t = 100$ to 160. The four displacement $(x(t))$ curves in Figure 1.15 result, each corresponding to the matching phase-plane portrait in Figure 1.14.

For $F_1 = 0.325$, the inverted spring responds periodically at exactly the driving frequency, the period being $T = 2\pi/\omega = 6.28$. If the period is written as $T = n(2\pi/\omega)$, then $n = 1$ for this case, and the motion is referred to as a *period-one* response.

For $F_2 = 0.35$, the spring has a repeat period that is twice that of the driving term, i.e., one has $n = 2$ and therefore a *period-two* response. Notice that now the system alternates each half-cycle between different maximum values of the displacement.

For $F_3 = 0.356$, the repeat period is four times as large, corresponding to *period four*. As F was increased, the period doubled from period one to period two to period four. As F is further increased, this *period doubling* will continue until the repeat period is so large that the motion appears to be chaotic. This is what has happened for $F_4 = 0.42$.

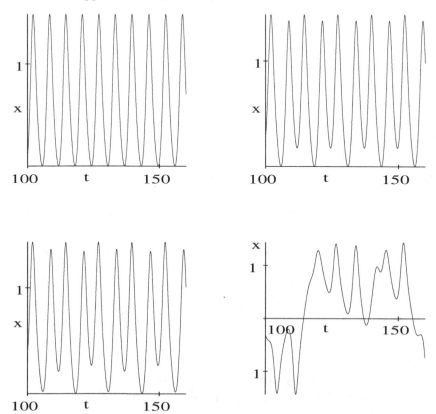

Figure 1.15: Displacement x versus time t for $F_1 = 0.325$ (top left), $F_2 = 0.35$ (top right), $F_3 = 0.356$ (bottom left), and $F_4 = 0.42$ (bottom right).

The scenario that Jennifer has outlined for the inverted Duffing equation is a commonly observed phenomenon for forced oscillator systems and is referred to in the literature as the period-doubling route to chaos.

PROBLEMS:

Problem 1-9: Unforced damped motion
In the text example, keep all equation coefficients the same (leaving $\gamma = 0.25$) but take the forcing amplitude F to be zero.

(a) If the inverted spring system is initially at rest in the right potential well with total energy $E = +\frac{1}{4}$, to what stationary point does it asymptotically evolve? Make a phase-plane portrait and a plot of x versus t.

(b) Identify the stationary point.

(c) How long does it take for the system to be within 1% of this point?

(d) If the spring system is initially at rest in the right potential well, what is the minimum value that the total energy must have so that it asymptotically approaches the stationary point in the left potential well?

Problem 1-10: Period 8
In the text example, determine an approximate F for which period 8 occurs.

Problem 1-11: A different response
In the text example, what effect does changing the driving frequency to $\omega = 2$ have on the four graphs? Identify the period response for each F value.

Problem 1-12: Varying frequency
With all other parameters the same as in the text, but with $F = 0.42$, study the response of the inverted Duffing system as ω is varied over the range between zero and one. Interpret the results in each case.

Problem 1-13: Varying the damping coefficient
With all other parameters the same as in the text, investigate the effect on the four graphs when the damping coefficient is reduced to $\gamma = 0.125$. Identify the period response for each F value. Repeat with $\gamma = 0.0625$.

Problem 1-14: Varying the force law
Execute the text recipe with the x^3 term in the force law replaced with x^5 and discuss how this change affects the results. Then try some larger F values (all other parameters remaining the same) and determine the period response of each solution.

Problem 1-15: Nonharmonic Duffing oscillator
For the nonharmonic Duffing oscillator with $\alpha = 0$, $\beta = 1$, $\gamma = 0.04$, $\omega = 1$, $F = 0.2$, $x(0) = 0.25$, and $y(0) \equiv \dot{x}(0) = 0$, determine the period response of the solution. Use both scene=[x,y] and scene=[t,x] before making your conclusion. What is the period response if $x(0) = 0.2$, all other parameter values remaining the same?

Problem 1-16: Another forced oscillator
Determine the period response of the forced-oscillator equation

$$\ddot{x} + 0.7\,\dot{x} + x^3 = 0.75\cos t,$$

subject to the initial condition $x(0) = \dot{x}(0) = 0$. Explore the change in period response of the solution as the force amplitude is varied. What type of Duffing equation is the above equation?

Problem 1-17: Three-dimensional plots
Instead of the planar plots presented in the text recipe, make use of the DEplot3d command to make three-dimensional plots in the t versus x versus y space. Plot the trajectory for each F value separately, choosing an orientation in each case that gives the best view.

1.2.2   The Oregonator

Everybody's youth is a dream, a form of chemical madness.
F. Scott Fitzgerald, American writer (1896–1940)

The Belousov–Zhabotinski (BZ) chemical reaction is now probably one of the best known of the chemical oscillators. However, because it was contrary to the then current belief that all solutions of reacting chemicals must go monotonically to equilibrium, Belousov could not initially get his chemical oscillator discovery published in any Soviet journal. Only years later, when his results were confirmed by Zhabotinski, was he given due recognition for his discovery. For his pioneering research work he was awarded, along with Zhabotinski, the Soviet Union's highest medal, but unfortunately for him, he had passed away 10 years earlier.

The BZ reaction may be achieved [Tys76] by dissolving 4.292 g of 0.28 M malonic acid[7] and 0.175 g of 0.002 M cerium ammonium nitrate in 150 ml of 1 M sulfuric acid and stirring well. The solution will initially be yellow, then turn clear after a few minutes. Then, on adding 1.415 g of 0.063 M sodium bromate, the solution will oscillate between yellow and clear with a period of about one minute. A more dramatic color change between red and blue can be achieved by adding a few ml of 0.025 M ferroin.

Field, Kőrös, and Noyes [FKN72][FN74] were able to measure the periodic oscillations in the Br^- concentration and the Ce^{4+}/Ce^{3+} ratio in the BZ reaction. They then isolated the five most important reactions in the complicated chemistry that was taking place and created a kinetic model called the *Oregonator*, the name reflecting the location of where the research was carried out.

Labeling the concentration of BrO_3^- as A, that of $HBrO_2$ as X, that of Br^- as Y, that of Ce^{4+} as Z, and all other less-important chemical species as $*$, the relevant chemical reactions in the Oregonator model are as follows:

$$
\begin{aligned}
A + Y + * & \xrightarrow{k_1} X + * \\
X + Y + * & \xrightarrow{k_2} * \\
A + X + * & \xrightarrow{k_3} 2X + 2Z + * \\
2X & \xrightarrow{k_4} A + * \\
Z + * & \xrightarrow{k_5} hY + *
\end{aligned}
\tag{1.13}
$$

Here the k_i denote the rates of reaction and h is a numerical "fudge factor" introduced because of the severe truncation of the full set of equations describing the complicated chemistry. How sensitive the results are to the value chosen for h will be left as a problem.

Ignoring the less-important chemical species and noting that the depletion of A can be neglected, i.e., A is constant, the rate equations for the production

---

[7]Belousov [Bel58] used citric acid, but malonic acid is now commonly substituted.

of X, Y, Z are

$$\dot{X} = k_1\,AY - k_2\,XY + k_3\,AX - k_4\,X^2,$$
$$\dot{Y} = -k_1\,AY - k_2\,XY + h\,k_5\,Z, \qquad (1.14)$$
$$\dot{Z} = 2\,k_3\,AX - k_5\,Z.$$

In writing down equations (1.14), use has been made of the following empirical rule: When two substances react to produce a third, the reaction rate is proportional to the product of the concentrations of the two substances. Thus, for example, the structure of the $\dot{X}$ equation can be easily understood. In the first chemical reaction, the rate of producing X is $+k_1\,AY$. In the second reaction, there is a decrease in X, the rate contribution being $-k_2\,XY$. The third reaction provides a positive contribution $+k_3\,AX$. Finally, noting that $2\,X$ in the fourth reaction is treated as $X + X$, this reaction provides a rate decrease given by $-k_4\,X^2$. The other rate equations may be similarly understood. The factor of 2 in the $\dot{Z}$ equation appears because in the third reaction, two of Z appear for each net $(2\,X - X)$ one of X.

The nonlinear rate equations (1.14) can be converted into a *normalized* form that reduces the number of parameters. Introducing a normalized time $\tau = (k_1\,A)\,t$ and concentrations

$$x = (k_2\,X)/(k_1\,A), \quad y = (k_2\,Y)/(k_3\,A), \quad z = (k_2\,k_5\,Z)/(2\,k_1\,k_3\,A^2),$$

and positive parameters,

$$\epsilon = k_1/k_3, \ \ p = (k_1\,A)/k_5, \ \ q = (k_1\,k_4)/(k_2\,k_3),$$

the Oregonator equations reduce to

$$\epsilon\,\dot{x}(\tau) = x + y - q\,x^2 - x\,y, \quad \dot{y}(\tau) = -y + 2\,h\,z - x\,y, \quad p\,\dot{z}(\tau) = x - z.$$

As an illustrative example, we will look at the onset of oscillations in the Oregonator model for $\epsilon = 0.03$, $p = 2$, $q = 0.006$, $h = 0.75$ and initial (normalized) concentrations $x(0) = 1$, $y(0) = 1$, and $z(0) = 20$.

After loading the plots package,

```
>   restart: with(plots):
```
the three governing differential equations are entered,
```
>   de1:=epsilon*diff(x(t),t)=x(t)+y(t)-q*x(t)^2-x(t)*y(t);
```

$$de1 := \epsilon\left(\frac{d}{dt}x(t)\right) = x(t) + y(t) - q\,x(t)^2 - x(t)\,y(t)$$

```
>   de2:=diff(y(t),t)=-y(t)+2*h*z(t)-x(t)*y(t);
```

$$de2 := \frac{d}{dt}y(t) = -y(t) + 2\,h\,z(t) - x(t)\,y(t)$$

```
>   de3:=p*diff(z(t),t)=x(t)-z(t);
```

$$de3 := p\left(\frac{d}{dt}z(t)\right) = x(t) - z(t)$$

along with the parameter values

>   epsilon:=0.03: p:=2: q:=0.006: h:=0.75:

and the initial condition.

>   ic:=x(0)=1,y(0)=1,z(0)=20:

The set of nonlinear ODEs cannot be solved analytically, so we seek a numerical solution for the set of three unknowns using the `dsolve` command with the `numeric` option. Unless otherwise specified, the default numerical scheme is the Runge–Kutta–Fehlberg 45 (rkf45) algorithm. See Burden and Faires. [BF89]

>   sol:=dsolve({de1,de2,de3,ic},{x(t),y(t),z(t)},numeric):

The 3-dimensional text plot command is used to place the blue-colored word "start" near the starting point of the trajectory. The three numbers in the list are the x-, y-, and z-coordinates where the word is to be placed.

>   tp:=textplot3d([1,1,21,"start"],color=blue):

The `odeplot` command enables us to plot the numerical solution over the time interval 0 to 20 time units. The minimum number of plotting points is taken to be 3000 in order to obtain a smooth curve. The default number is 50.

>   gr:=odeplot(sol,[x(t),y(t),z(t)],0..20,numpoints=3000,
        thickness=2,axes=frame,labels=["x","y","z"],
        tickmarks=[3,3,3]):

The above two plots are then displayed together, producing Figure 1.16.

>   display({tp,gr});

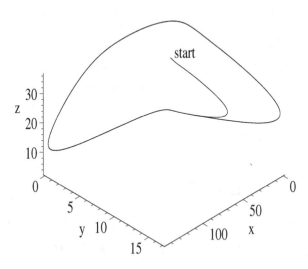

Figure 1.16: Evolution of Oregonator system onto limit cycle in phase space.

The trajectory in the 3-dimensional *phase space* evolves onto a closed loop, characteristic of an oscillatory solution. No matter what the starting point (initial condition), the trajectory will wind onto the same loop, indicating that the loop is a stable 3-dimensional limit cycle.

The `odeplot` command is used to plot the concentrations $x(t)$, $y(t)$, and $z(t)$ as red, blue, and green curves respectively, with a title indicating this included.

```
>   odeplot(sol,[[t,x(t),color=red],[t,y(t),color=blue],
    [t,z(t),color=green]],0..20,numpoints=3000,tickmarks=[3,3],
    thickness=2,title="red=x, blue=y, green=z");
```

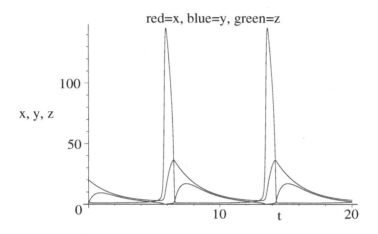

Figure 1.17: Oscillatory behavior of $HBrO_2$ (x), Br^- (y), and Ce^{4+} (z).

The black-and-white version is shown in Figure 1.17, the tallest curve being $x(t)$, the intermediate curve $z(t)$, and the shortest curve $y(t)$. Each curve reaches its maximum amplitude at a different time.

PROBLEMS: Problem 1-18: Oregonator limit cycle
Confirm that a limit cycle results in the Oregonator model, regardless of the initial (nonzero) concentrations.

Problem 1-19: Another chemical oscillator
The rate equations for a certain chemical oscillator are

$$A \xrightarrow{k_1} X$$
$$B + X \xrightarrow{k_2} Y + *$$
$$2X + Y \xrightarrow{k_3} 3X$$
$$X \xrightarrow{k_4} *$$

where the concentrations A and B of species A and B are held constant.

(a) Using the empirical rule for chemical reactions, write down the rate equations for X and Y.

(b) Convert the rate equations into a normalized form by setting $\tau = k_4 t$, $x = \sqrt{(k_3/k_4)}\, X$, $y = \sqrt{(k_3/k_4)}\, Y$, $a = \sqrt{(k_3/k_4)}\, (k_1/k_4)\, A$, $b = (k_2/k_4)\, B$.

(c) Taking $a = 1$, $b = 2.5$, $x(0) = y(0) = 0.1$, produce a 3-dimensional plot showing $x(t)$ vs. $y(t)$ vs. t. Choose an orientation that clearly shows a periodic orbit.

(d) Confirm the limit-cycle nature by trying a few different initial normalized concentrations.

Problem 1-20: Oregonator fudge factor
In the text recipe for the Oregonator model, the fudge factor was taken to be $h = 0.75$. Exploring the range $h = 0.1$ to $h = 1$, with all other conditions the same, determine whether a limit cycle occurs. Comment on the sensitivity of the model on h.

1.2.3   Rössler's Strange Attractor

Strange as it may seem, no amount of learning can cure stupidity, and formal education positively fortifies it.
Stephen Vizinczey, Hungarian novelist (1933–)

The following 3-dimensional nonlinear ODE system, due to Rössler [R76],

$$\dot{x} = -(y+z), \qquad \dot{y} = x + a\,y, \qquad \dot{z} = b + z\,(x-c), \qquad (1.15)$$

can display a variety of different trajectories in the x-y-z phase space, depending on the values assigned to the parameters a, b, and c and the initial condition. The following recipe produces one of the more interesting trajectories.

After loading the DEtools library package,

>   `restart: with(DEtools):`

the right-hand sides of the $\dot{x}$, $\dot{y}$, and $\dot{z}$ equations are entered and assigned the names P, Q, and R, respectively.

>   `P:=-(y+z); Q:=x+a*y; R:=b+z*(x-c);`

$$P := -y - z \qquad Q := x + a\,y \qquad R := b + z\,(x-c)$$

We take the parameter values to be $a = 0.2$, $b = 0.2$, and $c = 5.7$.

>   `a:=0.2: b:=0.2: c:=5.7:`

The number and locations of the fixed points are determined by solving the three equations $P = 0$, $Q = 0$, and $R = 0$ for x, y, and z. Lists are used so that the order x, y, z of the fixed-point coordinates is maintained in the output.

>   `points:=solve([P=0,Q=0,R=0],[x,y,z]);`

$$points := [\,[x = 0.007026204834,\ y = -0.03513102417,\ z = 0.03513102417],$$
$$[x = 5.692973795,\ y = -28.46486898,\ z = 28.46486898]\,]$$

There are two fixed points, one of which is near the origin. We take our initial condition to be $x(0) = 0.1$, $y(0) = 0.1$, $z = 0.1$, i.e., near this fixed point. The trajectory that results if the initial condition is near the other fixed point is left as a problem.

```
>  ic:=[x(0)=0.1,y(0)=0.1,z(0)=0.1]:
```

To enter the relevant ODEs, the dependent variables x, y, and z must be made time-dependent. This is done in the following command line.

```
>  vars:={x=x(t),y=y(t),z=z(t)}:
```

Substituting the variables into P, Q, and R, and equating to dx/dt, dy/dt, and dz/dt, yields the Rössler system of ODEs.

```
>  sys:=diff(x(t),t)=subs(vars,P),diff(y(t),t)=subs(vars,Q),
        diff(z(t),t)=subs(vars,R);
```

$$sys := \frac{d}{dt}x(t) = -y(t) - z(t), \ \frac{d}{dt}y(t) = x(t) + 0.2\,y(t),$$
$$\frac{d}{dt}z(t) = 0.2 + z(t)\,(x(t) - 5.7)$$

Choosing the option `scene=[x,y,z]` in the `DEplot3d` command, we plot the trajectory in x-y-z space over the time interval $t = 0$ to 150, subject to the given initial condition. The step size is taken to be 0.01 in order to obtain a smooth curve. The trajectory is colored with the `zhue` shading option, and a particular orientation of the viewing box is chosen.

```
>  DEplot3d([sys],[x(t),y(t),z(t)],t=0..150,[ic],scene=[x,y,z],
        stepsize=0.01,shading=zhue,orientation=[-120,60],
        tickmarks=[3,3,3],thickness=1);
```

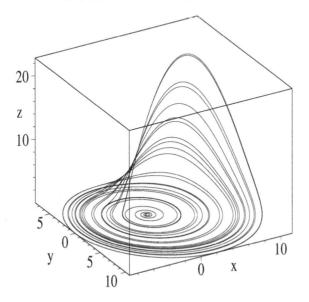

Figure 1.18: Rössler's strange attractor.

The resulting picture is shown in Figure 1.18. The trajectory unwinds in a spiral fashion from its starting point near the origin, indicating that the fixed point at the origin is an unstable focal point. As time progresses, the trajectory is attracted to a localized region of the phase space where it traces out a never-repeating (chaotic) path. This is an example of a *strange attractor*, the word strange being introduced historically because it was not like a "normal" attractor (e.g., a focal point). Strange attractors also have the property that they have noninteger, or *fractal*, dimensions. If you wish to learn more about strange attractors and fractal patterns, this topic is discussed at length in the *Introductory Guide*.

PROBLEMS: Problem 1-21: Second fixed point
Run the text recipe with an initial condition near the second fixed point. What is the probable nature of this fixed point? What is the nature of the resulting trajectory as time progresses?

Problem 1-22: Varying c
Holding all other parameters as in the text recipe, explore the behavior of the Rössler system as the coefficient c is varied. Interpret the results.

Chapter 2

Phase-Plane Analysis

The more important the subject and the closer it cuts to the bone of our hopes and needs, the more we are likely to err in establishing a framework for analysis.
Stephen Jay Gould, American paleontologist and science historian (1941–2002)

In the first chapter, the reader has seen examples of phase-plane portraits for two-dimensional autonomous ODE systems of the structure

$$\dot{x} = P(x, y), \quad \dot{y} = Q(x, y), \tag{2.1}$$

where P and Q were specified real functions. Given the mathematical forms of P and Q, the number and locations of the fixed points is easily established, either analytically or numerically. Quite generally, the topological nature of a fixed point can then be determined by examining the flow of tangent arrows in its vicinity and/or the temporal evolution of a nearby trajectory. For nonlinear systems, this was done with numerically based graphing commands. In this chapter, we will complement this approach by introducing *phase-plane analysis*, which involves analytically examining the nature of the trajectories at ordinary points lying near each fixed point. The method can be generalized [Hay64] to three-dimensional systems, but becomes considerably more complicated.

2.1 Phase-Plane Analysis

Consider an ordinary point (x, y) lying near a fixed point (x_0, y_0). From equation (2.1), the general expression for the slope of a trajectory at (x, y) is

$$dy/dx = Q(x, y)/P(x, y). \tag{2.2}$$

At a stationary point, $Q(x_0, y_0) = P(x_0, y_0) = 0$, while at ordinary points, although either Q or P may be zero (corresponding to zero slope or infinite slope), they are not zero simultaneously. For ordinary points close to a given fixed point, we can write $x = x_0 + u$, $y = y_0 + v$, where u and v are small, so that equation (2.2) becomes

$$\frac{dy}{dx} = \frac{Q(x_0 + u, y_0 + v)}{P(x_0 + u, y_0 + v)}. \tag{2.3}$$

Taylor expanding the right-hand side of (2.3) in powers of u and v yields

$$\frac{dy}{dx} = \frac{dv}{du} = \frac{c\,u + d\,v + c'\,u^2 + d'\,v^2 + f'\,u\,v + \cdots}{a\,u + b\,v + a'\,u^2 + b'\,v^2 + e'\,u\,v + \cdots}, \tag{2.4}$$

where the coefficients

$$a \equiv \left(\frac{\partial P}{\partial x}\right)_{x_0, y_0}, \quad b \equiv \left(\frac{\partial P}{\partial y}\right)_{x_0, y_0}, \quad c \equiv \left(\frac{\partial Q}{\partial x}\right)_{x_0, y_0}, \quad d \equiv \left(\frac{\partial Q}{\partial y}\right)_{x_0, y_0},$$

etc., are real since Q, P were assumed to be real.

For a linear model, only linear terms in u and v will be present in equation (2.4), the coefficients $a' \equiv (\partial^2 P/\partial x^2)_{x_0, y_0}$, $b' \equiv (\partial^2 P/\partial y^2)_{x_0, y_0}$, etc., being identically zero. On the other hand, for a nonlinear model, higher-order terms corresponding to some of these coefficients being nonzero must be present.

A *simple* stationary point for a nonlinear model is one in the neighborhood of which the qualitative behavior of the trajectories is correctly described by retaining only the linear terms in u and v in equation (2.4), so that

$$\frac{dv}{du} = \frac{c\,u + d\,v}{a\,u + b\,v}. \tag{2.5}$$

Clearly, if a, b, c, and d are nonzero and u and v are sufficiently small, then equation (2.5) should be a good approximation to equation (2.4), the higher-order terms in u and v making only small corrections that, except for the vortex,[1] do not qualitatively change the nature of the trajectories.

If, on the other hand, c and d (or a and b) both vanish, then higher-order terms should be kept in the numerator (or denominator). Even for a, b, c, and d all nonzero, one can have $a\,u + b\,v = 0$ and $c\,u + d\,v = 0$ for u and $v \neq 0$ (u, $v = 0$ corresponds to the fixed point (x_0, y_0) of interest), in which case higher-order terms should be kept in both the numerator and denominator. A nontrivial solution of $c\,u + d\,v = 0$, $a\,u + b\,v = 0$, can occur only if the system has zero determinant:

$$\begin{vmatrix} c & d \\ a & b \end{vmatrix} = b\,c - a\,d = 0. \tag{2.6}$$

If this occurs, the stationary point is no longer simple (i.e., it is not determined by linear terms in u and v alone). Since setting either c and d or a and b equal to zero also makes $b\,c - a\,d$ vanish, it follows that a simple fixed point can occur if $b\,c - a\,d \neq 0$. In the neighborhood of such a stationary point, the trajectories are described by equation (2.5), so their nature is completely determined by the four coefficients a, b, c, and d.

Next we shall establish that there are only four types of simple fixed points for the two-dimensional phase plane, namely the vortex, focal, nodal, and saddle points introduced in Chapter 1. The expression (2.5) for dv/du can be thought of as resulting from a pair of coupled first-order linear ODEs, viz.,

$$\dot{u} = a\,u + b\,v, \quad \dot{v} = c\,u + d\,v. \tag{2.7}$$

---

[1]Even the smallest corrections can change a vortex into a focal point.

Solving for v in the first equation and substituting into the second yields the second-order linear ODE

$$\ddot{u} + p\dot{u} + qu = 0, \quad \text{with } p \equiv -(a+d), \quad q \equiv ad - bc. \tag{2.8}$$

Since this ODE has constant coefficients, a solution of the form $u = e^{\lambda t}$ is sought. Substituting u into (2.8) yields the quadratic auxiliary equation

$$\lambda^2 + p\lambda + q = 0, \quad \text{with two roots, } \lambda_{1,2} = -\frac{p}{2} \pm \frac{1}{2}\sqrt{p^2 - 4q}. \tag{2.9}$$

These roots may be either real or complex. Since the general solution u is a linear combination of $e^{\lambda_1 t}$ and $e^{\lambda_2 t}$, it is clear that $u \to 0$ as $t \to \infty$ if the real parts of both λ_1 and λ_2 are negative and $u \to \infty$ if either one (or both) of the roots has a positive real part. For the former, the stationary point will be stable, while for the latter it will be unstable.

For simple fixed points, $ad - bc \equiv q \neq 0$, so a zero λ root is not possible. The case $q = 0$ corresponds to a *higher-order* stationary point, which will be illustrated later. The possible roots λ_1, λ_2, which dictate the topological nature of the fixed point, depend on the relative size and signs of p and q. First, let's consider $q > 0$ and $p \neq 0$. Three cases have to be examined:

- For $p^2 > 4q$, both the roots λ_1 and λ_2 are real and of the same sign, negative for $p > 0$ and positive for $p < 0$. For $p > 0$, the general solution for u is of the structure $u = Ae^{-|\lambda_1|t} + Be^{-|\lambda_2|t}$ (A, B are arbitrary constants), while for $p < 0$, $u = Ae^{|\lambda_1|t} + Be^{|\lambda_2|t}$. The mathematical form of these solutions is characteristic of trajectories in the neighborhood of a nodal point. Stable nodal points ($u \to 0$ as $t \to \infty$) occur for $p > 0$ and unstable nodal points for $p < 0$. The overdamped SHO, encountered earlier, is an example of a stable nodal point solution.

- For $p^2 - 4q = 0$, we have $\lambda_1 = \lambda_2 = -p/2$, so the roots are degenerate and obviously of the same sign. In this case, a second linearly independent solution $te^{\lambda t}$ must be introduced. Since this second solution is dominated by the exponential function for large t, nodal points must also occur along the parabolic curve $p^2 - 4q = 0$. An example of this case for $p > 0$ is the critically damped SHO.

- For $p^2 - 4q < 0$, the roots are $\lambda_{1,2} = -p/2 \pm (i/2)\sqrt{|4q - p^2|}$, i.e., are complex conjugate roots. Using the trigonometric identity

$$e^{\pm i\theta} = \cos(\theta) \pm i\sin(\theta),$$

the general solution for u is of the form

$$u = e^{-(p/2)t}(A\cos(\sqrt{|4q - p^2|}\, t) + B\sin(\sqrt{|4q - p^2|}\, t)).$$

These damped oscillatory solutions are characteristic of trajectories in the neighborhood of a focal point. Clearly, stable focal points occur for $p > 0$, and unstable focal points for $p < 0$. The underdamped SHO is an example of a stable focal point.

For $q > 0$, now consider the case $p = 0$. The two roots, λ_1 and λ_2, are now purely imaginary, viz., $\lambda_{1,2} = \pm i\,q$, and the general solution is of the undamped oscillatory form $u = A\cos(q\,t) + B\sin(q\,t)$. The solution is characteristic of trajectories in the vicinity of a vortex point.

Finally, we examine the situation in which $q < 0$. Independent of the value or sign of p, the roots λ_1 and λ_2 are real but of opposite signs. Because the roots are of opposite signs, the associated stationary points are always unstable. This situation corresponds to the occurrence of saddle points. Since all possible roots of λ have been examined, it follows that there are only four types of simple stationary points. The regions of p and q for which each type holds may be summarized by making a p-q *diagram* as illustrated in Figure 2.1.

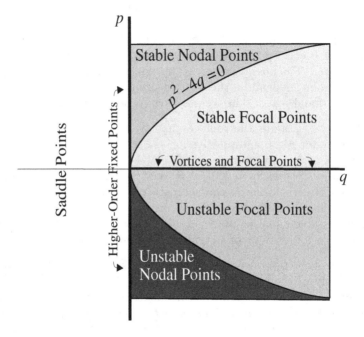

Figure 2.1: The p-q diagram for establishing types of simple stationary points.

In the figure, it should be noted that the line $p = 0$ for $q > 0$ has been labeled as vortices and focal points, rather than vortices alone. This is because the analysis for the vortices is not definitive for nonlinear models,[2] since we have kept only first-order terms in u and v in the Taylor expansion (2.4). Higher-order terms in the expansion may turn vortices into focal points. With the Taylor expansion option available in Maple, we could, of course, keep higher-order terms in an attempt to distinguish between the two types of fixed points. This can be done for individual cases, but it is difficult to make "global" statements that

---

[2]For linear ODE models, recall that higher-order terms are not present and one can have vortices only for $p = 0$ and $q > 0$.

apply to all nonlinear systems. A simple global theorem, which is left for the reader to prove, is due to Poincaré:

Suppose that for the system of equations $\dot{x} = P(x, y)$, $\dot{y} = Q(x, y)$, the functions $P(x, y)$, $Q(x, y)$ satisfy, in the neighborhood of the stationary point O, the conditions for O to be a vortex or a focus. If $P(x, y)$ and $Q(x, y)$ satisfy the conditions $P(x, -y) = -P(x, y)$, $Q(x, -y) = Q(x, y)$, then O is a vortex.

In some situations, P and Q may not satisfy the above conditions, yet O is a vortex. *Poincaré's theorem* represents a sufficient condition for the existence of a vortex, but is not a necessary condition.

2.1.1 Foxes Munch Rabbits

As for life, it is a battle ...
Marcus Aurelius Antonius, Roman emperor and philosopher (AD 121–180)

In mathematical biology there has been a great deal of interest in predator–prey systems in which certain animal species (the predator) survive by munching or crunching on one or more others (the prey). As a simple example, suppose that a species of fox survives by eating jackrabbits in the rolling hills of Rainbow County. The rabbits in turn subsist on the available vegetation, of which we shall assume there is an adequate supply. A model of this predator–prey interaction can be built up phenomenologically. Let's call $f(t)$ and $r(t)$ the fox and jackrabbit numbers per unit area (acre, hectare, or whatever) at time t.

If no foxes were present, the rabbit population would increase, the rate of increase assumed to be proportional to the number of rabbits present, i.e., $\dot{r}(t) = A_1 r(t)$, with the rate constant A_1 positive. On the other hand, if no rabbits were present, the foxes would starve to death and their numbers decrease, the rate equation being $\dot{f}(t) = -A_2 f(t)$, with $A_2 > 0$.

With both species present, the probability of an interaction will be proportional to the product $r(t) f(t)$ of the population numbers. For the foxes the interaction will be positive in nature, but negative for the rabbits. Thus, the simple phenomenological model takes the following form:

$$\dot{r} = A_1 r - B_1 r f, \quad \dot{f} = -A_2 f + B_2 r f, \tag{2.10}$$

with the interaction coefficients B_1 and B_2 positive. In practice, mathematical biologists create more realistic models with other factors taken into consideration and the coefficient values determined from observational data. Despite its simple appearance, this set of nonlinear ODEs cannot be solved analytically.

To begin the recipe, the DEtools library package is loaded because a phase-plane portrait will be constructed.

> `restart: with(DEtools):`

We enter the following numerical values: $A_1 = 2$, $A_2 = 1$, $B_1 = 3/100$, and $B_2 = 1/100$, for the coefficients. The coefficient numbers are strictly artificial, and you should feel free to experiment with different positive values.

>   A[1]:=2: A[2]:=1: B[1]:=3/100: B[2]:=1/100:

The nonlinear functions $P = A_1 r - B_1 r f$ and $Q = -A_2 f + B_2 r f$, corresponding to the right-hand sides of the ODEs in (2.10), are entered as "functional operators" using the "arrow notation."

>   P:=(r,f)->A[1]*r-B[1]*r*f;

$$P := (r, f) \rightarrow A_1 r - B_1 r f$$

>   Q:=(r,f)->-A[2]*f+B[2]*r*f;

$$Q := (r, f) \rightarrow -A_2 f + B_2 r f$$

The "arrow" (->) in the above inputs is formed on the keyboard by entering a "hyphen" followed by a "greater than" sign. The operation or "procedure" on the right-hand side of each arrow will be applied when the two[3] variables r and f on the left-hand side are supplied as arguments to P and Q. For example, let's take $r = 100$ and $f = 10$ and calculate $P(100, 10)$.

>   P(100,10); #example

$$170$$

The procedure on the rhs of P has been applied with the coefficient values automatically substituted, namely, $2 \times 100 - (3/100) \times 100 \times 10 = 170$.

To classify the stationary points, we need to evaluate a, b, c, and d, which respectively involve the partial derivatives $\partial P/\partial r$, $\partial P/\partial f$, $\partial Q/\partial r$, and $\partial Q/\partial f$ evaluated at the stationary points. A functional operator F is introduced to differentiate an arbitrary quantity $X(r, f)$ with respect to a variable v.

>   F:=(X,v)->diff(X(r,f),v):

Then F is used to calculate the four relevant partial derivatives of P and Q.

>   a:=F(P,r): b:=F(P,f): c:=F(Q,r): d:=F(Q,f):

The results have been labeled a, b, c, and d, but remember that the derivatives must still be evaluated at each stationary point. So let's locate the stationary points by solving $P(r, f) = 0$ and $Q(r, f) = 0$ for r and f.

>   sol:=solve({P(r,f)=0,Q(r,f)=0},{r,f});

$$sol := \{r = 0, \, f = 0\}, \left\{ f = \frac{200}{3}, \, r = 100 \right\}$$

There are two fixed points, one at the origin ($r = 0$, $f = 0$) and the second at $r = 100$, $f = 200/3 = 66\frac{2}{3}$. By choosing one of the fixed points and assigning the solution, the coordinates will automatically be substituted into the expressions that follow. Let's select the nonzero fixed point, which is the second solution here. As a check, the coordinates $r0$ and $f0$ of the fixed point are displayed.

>   assign(sol[2]); r0:=r; f0:=f;

$$r0 := 100 \qquad f0 := \frac{200}{3}$$

Then the values of a, b, c, and d are determined for the assigned fixed point.

---

[3] Any (finite) number of variables may be used.

```
>  a:=a; b:=b; c:=c; d:=d;
```

$$a := 0 \quad b := -3 \quad c := \frac{2}{3} \quad d := 0$$

The quantities $p = -(a + d)$ and $q = a\,d - b\,c$ are calculated.

```
>  p:=-(a+d); q:=a*d-b*c;
```

$$p := 0 \qquad q := 2$$

Referring to Figure 2.1, these values of p and q indicate that the stationary point must be either a vortex or a focal point. Instead of always referring back to the p-q picture in the text when tackling other examples, it is more convenient to create the picture directly in the code and place the (q, p) point on it. The range of q and p in the figure will be set to be from $R = -3$ to $+3$. For other stationary-point problems, this range may have to be adjusted. Further, the `display` command will be used to superimpose a number of graphs in the same figure. This command is found in the plots library package, which is now loaded.

```
>  R:=3: with(plots):
```

The first graph, assigned the name `gr1`, uses the `pointplot` command to plot the point (q, p) (entered as a list) as a size-20 blue box. The default size is 10.

```
>  gr1:=pointplot([q,p],symbol=box,symbolsize=20,color=blue):
```

The second graph, `gr2`, plots the two branches $p = \pm\sqrt{4\,q}$ of the parabola that divides the focal point and nodal point regions in the p-q diagram. Note that q is entered as `qq` (which ranges from $-R$ to R), because `q` has already been assigned a specific value. The resulting parabola is represented by a thick (the default thickness is 0) red curve on the computer screen. The minimum number of plotting points is taken to be 250, the default being 50.

```
>  gr2:=plot([sqrt(4*qq),-sqrt(4*qq)],qq=-R..R,
          numpoints=250,thickness=2,color=red):
```

The third graph, `gr3`, generates a thick green line along the $p=0$ axis between $q=0$ and $q=R$ and along the $q=0$ axis between $p=-R$ and $p=+R$.

```
>  gr3:=plot([[[0,0],[R,0]],[[0,-R],[0,R]]],style=line,
          color=green,thickness=3):
```

Using the `textplot` command, the fourth graph labels the various regions of the p-q diagram. Here `sFP` stands for stable Focal Point, `uFP` for unstable Focal Point, `sNP` for stable Nodal Point, `uNP` for unstable Nodal Point, `SP` for Saddle Point, and `V` and `FP` for Vortices and Focal Points. The two numbers in each list are determined by trial and error and indicate the horizontal and vertical locations of each string of letters.

```
>  gr4:=textplot([[1,0.9,"sFP"],[1,-0.9,"uFP"],
          [0.7,2.5,"sNP"],[0.7,-2.5,"uNP"],[-1,0.8,"SP"],
          [-1,-0.8,"SP"],[1,0.2,"V and FP "]]):
```

All four graphs are superimposed with the `display` command, the axes being labeled, and the minimum number of tickmarks along each axis specified.

```
>  display({gr1,gr2,gr3,gr4},labels=["q","p"],tickmarks=[3,3]);
```

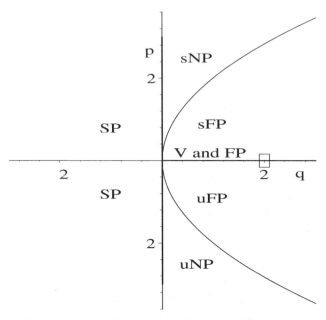

Figure 2.2: p-q diagram for the second fixed point.

The resulting p-q diagram is shown in Figure 2.2, the location of the box at $p = 0$, $q = 2$ indicating that the second stationary point is either a vortex or a focal point. As you may verify, assigning the first fixed point would move the box to the saddle-point region of the p-q diagram.

So, is the second fixed point a vortex or a focal point? Let's apply Poincaré's theorem. Forming $P(r0 + u, f0 - v) + P(r0 + u, f0 + v)$ and $Q(r0 + u, f0 - v) - Q(r0 + u, f0 + v)$ and expanding,

```
>   expand(P(r0+u,f0-v)+P(r0+u,f0+v));
```
$$0$$
```
>   expand(Q(r0+u,f0-v)-Q(r0+u,f0+v));
```
$$-\frac{u\,v}{50}$$

we see that the first relation yields zero, but the second doesn't. So Poincaré's theorem is inconclusive here. A phase-plane portrait that includes the two fixed points should settle this issue.

To proceed, the quantities r and f are unassigned from their previous values,

```
>   unassign('r','f'):
```

and the rabbits–foxes equations generated using the operators P and Q.

```
>   req:=diff(r(t),t)=P(r(t),f(t));
```
$$req := \frac{d}{dt}r(t) = 2\,r(t) - \frac{3}{100}\,r(t)\,f(t)$$

```
>  feq:=diff(f(t),t)=Q(r(t),f(t));
```

$$feq := \frac{d}{dt}f(t) = -f(t) + \frac{1}{100}r(t)f(t)$$

A phase-plane portrait operator PP is formed to generate a phase-plane portrait for the rabbits–foxes equations over the time interval $t = 0$ to 10 time units, given an initial population of 100 rabbits and 5 foxes (per unit area). The scene variables x and y and the linecolor c must be specified. Instead of the default color red, the tangent field arrows are colored blue.

```
>  PP:=(x,y,c)->phaseportrait([req,feq],[r(t),f(t)],
        t=0..10,[[r(0)=100,f(0)=5]],stepsize=0.01,scene=[x,y],
        color=blue,linecolor=c,arrows=MEDIUM,dirgrid=[25,25]):
```

The operator PP is used to plot a red-colored phase-plane trajectory in the f versus r plane, the resulting picture being shown in Figure 2.3.

```
>  PP(r,f,red);
```

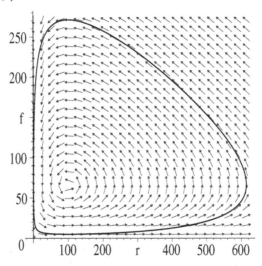

Figure 2.3: The phase-plane portrait of the rabbits–foxes interaction.

The phase-plane trajectory is a closed loop around the stationary point at $r_0 = 100$, $f_0 = 66\frac{2}{3}$, indicating that this stationary point is probably[4] a vortex, not a focal, point. The behavior of the trajectory and the appearance of the tangent field near the origin is consistent with the origin being a saddle point.

The temporal evolution of the rabbit and fox population numbers can be observed by entering PP(t,r,blue) and PP(t,f,red) and superimposing the blue and red curves using the display command. This produces Figure 2.4, the taller curve in this black-and-white rendition being for the rabbits.

---

[4]One can feel more confident about this conclusion by considering other initial conditions and taking longer time intervals.

```
>  display([PP(t,r,blue),PP(t,f,red)],labels=["t","r,f"]);
```

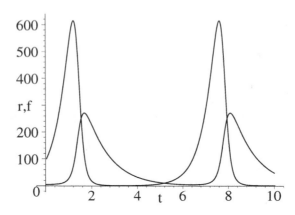

Figure 2.4: Periodic variation in the rabbits–foxes population numbers.

The periodic variation in population numbers is clearly seen, the two curves being slightly out of phase with each other as one would intuitively expect. You might argue that this cyclic result depends on the particular values chosen for the coefficients in the model. As you may confirm, choosing other positive values shifts the vortex-point location but doesn't destroy the periodicity.

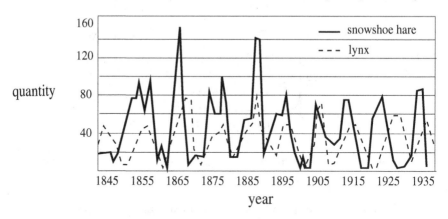

Figure 2.5: Trading records of fur catches for the Hudson's Bay Company.

Although the above predator–prey model is a phenomenological and over-simplified model of reality, the cyclic variations in population numbers that it predicts is a feature that has been observed in nature for different predator–prey interactions. For example, this can be seen in Figure 2.5, which shows the trading records for the period 1845 to 1935 of fur catches by trappers working in the Canadian north for the Hudson's Bay Company. In this case the lynx were the predators and the snowshoe hares the prey. Of course, the periodic variations observed in the lynx and snowshoe hare data curves are not as smooth and regular as in our idealized mathematical model.

PROBLEMS:

Problem 2-1: Iron core inductor

Consider the simple circuit shown in Figure 2.6, consisting of a charged capacitor C connected to a coil of N turns wrapped around an iron core. The current i

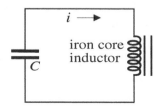

Figure 2.6: Iron core inductor circuit.

versus flux Φ relation for the iron core inductor has the form $i = N\,\Phi/L_0 + A\,\Phi^3$, where L_0 is the self-inductance of the coil, Φ is the flux threading through one turn of the coil, and $A > 0$.

(a) Using Kirchhoff's voltage rule, show that the governing ODE is given by

$$\ddot{\Phi} + \alpha\,\Phi + \beta\,\Phi^3 = 0,$$

where α and β are left for you to identify.

(b) Reexpress the ODE in a dimensionless form with α and β scaled out.

(c) Analytically show that the origin of the phase plane is a vortex. Confirm with a phase-plane ($\dot{\Phi}$ versus Φ) portrait containing a representative orbit.

(d) Use the scene option to plot $\Phi(t)$.

Problem 2-2: Competing armies

The armies of two warring countries are modeled by the following equations:

$$\dot{C}_1 = \alpha\,C_1 - \beta\,C_1\,C_2, \qquad \dot{C}_2 = (\alpha + 1)\,C_2 - \gamma\,\beta\,C_1\,C_2,$$

with α and β both positive and $\gamma > 1$. Here C_1 and C_2 are the numbers of individuals in the armies of countries 1 and 2.

(a) Discuss the model equations and how the model could be improved.

(b) Analytically locate and identify all the stationary points.

(c) Taking $\alpha = 5$, $\gamma = 1.15$, and $\beta = 1/2500$, make a tangent field plot that includes all stationary points and some representative trajectories. Discuss possible outcomes on the basis of this plot.

(d) Using appropriate scene options, create plots of $C_1(t)$ and $C_2(t)$.

Problem 2-3: Vortex or focal point?
You are told that the following system has either a vortex or a focal point at
the origin. Analytically determine which it is and support your conclusion by
creating a suitable phase-plane portrait.

$$\dot{x} = y + x\left(x^2 + y^2\right), \quad \dot{y} = -x + y\left(x^2 + y^2\right)$$

2.1.2   The Mona Lisa of Nonlinear Science

*Opinion is like a pendulum .... If it goes past the center of gravity on
one side, it must go a like distance on the other.*
Arthur Schopenhauer, German philosopher (1788–1860)

In the world of art, one of the most famous portraits ever painted is that of
a woman with an enigmatic smile, referred to as the Mona Lisa. The artist
was the Italian Leonardo da Vinci (1452–1519), who was not only a painter,
but also a sculptor, architect, musician, and scientist. If Leonardo were alive
today, he would probably appreciate on esthetic, as well as scientific, grounds
one of the most important phase-plane portraits of nonlinear science, that of
the simple plane pendulum. If we may make a puny pun, the plane pendulum
is not so plain as its name implies.

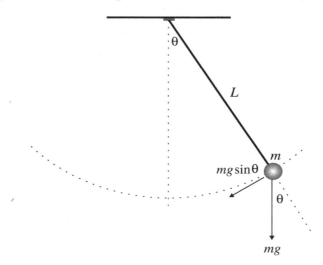

Figure 2.7: A simple plane pendulum.

What is a simple plane pendulum? It can be modeled as a small mass m
attached to the end of a very light rigid rod of length L that is allowed to
swing freely in a circular arc in the vertical plane as shown in Figure 2.7. Our

goal is to derive the equation of motion, locate and classify all of the possible stationary points, and determine all possible solutions of the simple pendulum.

The necessary packages to carry out this program are now loaded.

> restart: with(plots): with(DEtools): with(PDEtools):

The plots library package contains the display command for superimposing plots and the textplot command for placing text on a figure. The DEtools package is needed in order to use the DEplot command to produce a phase-plane portrait. Finally, the PDEtools package contains the dchange command, which will enable us to make variable changes in the equation of motion.

Since the coefficient γ will be introduced as a normalized damping coefficient, it must be unprotected from its Maple assignment as Euler's constant.

> unprotect(gamma);

If $\theta(t)$ is the angle in radians that the pendulum rod makes with the vertical at time t, g is the acceleration due to gravity, and the air resistance is assumed to be proportional to the angular velocity $d\theta/dt$ with damping coefficient Γ, Newton's second law, applied in the direction tangent to the circular arc, yields

> eq1:=m*L*diff(theta(t),t,t)=-m*g*sin(theta(t))
      -Gamma*diff(theta(t),t);

$$eq1 := m L \left(\frac{d^2}{dt^2}\theta(t) \right) = -m g \sin(\theta(t)) - \Gamma \left(\frac{d}{dt}\theta(t) \right)$$

On dividing eq1 by mg and expanding, eq2 results.

> eq2:=expand(eq1/(m*g));

$$eq2 := \frac{L \left(\frac{d^2}{dt^2}\theta(t) \right)}{g} = -\sin(\theta(t)) - \frac{\Gamma \left(\frac{d}{dt}\theta(t) \right)}{m g}$$

Noting that the radian is actually a dimensionless unit, it is clear from the first term in eq2 that $\sqrt{L/g}$ has the units of time. The reciprocal of this quantity is the characteristic frequency ω of the plane pendulum, which is entered.

> omega:=sqrt(g/L);

$$\omega := \sqrt{\frac{g}{L}}$$

A new dimensionless time variable $\tau = \omega t$ is introduced in the following transformation tr. Since the angle θ is already dimensionless, the transformation of eq2 into a completely dimensionless form is quite trivial. If Maple is used, however, the dependent variable $\theta(t)$ must be replaced by a new symbol, say, $\Theta(\tau)$, in the variable transformation. The "old" variables are placed on the left of the transformation set, the "new" variables on the right.

> tr:={t=tau/omega,theta(t)=Theta(tau)};

$$tr := \left\{ \theta(t) = \Theta(\tau), \ t = \frac{\tau}{\sqrt{\frac{g}{L}}} \right\}$$

The variable transformation of *eq2* is implemented in *eq3* with the `dchange` command. The first argument is the variable transformation *tr*, the second argument the equation to be transformed, the third argument the new variables, and the last optional argument is to simplify the result.

> eq3:=dchange(tr,eq2,[Theta(tau),tau],simplify);

$$eq3 := \frac{d^2}{d\tau^2}\Theta(\tau) = -\frac{\sin(\Theta(\tau))\, m\, g + \Gamma\left(\dfrac{d}{d\tau}\Theta(\tau)\right)\sqrt{\dfrac{g}{L}}}{m\, g}$$

The equation is further simplified by defining a dimensionless damping coefficient γ through the relation $\Gamma = 2\gamma m g/\omega$. Inclusion of the factor 2 is a matter of taste, a choice often made for damped harmonic oscillator systems. Substituting this expression for Γ into *eq3* and expanding,

> eq4:=expand(subs(Gamma=2*gamma*m*g/omega,eq3));

$$eq4 := \frac{d^2}{d\tau^2}\Theta(\tau) = -\sin(\Theta(\tau)) - 2\gamma\left(\frac{d}{d\tau}\Theta(\tau)\right)$$

yields *eq4*, the dimensionless equation of motion for the plane pendulum.

To make a phase-plane portrait, this second-order nonlinear ODE is re-expressed as two coupled first-order equations by setting the angular velocity $d\Theta/d\tau = V(\tau)$ in *eq5*, and substituting *eq5* into *eq4*.

> eq5:=diff(Theta(tau),tau)=V(tau);

$$eq5 := \frac{d}{d\tau}\Theta(\tau) = V(\tau)$$

> eq6:=subs(eq5,eq4);

$$eq6 := \frac{d}{d\tau}V(\tau) = -\sin(\Theta(\tau)) - 2\gamma V(\tau)$$

From the right-hand sides of *eq5* and *eq6*, the forms of P and Q needed to analyze the possible stationary points are easily identified and now entered.

> P:=V; Q:=-sin(Theta)-2*gamma*V;

$$P := V \qquad Q := -\sin(\Theta) - 2\gamma V$$

The derivatives $\partial P/\partial\Theta$, $\partial P/\partial V$, $\partial Q/\partial\Theta$, $\partial Q/\partial V$ are calculated in a, b, c, d.

> a:=diff(P,Theta): b:=diff(P,V): c:=diff(Q,Theta): d:=diff(Q,V):

The quantities a, b, c, and d still have to be evaluated at the stationary points, which are found by solving $P = 0$, $Q = 0$ for the unknowns Θ and V.

> sol:=solve({P,Q},{Theta,V});

$$sol := \{\Theta = 0,\ V = 0\}$$

There is a fixed point at the origin of the phase plane, corresponding to the pendulum at rest with the supporting rod oriented vertically downward ($\theta = 0$ in Figure 2.7). This is obvious on physical grounds, since the fixed point arises because there is a zero net force on the mass m in this position. The downward pull of gravity on m is balanced by the upward tension in the supporting rod.

But the net force on m would also be zero if the pendulum were oriented vertically upward ($\theta = \pi$). If the mass also has zero initial angular velocity, it will remain at rest, so the point ($\theta = \pi$, $V = 0$) must be another stationary point in the phase plane. Clearly, this second fixed point is one of unstable equilibrium, since the slightest nudge would cause the pendulum to move away from $\theta = \pi$. So, why didn't the `solve` command produce a second solution to the equations $P = 0$, $Q = 0$? The reason is that the sine function is a transcendental function, so that in order to obtain the stationary points, one must solve an inverse transcendental function. To return the entire set of stationary points, we must make use of the following command line.

```
> _EnvAllSolutions:=true:
```

On applying the `solve` command a second time to the equations $P = 0$, $Q = 0$,

```
> Sol:=solve({P,Q},{Theta,V}); assign(Sol);
```

$$Sol := \{\Theta = \pi\_Z1, V = 0\}$$

and noting that the symbol _Z1 in the output stands for integer value, the complete family of stationary points is obtained for the simple pendulum. This new solution, *Sol*, is assigned.

For _Z1 $= 0$ and _Z1 $= 1$, the fixed points ($\theta = 0, V = 0$) and ($\theta = \pi, V = 0$) result. The other stationary points corresponding to _Z1 $= 2, 3, 4, \ldots$ and _Z1 $= -1, -2, -3, \ldots$ reflect the mathematical periodicity of the sine function, which has a period 2π. Physically, for example, the angular position $\theta = 2\pi$ is the same as $\theta = 0$. In our analysis of the types of stationary points, we shall therefore concentrate on the points ($\theta = 0, V = 0$) and ($\theta = \pi, V = 0$).

Continuing, the quantities $p = -(a + d)$ and $q = ad - bc$ are evaluated,

```
> p:=-(a+d); q:=a*d-b*c;
```

$$p := 2\gamma \qquad q := (-1)^{-Z1}$$

and found to be $p = 2\gamma$, while $q = +1$ for _Z1 $= 0$ (or any even integer) and $q = -1$ for _Z1 $= 1$ (or any odd integer). Let's label the former q value as $q1$ and the latter as $q2$.

```
> q1:=1: q2:=-1:
```

Three different γ values, $\gamma_1 = 0$, $\gamma_2 = 0.1$, and $\gamma_3 = 1.25$, are entered, the numbers selected to reveal different physical behaviors of the pendulum.

```
> gamma[1]:=0: gamma[2]:=0.1: gamma[3]:=1.25:
```

In the following do loop,

```
> for i from 1 to 3 do
```

two plotting points, corresponding to the stationary points at ($\theta = 0, V = 0$) and at ($\theta = \pi, V = 0$), are calculated for each γ value,

```
> pt1:=[q1,subs(gamma=gamma[i],p)];
> pt2:=[q2,subs(gamma=gamma[i],p)];
```

and three plots formed, the fixed points represented by size-20 blue circles.

```
> pl[i]:=plot({pt1,pt2},style=point,symbol=circle,
             symbolsize=20,color=blue):
```

> end do:

Now the rest of the p-q diagram is created. The parabola $p^2 = 4q$ is plotted as a thick red line over the range $q = -2$ to $+2$. Note that the "dummy variable" r is used below instead of q as the latter has already been assigned.

```
>  pl[4]:=plot([sqrt(4*r),-sqrt(4*r)],r=-2..2,
            numpoints=250,thickness=2,color=red):
```

The segments $q=0$ to $+2$ along the q-axis, and $p=-3$ to $+3$ along the p-axis, are plotted as thick green lines.

```
>  pl[5]:=plot([[[0,0],[2,0]],[[0,-3],[0,3]]],style=line,
            color=green,thickness=3):
```

The various regions of the p-q plane are labeled in the same manner as the previous recipe.

```
>  pl[6]:=textplot([[1,0.8,"sFP"],[1,-0.8,"uFP"],
            [0.7,2.5,"sNP"],[0.7,-2.5,"uNP"],[-1,0.8,"SP"],
            [-1,-0.8,"SP"],[1.7,0.2,"V/FP"]]):
```

The six plots are superimposed to produce the entire p-q diagram in Figure 2.8.

```
> display({seq(pl[i],i=1..6},labels=["q","p"],tickmarks=[2,3]);
```

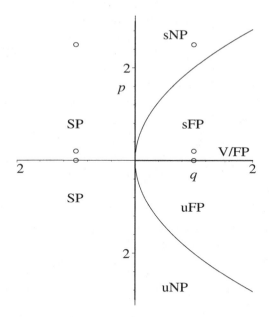

Figure 2.8: p-q diagram for a simple pendulum.

From bottom to top, the pairs of points in Figure 2.8 correspond to $\gamma = 0$, $\gamma = 0.1$, and $\gamma = 1.25$. In all three cases, the point on the left, which is associated with the fixed point ($\theta = \pi$, $V = 0$), is an unstable saddle point. Because of the mathematical periodicity of the sine function, saddle points will also occur at $(\theta = -\pi, \pm 3\pi, \ldots, V = 0)$. The fixed points on the right correspond to $(\theta = 0, V = 0)$ as well as $(\theta = \pm 2\pi, \ldots, V = 0)$. For $\gamma = 0$, the fixed point is a vortex or a focal point. Applying Poincaré's theorem, the reader will be able to confirm that it is a vortex. For $\gamma = 0.1$, the damping coefficient is sufficiently small that $(\theta = 0, V = 0)$ is a stable focal point, while for $\gamma = 1.25$ the damping is sufficiently large that a stable nodal point results.

With the stationary point analysis completed, a phase-plane portrait is now constructed for each of the γ values. First, the values of Θ and V are unassigned.

>   unassign('Theta','V'):

Two initial conditions are chosen, both corresponding to pulling the pendulum to the left of the vertical so that the initial angle is -2 radians (about $-115°$). For *ic1* the pendulum is given an initial angular velocity of 2.75 radians per second, while for *ic2* it is released from rest.

>   ic1:=Theta(0)=-2,V(0)=2.75;

$$ic1 := \Theta(0) = -2, \ V(0) = 2.75$$

>   ic2:=Theta(0)=-2,V(0)=0;

$$ic2 := \Theta(0) = -2, \ V(0) = 0$$

A functional operator de is formed to evaluate the differential equation *eq6* for the *i*th γ value.

>   de:=i->eval(eq6,gamma=gamma[i]):

Another arrow operator, G, is introduced which uses the DEplot command to produce a phase-plane portrait for the two initial conditions. Using the linecolor option, the trajectories are colored red. A tangent field of full-headed blue-colored arrows is also included. The argument $i = 1$, 2, or 3 must be provided for the operator G.

>   G:=i->DEplot([eq5,de(i)],[Theta(tau),V(tau)],tau=0..25,
          [[ic1],[ic2]],Theta=-2..8.5,V=-2..3.5,stepsize=0.05,
          color=blue,linecolor=red,arrows=MEDIUM,
          dirgrid=[30,30]):

Entering G(1) and G(2),

>   G(1); G(2);

produces Figures 2.9 and 2.10, corresponding to $\gamma = 0$ and $\gamma = 0.1$, respectively.

In Figure 2.9, the vortex fixed points at $V = 0$ and $\Theta = 0$ and 2π can be clearly seen, as well as the saddle point at $V = 0$, $\Theta = \pi$.

The closed curve encircling the vortex point at the origin is for the second initial condition (zero initial velocity). In this case, the mass has insufficient initial energy to swing over the top, i.e., past the saddle point. The direction of increasing time is indicated by the tangent-field arrows.

For the first initial condition (nonzero initial velocity), on the other hand, the initial energy is sufficiently large that the pendulum can swing over the top. Since no damping is present, the pendulum continues to move in the positive θ-direction, alternately speeding up and slowing down, but never approaching a stationary point.

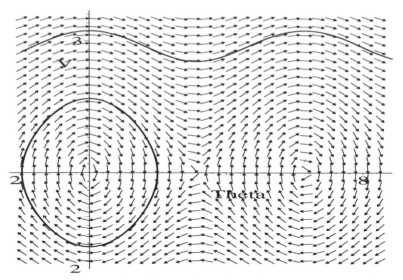

Figure 2.9: Phase-plane diagram for $\gamma = 0$.

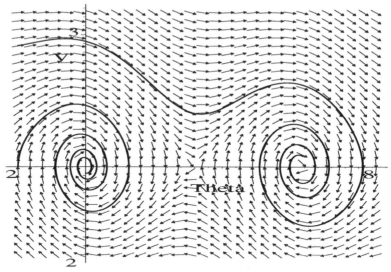

Figure 2.10: Phase-plane diagram for $\gamma = 0.1$.

Figure 2.10, corresponding to $\gamma = 0.1$, displays a completely different behavior. Due to the presence of small damping, the vortices have turned into stable focal points. For $ic2$, the pendulum initially overshoots $\theta = 0$ but eventually winds onto the origin. For $ic1$, the pendulum makes it over the top once, before winding onto the stationary point ($\theta = 2\pi$, $V = 0$). Physically, of course, this point corresponds to the vertically downward position of the pendulum rod.

The third graph, for $\gamma = 1.25$, is shown in Figure 2.11. Because the trajectories turn out to be confined to a limited region of the phase plane, we use the `display` command to better control the viewing range of `G(3)` and effectively magnify the region of interest. Care must be taken in doing this, since the number of tangent-field arrows is correspondingly reduced and their directions may not accurately reflect the directions of the trajectories.

```
>   display(G(3),view=[-2..1,-0.5..3],tickmarks=[3,3]);
```

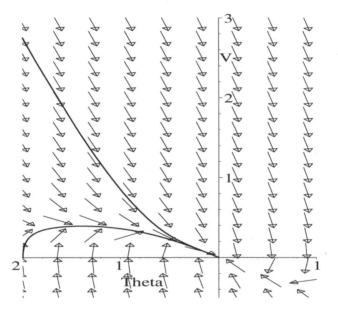

Figure 2.11: Phase-plane diagram for $\gamma = 1.25$.

The damping is now sufficiently large that both trajectories approach the stable nodal point at the origin. One would have to increase the initial angular velocity to see any over-the-top behavior before things settle down to a stable nodal point. Here, the two trajectories merge along a common path for large times.

From an artistic, as well as a scientific, viewpoint Leonardo da Vinci would undoubtedly have appreciated the many different faces revealed by the, perhaps not so plain, plane pendulum.

PROBLEMS:

Preamble: For each of the nonlinear ODE systems in the following set of problems, carry out the following steps:

(1) Use the p-q diagram approach to locate and identify the stationary points of the system, using Poincaré's theorem where necessary.

(2) In each case make a phase-plane portrait with all the fixed points included as well as the tangent field and some representative trajectories.

(3) Plot the temporal evolution of the dependent variable(s) for the above trajectories, using the appropriate scene option.

(4) Answer any additional questions posed for the system.

Problem 2-4: Hard and soft springs

For a hard (soft) spring, the displacement x from equilibrium is described by
$$\ddot{x} + \omega^2 \left(1 \pm \alpha^2 x^2\right) x = 0, \quad \text{with} \quad \omega > 0, \ \alpha > 0.$$

The plus sign is for the hard spring, the minus sign for the soft spring. Carry out the steps listed in the preamble for each spring type. Discuss the origin and nature of the restoring force leading to each equation.

Problem 2-5: Eardrum equation

The displacement x of an eardrum is described by the model equation
$$\ddot{x} + x - x^2/2 = 0.$$

Carry out the steps listed in the preamble.

Problem 2-6: Some nonlinear systems

Carry out the steps listed in the preamble for each of the following systems:

(a) $\dot{x} = x^2 - y^3, \quad \dot{y} = 2x\left(x^2 - y^2\right);$

(b) $\dot{x} = -x, \quad \dot{y} = 1 - x^2 - y^2;$

(c) $\dot{x} = x\left(1 - x^2 - 6y^2\right), \quad \dot{y} = y\left(1 - 3x^2 - 3y^2\right).$

Problem 2-7: SIR model of infectious diseases

The study of disease occurrence is called *epidemiology*. There are basically three types of deterministic models for infectious diseases that are spread by direct person-to-person contact. One of these models is referred to as the *SIR model*, the name being an acronym for the three population categories in the model. The S refers to the number of susceptibles who have not yet caught the disease, the I to the number of infectibles who have become infected with the disease, and R to the number of removables who have had the disease and are immune to catching that disease again. The disease being considered is such that very few people die from it. The SIR model equations are given by
$$\dot{S} = b - aIS, \quad \dot{I} = aIS - cI, \quad \dot{R} = cI,$$

where $b = 1$ is the constant birth rate, $a = 0.001$ is the interaction coefficient between susceptibles and infectibles, and $c = 0.1$ is the rate of increase of the removables. More generally, the model will have natural deaths in each category. The initial conditions are $S(0) = 100$, $I(0) = 1$, and $R(0) = 0$.

Carry out the steps listed in the preamble. Note: To use the symbol I, first enter the command `interface(imaginaryunit=j)`, which assigns j to stand for $\sqrt{-1}$, rather than `I`. Discuss problems with the SIR model.

2.1.3   Mike Creates a Higher-Order Fixed Point

I never learn anything talking.
I only learn things when I ask questions.
Lou Holtz, American football coach

Echoing Lou Holtz, Vectoria asks her mathematician boyfriend Mike, "Can you give me a simple physical example of a higher-order stationary point?"

"Sure," Mike replies, "I can build up a phenomenological force law model for a nonlinear spring system that will display a higher-order fixed point. To do this, let's consider a light (weightless) spring suspended from one end with a unit mass attached to the other. Since you are into using computer algebra, I will do so as well. Let me load the following Maple library packages, that I am sure we will need, onto your computer.

```
>   restart: with(DEtools): with(plots):
```
If the spring is stretched by only a small amount x from the equilibrium position, the force F required to deflect the unit mass is given by Hooke's law $F = kx$ with the spring constant k being positive. The restoring force is, of course, of the opposite sign. As you already know, the equation of motion is then just the SHO equation, which has periodic solutions. There is only a single stationary point, namely a vortex, located at the origin of the velocity–displacement phase plane. Physically, the stationary point corresponds to the situation that both the velocity and the force F (and therefore the acceleration) are equal to zero. By integrating F with respect to x, the potential energy V is given by the parabolic curve $V = \frac{1}{2} k x^2$. The stationary point is at $x = 0$, the bottom of the potential well.

If we stretch the spring even more, the displacement can be sufficiently large that higher-order terms in a Taylor expansion of $F(x)$ should be included, thus leading to a nonlinear equation of motion. If x is not too large, we need keep terms only up to third order in the Taylor expansion of $F(x)$. I will write the polynomial force law in the form $F = k x - g x^2 + h x^3$ with k, g, and h all positive.

```
>   F:=k*x-g*x^2+h*x^3; #deflecting force
```
$$F := k x - g x^2 + h x^3$$
The associated potential energy V is again easily obtained by integrating F."

```
>   V:=int(F,x);
```
$$V := \frac{1}{2} k x^2 - \frac{1}{3} g x^3 + \frac{1}{4} h x^4$$
"Hold on a moment, Mike. I don't see why you went to third order in x in the force law and why you took the quadratic term to be negative?"

"Let me answer your last question first. If all three terms in F were positive, then $F(x)$ would have only one real root at $x = 0$, which clearly still corresponds to a vortex. A higher-order stationary point is not possible in this case, no matter what the size of the coefficients. Now let's make the quadratic term negative. If the coefficient g is sufficiently large, then as x is increased from zero, F will grow linearly at small x due to the $k\,x$ term, then begin to decrease at intermediate x due to the $-g\,x^2$ term, and again increase at larger x due to the $h\,x^3$ term. Since F is a cubic polynomial, there exists the possibility of $F = 0$ having three real x roots and therefore three stationary values. Since we can still look at very small vibrations around the origin where the Hooke's law contribution predominates, there will still be a vortex at the origin of the phase plane. So this still leaves two other possible nonzero stationary points corresponding to the nonzero x roots of $F = 0$. If you substitute numbers and play around with the cubic polynomial, the other two stationary points will be simple stationary points unless the coefficients are such that the two nonzero roots coalesce into a single degenerate root. When coalescence takes place, the F curve will just be tangent to the x-axis at the location of the degenerate root. So we must impose the condition that $dF(x_0)/dx = 0$, as well as $F(x_0) = 0$, in order to adjust the coefficients to give a degenerate root x_0. It is this degenerate situation that gives rise to a higher-order fixed point, as I will now show you. We would not have obtained a degenerate root and therefore a higher-order stationary point if I had not kept the cubic term in F.

Let's calculate the derivative of the force F with respect to x,

```
>   derF:=diff(F,x);
```

$$derF := k - 2\,g\,x + 3\,h\,x^2$$

and substitute $x = x_0$ into both $F = 0$ and $derF = 0$.

```
>   eq1:=subs(x=x[0],{F=0,derF=0});
```

$$eq1 := \{k - 2\,g\,x_0 + 3\,h\,x_0^2 = 0,\ k\,x_0 - g\,x_0^2 + h\,x_0^3 = 0\}$$

Solving the set of equations contained in $eq1$ for the coefficients g and h,

```
>   sol:=solve(eq1,{g,h});
```

$$sol := \left\{ g = 2\,\frac{k}{x_0},\ h = \frac{k}{x_0^2} \right\}$$

yields $g = 2\,k/x_0$ and $h = k/x_0^2$. Assigning the solution and collecting the coefficients of k in F yields the force law necessary for coalescence of the nonzero roots to occur.

```
>   assign(sol): F:=collect(F,k);
```

$$F := \left(x - \frac{2\,x^2}{x_0} + \frac{x^3}{x_0^2} \right) k$$

In order to plot both the force and potential energy curves as a function of x, I will choose some specific values for x_0 and k. What do you suggest?"

"Oh, I think that we will get a good idea of the physical behavior if we take $x_0 = 1$ and $k = 1$."

"OK," Mike replies, "I will evaluate F and V with your suggested numbers,

```
>   F:=eval(F,{x[0]=1,k=1});
```

$$F := x - 2\,x^2 + x^3$$

> `V:=eval(V,{x[0]=1,k=1});`

$$V := \frac{1}{2}x^2 - \frac{2}{3}x^3 + \frac{1}{4}x^4$$

and now plot the force and potential energy in the same figure. To distinguish the two curves on the computer screen and in a black-and-white text rendition, I will use different colors and line styles for the two curves. Using a list format to preserve order, the following plot command will generate a dashed red curve for F and a solid blue curve for V."

> `plot([F,V],x=-0.25..1.5,color=[red,blue],`
> `linestyle=[DASH,SOLID],labels=["x","F,V"]);`

For the reader's benefit, the resulting picture observed by Mike and Vectoria on executing the `plot` command is reproduced in Figure 2.12.

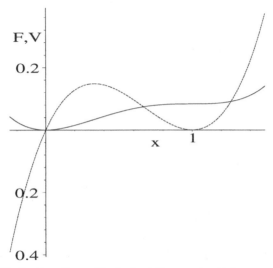

Figure 2.12: Deflecting force F, dashed line; potential energy V, solid line.

Mike continues, "At the degenerate root location, $x_0 = 1$, the (solid) potential energy curve is horizontal, corresponding to the (dashed) force curve touching zero. The equation of motion takes the form

$$\ddot{x} = -F = -x + 2x^2 - x^3, \tag{2.11}$$

so that on setting the velocity $\dot{x}$ equal to y, we can identify the functions P and Q needed for stationary point analysis.

> `P:=-F; Q:=y;`

$$P := -x + 2x^2 - x^3 \qquad Q := y$$

To evaluate a, b, c, and d, the following partial derivatives are calculated.

> `a:=diff(P,x): b:=diff(P,y): c:=diff(Q,x): d:=diff(Q,y):`

Solving the simultaneous equations $P = 0$, $Q = 0$,

> `sol:=solve({P,Q},{x,y});`

$$sol := \{x = 0, \, y = 0\}, \, \{y = 0, \, x = 1\}, \, \{y = 0, \, x = 1\}$$

yields the expected root at the origin and the twofold degenerate root ($x = 1$, $y = 0$). Since we already know that the former is a vortex, let's select the second (or third) solution and assign it for later use.

> `assign(sol[2]);`

Then, calculating p and q using the standard formulas,

> `p:=-(a+d); q:=a*d-b*c;`

$$p := -1 \qquad q := 0$$

we obtain $p = -1$ and $q = 0$. Since q is zero, the degenerate root is a higher-order stationary point as I predicted."

"OK, I understand your construction of a higher-order fixed point. But what do the trajectories look like in the phase plane, and how does the displacement $x(t)$ behave in this case?"

"To answer your questions, I first have to unassign x and y from their stationary-point values,

> `unassign('x','y');`

and then enter the associated first-order ODEs.

> `ODEs:={diff(x(t),t)=y(t),diff(y(t),t)=-x(t)+2*x(t)^2-x(t)^3};`

$$ODEs := \left\{ \frac{d}{dt} x(t) = y(t), \, \frac{d}{dt} y(t) = -x(t) + 2\,x(t)^2 - x(t)^3 \right\}$$

In order to put the stationary points in the phase-plane diagram, let's create a plot for the two points, using size-20 red circles to represent their locations.

> `gr1:=plot([[0,0],[1,0]],style=point,symbol=circle,`
>   `symbolsize=20,color=red):`

Employing the `DEplot` command, we can produce a phase-plane portrait for different initial conditions."

After some experimentation, Mike comes up with four different initial conditions, which are included in the following `DEplot` command line. He chooses blue arrows with two barbs on the head (using `arrows=MEDIUM`) for the tangent field and colors the trajectories red with the `linecolor` option. The time range is taken from $t = 0$ to 20 and the time step size equal to 0.05.

> `gr2:=DEplot(ODEs,[x(t),y(t)],t=0..20,x(t)=-1..2.5,`
>   `y(t)=-1.5..1.5,[[x(0)=-0.4,y(0)=1],[x(0)=-0.35,y(0)=0],`
>   `[x(0)=-0.3,y(0)=0],[x(0)=-0.2,y(0)=0]],stepsize=0.05,`
>   `color=blue,linecolor=red,arrows=MEDIUM,dirgrid=[20,20]):`

Putting the two graphs `gr1` and `gr2` together with the `display` command

> `display({gr1,gr2},tickmarks=[2,3]);`

and controlling the tick marks on the coordinate axes results in the phase portrait shown in Figure 2.13. The four trajectories, corresponding to the four initial conditions, are clearly seen.

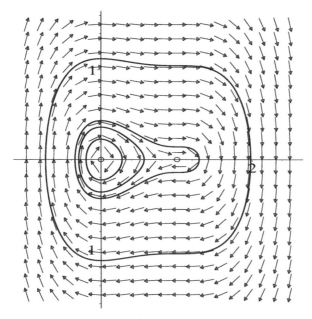

Figure 2.13: Phase-plane diagram for the nonlinear spring system.

"That's a nice plot, Mike. I can see the two stationary points and the change in shape of the trajectories from the inner one to the outer one. Close to the origin, the tangent field is characteristic of a vortex. The inner closed trajectory for the initial condition $x(0) = -0.2$ and $y = 0$ is a distorted circle cycling around this vortex point. As the loops grow larger, corresponding to the other initial conditions, the higher-order terms in the polynomial force law become more important and there is even more distortion of the closed loops.

For $x(0) = -0.35$ and $y(0) = 0$, the third-largest loop encloses both station-ary points. For the degenerate stationary point at $x = 1$, $y = 0$, the tangent field to the left of the point looks like that for a saddle point, while on the right it looks like that for a vortex. Does this type of higher-order fixed point have a name?"

"This hybrid stationary point is, not surprisingly, referred to as a *saddle-vortex* fixed point. For other examples involving higher-order stationary points, different hybrid combinations are possible, and are best studied on a case-by-case basis.

As the trajectory moves further away from both stationary points, as in the case of the outer loop in Figure 2.13, generated by the initial condition $x(0) = -0.4$ and $y(0) = 1$, the positive cubic term in the force law predominates and periodic motion about both stationary points still occurs.

You also asked about what $x(t)$ looks like. The $x(0) = -0.35$, $y = 0$ trajectory looks as if it could be interesting. To see $x(t)$, we can use the DEplot command again, but this time with the `scene=[t,x]` option."

```
>   DEplot(ODEs,[x(t),y(t)],t=0..50,[[x(0)=-0.35,y(0)=0]],scene
    =[t,x],stepsize=.05,color=blue,linecolor=red,axes=normal);
```

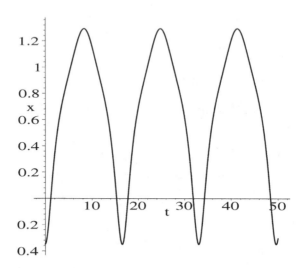

Figure 2.14: Displacement of the nonlinear spring as a function of time.

Again, we have reproduced what Mike and Vectoria see on the computer screen in Figure 2.14.

Mike continues, "Two features displayed by the $x(t)$ curve leap out at me, Vectoria. The shape is not sinusoidal and it is highly asymmetric about the origin. The deviation away from sinusoidal behavior is a signal that nonlinear terms are present in the force law. The asymmetry arises specifically from the quadratic term, which doesn't reverse sign as the spring system passes through the origin."

"That was great, Mike. I have learned a lot from this example. It's Friday and I feel like wrapping the week up by going to that little Greek restaurant on West 4th for supper. How about you?"

"With a glass or two of Tsantali to drink, that sounds good. Save the file, shut down your computer, and let's be off."

PROBLEMS:

Problem 2-8: Verhulst predator–prey equations

The population densities of prey (variable x) and predators (variable y) are governed by the following nonlinear ODEs:

$$\dot{x} = x - A x^2 - B x y, \quad \dot{y} = y - C y^2 + D x y,$$

with the coefficients A, B, C, and D all positive.

(a) What do the terms involving A and C represent physically?

(b) Show that the nature of the stationary points depends on whether $C > B$, $C = B$, or $C < B$.

(c) For $C > B$ and $C < B$, show that in each case four simple stationary points occur, and locate and identify them.

(d) Choosing appropriate numerical values for the parameters, make phase-plane portraits for each case in part (c), superimposing representative trajectories on the tangent field. Represent the stationary points in the plot by colored circles.

(e) For $C = B$, show that two of the four stationary points are higher-order fixed points. What are the other two stationary points?

(f) Make phase-plane portraits for part (e), superimposing representative trajectories on the tangent field. Represent the stationary points in the plot by colored circles.

Problem 2-9: Variation on the Verhulst problem
Suppose that in the preceding problem $D = -B$, so that the interaction is disadvantageous to both species. Taking $A = B = C = 1$, find the stationary points of the new system and identify them. Confirm the stationary-point analysis by producing appropriate phase-plane portraits.

2.1.4 The Gnus and Sung of Erehwon

These are the voyages of the starship Enterprise. Its five-year mission ... to boldly go where no man has gone before.
Gene Roddenberry, *Star Trek* television series, 1966

On the terraformed[5] planet of Erehwon, there coexist two related species of animals, the gnus and their somewhat backward relatives the sung. If the density of gnus per unit area at time t is $g(t)$ and the density of sung is $s(t)$, the temporal evolution of the densities is given by[6]

$$\dot{g}(t) = a_g \, g(t) - b_g \, g(t) \, s(t) - c_g \, g(t)^6 \, s(t)^7,$$

$$\dot{s}(t) = -a_s \, s(t) + b_s \, s(t) \, g(t) + c_s \, g(t)^5 \, e^{-s(t)},$$

$$(2.12)$$

with $a_g = 2$, $a_s = 1$, $b_g = 0.03$, $b_s = 0.01$, $c_g = 0.1$, and $c_s = 0.1$. The question facing the Gnus–Sung Preservation Society is, what is the long-term prognosis for the two animal populations given that the current population densities are

---

[5]This word, which is commonly used in science fiction novels, refers to the attempt by colonizers from Earth to alter a planet's atmosphere and other physical characteristics to be more Earth-like.

[6]Reference: page 454, *Encyclopedia Erehwonia*, 44th edition, Springer Intergalactic, Inc.

$g(0) = 0.5$ and $s(0) = 1$? Can the gnus and sung get along sufficiently well that both groups will survive, or is the interaction such that only one will survive?

Although this competition problem is similar to that for the rabbits and foxes, it is complicated by the appearance of higher-order polynomial and transcendental terms on the right-hand sides of the evolution equations. This makes locating the stationary points more of a challenge than in the earlier examples.

To answer the question, the society has called on the services of the pre-eminent scientist and conservationist Dr. Eiram Eiruc. Dr. Eiruc begins her computer analysis by loading the plots and DEtools packages,

```
>   restart: with(plots): with(DEtools):
```
as well as the given coefficient values.

```
>   a[g]:=2: a[s]:=1: b[g]:=0.03: b[s]:=0.01: c[g]:=0.1: c[s]:=0.1:
```
From equations (2.12), the functions P and Q are identified and entered.

```
>   P:=a[g]*g-b[g]*g*s-c[g]*g^6*s^7;
```

$$P := 2\,g - 0.03\,g\,s - 0.1\,g^6\,s^7$$

```
>   Q:=-a[s]*s+b[s]*g*s+c[s]*g^5*exp(-s);
```

$$Q := -s + 0.01\,g\,s + 0.1\,g^5\,e^{(-s)}$$

The stationary points can be obtained by setting $P = Q = 0$. However, the simultaneous solution of these equations is nontrivial because of their complexity. Before carrying out a numerical search for the roots, Eiram uses the `implicitplot` command with boxed axes to graph the functions $P = 0, Q = 0$. The intersection points will then correspond to the stationary points. The ranges of g and s are determined by trial and error.

```
>   implicitplot({Q=0,P=0},g=-6..6,s=-10..10,grid=[60,60],
    numpoints=5000,tickmarks=[4,2],axes=box);
```

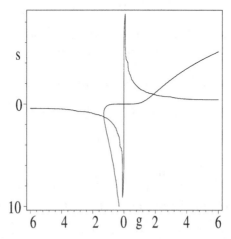

Figure 2.15: Graphically solving $P = 0, Q = 0$ to locate stationary points.

The result is shown in Figure 2.15. Since the population densities cannot be negative, the only stationary points of interest to Eiram are the one that appears to be at the origin and the one in the $g > 0$, $s > 0$ quadrant. Eiram could, of course, obtain approximate values by clicking the mouse arrow on the intersection points in the computer screen plot. More accurate numbers are found by using the floating point solve command. However, since the equations $P = 0$ and $Q = 0$ are quite nonlinear, the fsolve command will yield only one real root if no range is specified for the variables. When more than one root is present, a range must be given that includes the root of interest.

For example, by taking $g = -1$ to 1 in the fsolve command,

```
>   sol[1]:=fsolve({P,Q},{g,s},g=-1..1);
```
$$sol_1 := \{g = 0., \ s = 0.\}$$
Eiram confirms that there is indeed a stationary point at the origin. Referring to Figure 2.15, she then chooses the range $g = 1$ to 3,

```
>   sol[2]:=fsolve({P,Q},{g,s},g=1..3);
```
$$sol_2 := \{g = 1.902803669, \ s = 0.9669140476\}$$
and finds that the second stationary point is at $g \approx 1.90$ gnus and $s \approx 0.97$ sung (per unit area).

The relevant partial derivatives of P and Q for identifying the stationary points are calculated.

```
>   a:=diff(P,g): b:=diff(P,s): c:=diff(Q,g): d:=diff(Q,s):
```
A do loop is formed that will enable Eiram to carry out the identification process for each of the two relevant stationary points.

```
>   for i from 1 to 2 do
```
The "coordinates" of the ith stationary point are assigned.

```
>   assign(sol[i]):
```
The quantities $p = -(a + d)$, $q = ad - bc$, and $r = p^2 - 4q$ are calculated for the ith stationary point.

```
>   p[i]:=-(a+d); q[i]:=a*d-b*c; r[i]:=p[i]^2-4*q[i];
```
Then g and s are unassigned,

```
>   unassign('g','s');
```
and the do loop ended,

```
>   end do;
```
yielding the following results:
$$p_1 := -1. \qquad q_1 := -2. \qquad r_1 := 9.$$
$$p_2 := 11.78445045 \qquad q_2 := 87.09248243 \qquad r_2 := -209.4966573$$
Having the Springer Intergalactic edition of this text, Eiram refers to the p-q diagram given in Figure 2.1. Noting that $q_1 < 0$, she concludes that the stationary point at the origin is a saddle point. For the second nonzero stationary point, $p_2 > 0$ and $q_2 > 0$, so this fixed point is either a stable focal point or a stable nodal point. But $r_2 < 0$, so it is actually a stable focal point.

Eiram will confirm this analysis by making a phase-plane portrait. She forms the gnus and sung ODEs by inserting the g and s time-dependence in the P, Q functions and entering $dg(t)/dt = P(g(t), s(t))$ and $ds(t)/dt = Q(g(t), s(t))$.

```
> gnus:=diff(g(t),t)=subs({g=g(t),s=s(t)},P);
```

$$gnus := \frac{d}{dt} g(t) = 2\, g(t) - 0.03\, g(t)\, s(t) - 0.1\, g(t)^6\, s(t)^7$$

```
> sung:=diff(s(t),t)=subs({g=g(t),s=s(t)},Q);
```

$$sung := \frac{d}{dt} s(t) = -s(t) + 0.01\, g(t)\, s(t) + 0.1\, g(t)^5\, e^{(-s(t))}$$

A functional operator F is formed to produce a phase-plane portrait for the gnus and sung, given that initially there were 0.5 gnus and 1 sung (per unit area), and looking at a time span of 10 years. The scene variables x and y, and the line color c, must be supplied as arguments.

```
> F:=(x,y,c)->phaseportrait([gnus,sung],[g(t),s(t)],t=0..10,
      [[g(0)=0.5,s(0)=1]],scene=[x,y],g=0..3,s=0..2,stepsize
      =0.01,color=blue,linecolor=c,arrows=MEDIUM,
      dirgrid=[20,20]):
```

Then, entering F(g,s,red) produces the phase-plane portrait shown in Figure 2.16, the tangent arrows being colored blue and the trajectory red on the computer screen.

```
> F(g,s,red);
```

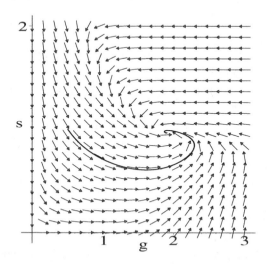

Figure 2.16: Phase-plane portrait for the gnus–sung interaction.

As expected, the trajectory winds onto the stable focal point. The tangent arrows near the origin are indicative of a saddle point.

Next, Eiram will show the time evolution of the two animal populations. First she uses the `textplot` command to create name labels to be added to the final figure.

```
>   tp:=textplot([[1.2,1.1,"sung"],[1.2,2,"gnus"]]):
```

She then uses the `display` command and the functional operator `F` to produce Figure 2.17, showing the temporal evolution of the gnus and sung populations for a period of two years. On the computer screen the gnus' population density is colored red, the sung's population density blue.

```
>   display({F(t,g,red),F(t,s,blue),tp},labels=["t",""],
    tickmarks=[3,3],view=[0..2,0..3]);
```

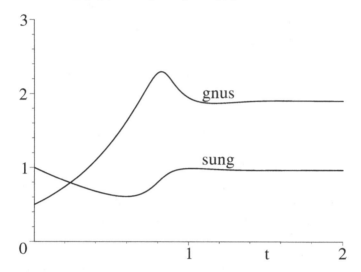

Figure 2.17: Temporal evolution of gnus and sung.

So Eiram concludes that if all conditions remain the same, then the two groups will ultimately live in relative harmony with each other, since both gnus and sung survive. However, she notes that the gnus will gain the upper hand in terms of population density over their more backward relatives, even though initially the density of sung was twice that of the gnus.

PROBLEMS:

Problem 2-10: Create your own model

Create your own complicated model of the gnus–sung interaction by modifying the last term in each of the equations. Follow the procedure in the text recipe and determine the fate of the two populations in your model. Feel free to experiment with parameter values and initial conditions.

2.1.5   A Plethora of Points

The sure conviction that we could if we wanted to
is the reason so many good minds are idle.
G. C. Lichtenberg, German physicist, philosopher (1742–1799)

We have asked Jennifer, the MIT mathematician, to provide us with a recipe
that locates and identifies all the fixed points of the nonlinear ODE system

$$\dot{x} = x\,(1 - x^2 - 6\,y^2), \quad \dot{y} = y\,(1 - 3\,x^2 - 3\,y^2),$$

making use of conditional logic statements, and to support the identification
with a tangent-field plot containing all the stationary points. Here is her recipe.
    After loading the required library packages,

>   `restart: with(plots): with(DEtools):`

Jennifer identifies the rhs of the $\dot{x}$ and $\dot{y}$ equations as the functions P and Q
needed for the fixed-point analysis. She then enters P and Q.

>   `P:=x*(1-x^2-6*y^2); Q:=y*(1-3*x^2-3*y^2);`

$$P := x\,(1 - x^2 - 6\,y^2) \quad Q := y\,(1 - 3\,x^2 - 3\,y^2)$$

Jennifer introduces a functional operator f for differentiating a given function
X with respect to an arbitrary variable v.

>   `f:=(X,v)->diff(X,v):`

Making use of the functional operator f, the partial derivatives $a = \partial P/\partial x$,
$b = \partial P/\partial y$, $c = \partial Q/\partial x$, and $d = \partial Q/\partial y$ are calculated.

>   `a:=f(P,x): b:=f(P,y): c:=f(Q,x): d:=f(Q,y):`

The partial derivatives must be evaluated at each fixed point. To locate these
points, the equations $P = 0$ and $Q = 0$ are solved for x and y. Lists are used
here, because the order of x and y in *sol* must be preserved for later use.

>   `sol:=solve([P=0,Q=0],[x,y]);`

$$sol := [\,[x = 0,\, y = 0],\, [x = 0,\, y = \text{RootOf}(-1 + 3\_Z^2,\, label = \_L1)],$$
$$[x = 1,\, y = 0],\, [x = -1,\, y = 0],$$
$$[x = \text{RootOf}(5\_Z^2 - 1),\, y = \text{RootOf}(-2 + 15\_Z^2)]\,]$$

In the solution, RootOf( ) is a placeholder for the roots of the polynomial
function given in each included argument. Jennifer wants to extract all the
roots explicitly. To this end, she first determines the number of operands (here
the number of lists) in *sol* using the nops command.

>   `N:=nops(sol);`

$$N := 5$$

The number of operands is 5, which is easily confirmed by visual inspection of
sol. The roots can be explicitly obtained by applying the allvalues command
to the *i*th entry in *sol*, and using the seq command (running from $i = 1$ to N)
to generate the sequence of roots. The results are again put into a list, so the
number of operands in the new list of lists can be extracted.

```
> sol2:=[seq(allvalues(sol[i]),i=1..N)];
```

$$sol2 := \left[\; [x = 0, \, y = 0], \; \left[x = 0, \, y = \frac{\sqrt{3}}{3} \right], \; \left[x = 0, \, y = -\frac{\sqrt{3}}{3} \right], \right.$$

$$[x = 1, \, y = 0], \; [x = -1, \, y = 0], \; \left[x = \frac{\sqrt{5}}{5}, \, y = \frac{\sqrt{30}}{15} \right],$$

$$\left. \left[x = \frac{\sqrt{5}}{5}, \, y = -\frac{\sqrt{30}}{15} \right], \; \left[x = -\frac{\sqrt{5}}{5}, \, y = \frac{\sqrt{30}}{15} \right], \; \left[x = -\frac{\sqrt{5}}{5}, \, y = -\frac{\sqrt{30}}{15} \right] \; \right]$$

In the *sol2* list of lists, there are clearly nine lists of x and y values, corresponding to the coordinates of nine stationary points. There is indeed a plethora of fixed points in this example. The number of operands (lists here) will be needed, so this number is extracted with the **nops** command.

```
> N2:=nops(sol2);
```

$$N2 := 9$$

Functional operators are formed to numerically evaluate $p = -(a + d)$ and $q = a \, d - b \, c$ for the ith entry in *sol2*.

```
> p:=i->evalf(eval(-(a+d),sol2[i])):
```

```
> q:=i->evalf(eval(a*d-b*c,sol2[i])):
```

Next, Jennifer creates a do loop, running from $i = 1$ to $N2 = 9$, to identify the nature of each fixed point and to plot each point.

```
> for i from 1 to N2 do
```

The ith entry (list of x and y coordinates) of *sol2* is entered, and then the x and y coordinates are extracted separately. For example, taking $i = 1$, sol2[1,1] extracts the first entry ($x = 0$) in the first list of *sol2*, while sol2[1,2] produces the second entry ($y = 0$). The right-hand sides are extracted to give the x and y coordinates of the fixed point for plotting purposes.

```
> sol2[i]; X[i]:=rhs(sol2[i,1]); Y[i]:=rhs(sol2[i,2]);
```

The values of p, q, and $r = p^2 - 4 q$ are determined for the ith fixed point. Jennifer uses the concatenation operator || to attach numbers to these quantities. For example, the outputs of p||1, p||2, etc., would be of the form $p1$, $p2$, etc. Concatenation is a useful alternative way of numbering quantities.

```
> p||i:=p(i); q||i:=q(i); r||i:=simplify(p||i^2-4*q||i);
```

A conditional statement is now entered that will classify the ith fixed point according to the region of the p-q diagram in which it is located. The general syntax for a conditional statement is

if <*conditional expression*> *then* <*statement sequence*>
elif <*conditional expression*> *then* <*statement sequence*>
else <*statement sequence*>
end if

where the *elif* (else if) and *else* phrases are optional.

```
>  if q||i<0 then s||i:=saddle;
   elif q||i>0 and p||i>0 and r||i<0 then s||i:=stablefocal;
   elif q||i>0 and p||i>0 and r||i>=0 then s||i:=stablenodal;
   elif q||i>0 and p||i<0 and r||i<0 then s||i:=unstablefocal;
   elif q||i>0 and p||i<0 and r||i>=0 then s||i:=unstablenodal;
   elif q||i>0 and p||i=0 then s||i:=vortexorfocal;
   else s||i:=higherorder;
   end if:
```

The ith stationary point is plotted as a size-20 blue circle,

```
>  gr[i]:=pointplot([[X[i],Y[i]]],symbol=circle,
             symbolsize=20,color=blue);
```

and the conditional statement ended, a colon being used to suppress the output.

```
>  end do:
```

Using the sequence command, the x and y fixed-point coordinates and fixed-point type are now displayed for all nine fixed points. For example, the fixed point at $x = 0$, $y = 0$, is an unstable nodal point.

```
>  seq([X[i],Y[i],s||i],i=1..N2);
```

$$[0,\, 0,\, unstablenodal],\ \left[0,\, \frac{\sqrt{3}}{3},\, stablenodal\right],\ \left[0,\, -\frac{\sqrt{3}}{3},\, stablenodal\right],$$

$$[1,\, 0,\, stablenodal]\,,\, [-1,\, 0,\, stablenodal],\ \left[\frac{\sqrt{5}}{5},\, \frac{\sqrt{30}}{15},\, saddle\right],$$

$$\left[\frac{\sqrt{5}}{5},\, -\frac{\sqrt{30}}{15},\, saddle\right],\, \left[-\frac{\sqrt{5}}{5},\, \frac{\sqrt{30}}{15},\, saddle\right],\, \left[-\frac{\sqrt{5}}{5},\, -\frac{\sqrt{30}}{15},\, saddle\right]$$

The two relevant ODEs are entered, with the time dependence of x and y being substituted into P and Q.

```
>  xeq:=diff(x(t),t)=subs({x=x(t),y=y(t)},P);
```

$$xeq := \frac{d}{dt}x(t) = x(t)\left(1 - x(t)^2 - 6\,y(t)^2\right)$$

```
>  yeq:=diff(y(t),t)=subs({x=x(t),y=y(t)},Q);
```

$$yeq := \frac{d}{dt}y(t) = y(t)\left(1 - 3\,x(t)^2 - 3\,y(t)^2\right)$$

The **dfieldplot** command is used to plot the tangent field for this ODE system, the resulting graph being assigned the name **gr[N2+1]**.

```
>  gr[N2+1]:=dfieldplot([xeq,yeq],[x(t),y(t)],t=0..10,
        x=-1.2..1.2,y=-1.2..1.2,dirgrid=[26,26],arrows=MEDIUM):
```

The above graph assignment allows Jennifer to use the sequence command to plot all the stationary points and the tangent field in the same figure.

```
>  display({seq(gr[i],i=1..N2+1)},tickmarks=[3,3]);
```

The resulting picture is shown in Figure 2.18. The plot confirms the identification of the fixed points, e.g., the unstable nodal point at the origin.

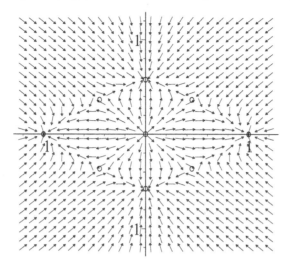

Figure 2.18: A plethora of fixed points.

PROBLEMS:

Problem 2-11: Lots of fixed points

Consider the coupled ODE system

$$\dot{x} = y\,(1 + x - y^2), \quad \dot{y} = x\,(1 + y - x^2).$$

Using conditional logic statements, locate and identify all the fixed points. Make a tangent-field plot containing all these stationary points.

Problem 2-12: Not so many fixed points

Locate and identify all the fixed points of the ODE system

$$\dot{x} = 16\,x^2 + 9\,y^2 - 25, \quad \dot{y} = 16\,x^2 - 16\,y^2,$$

making use of conditional logic statements. Support the identification with a tangent-field plot containing all the stationary points.

Problem 2-13: Squid and herring

The major food source for squid is herring. If S and H are the numbers of squid and herring, respectively, per acre of seabed, the interaction between the two species can be modeled [Sco87] by the system (with time in years)

$$\dot{H} = k_1\,H - k_2\,H^2 - k_3\,H\,S, \quad \dot{S} = -k_4\,S - k_5\,S^2 + b\,k_3\,H\,S,$$

with $k_1 = 1.1$, $k_2 = 10^{-5}$, $k_3 = 10^{-3}$, $k_4 = 0.9$, $k_5 = 10^{-4}$, and $b = 0.02$.

(a) Using conditional logic statements, locate and identify all the stationary points of the squid–herring system.

(b) Make a phase-plane portrait that shows all the stationary points and includes some representative trajectories. Discuss possible outcomes for different ranges of initial populations.

(c) Suppose that every last squid were removed from the area occupied by the herring and from all surrounding areas. Would the herring population increase indefinitely without bound or would there be an upper limit on the number of herring per unit area? If you believe the latter would occur, what is that number?

(d) If the squid-free situation just described had persisted for many years, how many squid would there be two years later if a pair of fertile squid were introduced into the area?

2.2   Three-Dimensional Autonomous Systems

The analysis of the general autonomous 3-dimensionsal system,

$$\dot{X} = P(X,Y,Z), \quad \dot{Y} = Q(X,Y,Z), \quad \dot{Z} = R(X,Y,Z), \qquad (2.13)$$

where P, Q, and R may be nonlinear functions of the dependent variables X, Y, and Z, can be tackled in a manner similar to the 2-dimensional case. However, the analytic identification of the fixed points is nontrivial, and we refer the interested reader to the texts by Jackson [Jac90] and Hayashi [Hay64].

We shall be content to look at one specific example, locating the fixed points, establishing their stability, and using a graphical approach to identify their nature. What better example is there than Lorenz's famous chaotic "butterfly"?

2.2.1   Lorenz's Butterfly

The butterfly's attractiveness derives not only from colors
and symmetry: deeper motives contribute to it.
Primo Levi, Italian chemist, author, *Other Peoples Trades, "Butterflies,"* 1985

In a famous paper, Edward Lorenz [Lor63] discussed the possibility of very-long-range weather forecasting. Starting with the nonlinear Navier–Stokes PDEs of fluid mechanics, Lorenz attempted to model thermally driven convection in the earth's atmosphere. The atmosphere was treated as a flat fluid layer that is heated from below by the surface of the earth and cooled from above by heat radiation into outer space. Lorenz managed to reduce the original set of PDEs to the following set of three nonlinear ODEs:

$$\dot{x} = \sigma\,(y - x), \quad \dot{y} = r\,x - y - x\,z, \quad \dot{z} = x\,y - b\,z, \qquad (2.14)$$

where x is proportional to the convective velocity, y to the temperature difference between ascending and descending flows, and z to the mean convective heat flow. The positive coefficients σ and r are the Prandtl and reduced Rayleigh numbers, respectively, and $b > 0$ is related to the wave number.

The Lorenz equations (2.14) must be solved numerically. On doing so, Lorenz discovered that a very small change in the initial condition could lead to dramatically different long-term behavior of the numerical solutions. This could have implications for very-long-range weather forecasting, because present conditions are never known exactly due to the inevitable inaccuracy and incompleteness of weather observations. One might argue that this conclusion was model dependent, particularly because the original PDEs were so severely simplified, but in fact Lorenz found that this sensitivity to initial conditions was a general feature of nonlinear systems.

In addition to the sensitivity feature, Lorenz's ODE system is well known because its trajectory in *x-y-z* phase space is attracted to a localized region where it traces out a never-repeating (chaotic) path whose shape somewhat resembles the two wings of a butterfly. This is another example of a strange attractor, the attractor differing in appearance from the Rössler strange attractor seen in Chapter 1. Lorenz's *butterfly attractor* is illustrated in the following recipe, along with a stability analysis of the fixed points.

The DEtools and LinearAlgebra packages are loaded. The former is needed because we shall be using the `DEplot` and `DEplot3d` commands to plot $x(t)$ and the butterfly attractor, respectively. The LinearAlgebra package contains commands to construct and manipulate matrices and vectors and solve linear algebra problems. Various commands (e.g., `GenerateMatrix` and `Eigenvalues`) in this package will be used for the stability analysis.

```
> restart: with(DEtools): with(LinearAlgebra):
```

The right-hand sides of the three Lorenz equations are entered and assigned the names P, Q, and R, to be consistent with equations (2.13).

```
> P:=sigma*(y-x); Q:=r*x-y-x*z; R:=x*y-b*z;
```

$$P := \sigma\,(y - x) \qquad Q := r\,x - y - x\,z \qquad R := x\,y - b\,z$$

To produce Lorenz's butterfly, we take $\sigma = 10$, $b = 8/3$, and $r = 28$.

```
> sigma:=10: b:=8/3: r:=28:
```

To locate the fixed points, the equations $P = 0$, $Q = 0$, and $R = 0$ are solved for x, y, and z.

```
> sol:=solve([P,Q,R],[x,y,z]);
```

$$sol := [\,[x = 0,\, y = 0,\, z = 0],\, [\,x = 6\,\mathrm{RootOf}(-2 + \_Z^2,\, label = \_L3),$$
$$y = 6\,\mathrm{RootOf}(-2 + \_Z^2,\, label = \_L3),\, z = 27]\,]$$

The number N of operands in *sol* is determined.

```
> N:=nops(sol);
```

$$N := 2$$

The `allvalues` command is used to remove the RootOf that appeared in *sol*.

```
> sol2:=[seq(allvalues(sol[i]),i=1..N)];
```

$$sol2 := [\,[x = 0,\, y = 0,\, z = 0],\, [x = 6\sqrt{2},\, y = 6\sqrt{2},\, z = 27],$$
$$[x = -6\sqrt{2},\, y = -6\sqrt{2},\, z = 27]\,]$$

The number of fixed points is equal to the number $N2$ of operands in $sol2$.

```
>   N2:=nops(sol2);
```

$$N2 := 3$$

So the Lorenz system has three fixed points. The stability of these fixed points will now be determined by extending the procedure used in the phase-plane analysis. A functional operator f is introduced to substitute the coordinates $x + u$, $y + v$, $z + w$ of an ordinary point, near a fixed point (x, y, z), into a dependent variable V, where V will be taken to be P, Q, and R.

```
>   f:=V->simplify(subs({x=x+u,y=y+v,z=z+w},V)):
```

Assuming that u, v, and w are sufficiently small, only linear terms in these variables will be retained in P, Q, and R. Since P is already linear, entering f(P) produces the linear algebraic form shown in $eq1$. However, on entering f(Q) the nonlinear term $u\,w$ will be present. This term can be removed with the **remove** command, which is done in $eq2$. Similarly, the nonlinear term $u\,v$ is removed from f(R) in $eq3$.

```
>   eq1:=f(P); eq2:=remove(has,f(Q),u*w);
    eq3:=remove(has,f(R),u*v):
```

$$eq1 := 10\,y + 10\,v - 10\,x - 10\,u$$
$$eq2 := 28\,x + 28\,u - y - v - x\,z - x\,w - u\,z$$
$$eq3 := x\,y + x\,v + u\,y - \frac{8\,z}{3} - \frac{8\,w}{3}$$

Let's take $i = 1$, so that assigning the ith operand in $sol2$ corresponds to taking the fixed point coordinates to be $x = 0$, $y = 0$, $z = 0$. The stability of the other two fixed points may be examined by choosing $i = 2$ or $i = 3$.

```
>    i:=1: assign(sol2[i]): #choose i
```

For $i = 1$, $eq1$, $eq2$, and $eq3$ reduce to the following linear forms:

```
>   eq1:=eq1; eq2:=eq2; eq3:=eq3;
```

$$eq1 := 10\,v - 10\,u \qquad eq2 := 28\,u - v \qquad eq3 := -\frac{8\,w}{3}$$

So, the linearized ODEs at an ordinary point near the fixed point at the origin are

$$\frac{du}{dt} = 10\,v - 10\,u, \qquad \frac{dv}{dt} = 28\,u - v, \qquad \frac{dw}{dt} = -\frac{8\,w}{3}.$$

The stability is determined by assuming that u, v, $w \simeq e^{\lambda t}$. For the ODEs to have a nontrivial solution, a determinantal equation must be satisfied that when expanded produces a cubic equation in λ with three roots. To have a stable fixed point, the real part of λ must be negative for all three roots. If any root has a real part that is positive, the solution will grow with time, leading to an unstable situation.

Let's now carry out this analysis using a matrix approach. Equating *eq1* to λu, *eq2* to λv, and *eq3* to λw, the `GenerateMatrix` command

```
>  (A,B):=GenerateMatrix([eq1=lambda*u,eq2=lambda*v,
      eq3=lambda*w],[u,v,w]);
```

$$A, B := \begin{bmatrix} -10-\lambda & 10 & 0 \\ 28 & -1-\lambda & 0 \\ 0 & 0 & \dfrac{-8}{3}-\lambda \end{bmatrix}, \begin{bmatrix} 0 \\ 0 \\ 0 \end{bmatrix}$$

yields the relevant 3×3 matrix A and zero column vector B such that

$$A \begin{bmatrix} u \\ v \\ w \end{bmatrix} = B \equiv \begin{bmatrix} 0 \\ 0 \\ 0 \end{bmatrix}.$$

The determinant of A is computed, yielding a cubic polynomial in λ.

```
>  det:=Determinant(A);
```

$$det := 720 + \frac{722}{3}\lambda - \frac{41}{3}\lambda^2 - \lambda^3$$

The determinant is set equal to zero, solved for λ, and the answer put in floating-point form.

```
>  lambdas:=evalf(solve(det=0,lambda));
```

$$lambdas := -2.666666667, \ 11.82772345, \ -22.82772345$$

Three real roots are obtained, with two of them being negative, but the third one positive. Thus, the fixed point at the origin is unstable. It should be noted that the same result can be obtained by setting $\lambda = 0$ in A and obtaining the eigenvalues by using the `Eigenvalues` command. In this case, the λ values are expressed as a column matrix or vector.

```
>  eiv:=evalf(Eigenvalues(eval(A,lambda=0)));
```

$$eiv := \begin{bmatrix} -2.666666667 \\ 11.82772345 \\ -22.82772345 \end{bmatrix}$$

To see how the trajectory evolves from the vicinity of the fixed point, the variables x, y, z are unassigned from their previous values,

```
>  unassign('x','y','z'):
```

and an initial condition near the fixed point (the origin for $i = 1$) chosen.

```
>  ic:=[x(0)=rhs(sol2[i,1])+0.1,y(0)=rhs(sol2[i,2])+0.1,
      z(0)=rhs(sol2[i,3])+0.1]:
```

$$ic := [x(0) = 0.1, y(0) = 0.1, z(0) = 0.1]$$

The system of Lorenz ODEs is entered by introducing time dependence into the x, y, z variables,

```
>  var:=(x=x(t),y=y(t),z=z(t)):
```

and substituting these variables into P, Q, and R, which are equated to dx/dt, dy/dt, and dz/dt, respectively.

>   sys:=diff(x(t),t)=subs(var,P),diff(y(t),t)=subs(var,Q),
>   diff(z(t),t)=subs(var,R);

$$sys := \frac{d}{dt}x(t) = 10\,y(t) - 10\,x(t),\ \frac{d}{dt}y(t) = 28\,x(t) - y(t) - x(t)\,z(t),$$

$$\frac{d}{dt}z(t) = x(t)\,y(t) - \frac{8}{3}\,z(t)$$

The DEplot command is used with the scene option scene=[t,x] to plot x versus t for the system of ODEs (subject to the initial condition), the line color being allowed to change with time as the time increases.

>   DEplot([sys],[x(t),y(t),z(t)],t=0..60,[ic],scene=[t,x],
>   stepsize=0.01,linecolor=t,thickness=1);

The black-and-white version of the plot is shown in Figure 2.19.

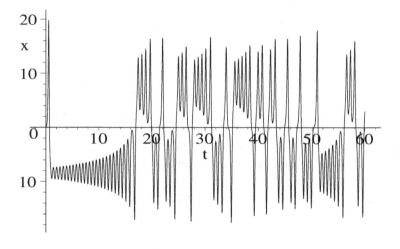

Figure 2.19: x versus t for an initial condition near the origin.

There is no apparent repeat pattern as x increases with t, the motion being chaotic. Longer times can be taken to confirm this. Sensitivity to the initial condition can be checked by changing the initial condition slightly and observing that the x versus t curve changes dramatically. Although the ODEs are deterministic, one's intuition is inadequate to predict what the value of x will be at large times if the initial condition is changed slightly. This is in contrast to a linear ODE system, where a slight change in initial condition will produce only a small change in the solution.

The DEplot3d command is now used to plot the trajectory in the x-y-z space, the line color again being allowed to change with t.

```
> DEplot3d([sys],[x(t),y(t),z(t)],t=0..50,[ic],scene=[x,y,z],
   stepsize=0.01,linecolor=t,orientation=[-23,68]):
```

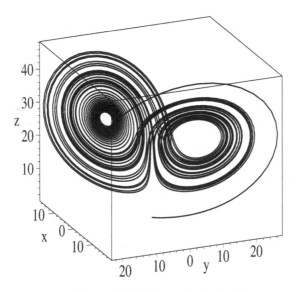

Figure 2.20: Lorenz's butterfly.

Qualitatively, the trajectory leaves the origin in a fixed direction, implying that the fixed point is an unstable nodal point. The trajectory then evolves into a localized region of phase space, producing a nonrepeating path that fills in the region so that the pattern resembles (to an active imagination, especially when seen in color on the computer screen) the wings of a butterfly.

By choosing $i = 2$ and 3, you can use the recipe to determine the nature of the other two fixed points and observe the strange attractor that results.

PROBLEMS:

Problem 2-14: Other fixed points
Use the text recipe to investigate the other two fixed points for the Lorenz system. Identify their probable nature and generate x versus t as well as the phase-space trajectory for an initial condition near each fixed point.

Problem 2-15: Sensitivity
Use the text recipe to explore the sensitivity to initial conditions for the Lorenz system.

Problem 2-16: The Oregonator
Consider the Oregonator system
$$\epsilon\dot{x} = x + y - q\,x^2 - x\,y, \quad \dot{y} = -y + 2\,h\,z - x\,y, \quad p\dot{z} = x - z,$$
with $\epsilon = 0.03$, $p = 2$, $q = 0.006$, $h = 0.75$. Show analytically that the origin is an unstable fixed point.

Problem 2-17: The Rössler system

Consider the Rössler system

$$\dot{x} = -(y+z), \quad \dot{y} = x + a\,y, \quad \dot{z} = b + z\,(x-c),$$

with a, b, $c > 0$. Analytically show that there are no stationary points for $c < \sqrt{4\,a\,b}$ and two fixed points for $c > \sqrt{4\,a\,b}$. Find analytic expressions for the latter points. Linearizing the system in the vicinity of these fixed points, find the cubic equation for the roots λ. Taking $a = b = 0.2$ and allowing c to take on the values $c = 2.4$, 3.5, 4.0, 5.0, 8.0, solve the cubic equation for the roots and determine the stability in each case.

Taking $x(0) = 0.1$, $y(0) = 0.1$, $z(0) = 0$, show by plotting the trajectories in phase space that each c value leads to a qualitatively different behavior of the solution. Identify the behavior by plotting $y(t)$.

Problem 2-18: Multiple stationary points

For the following 3-dimensional system,

$$\dot{x} = (1-z)\left[(4-z^2)(x^2+y^2-2\,x+y) + 4\,(2\,x-y) - 4\right],$$

$$\dot{y} = (1-z)\left[(4-z^2)(x\,y - x - z\,y) + 4\,(x+z\,y) - 2\,z\right],$$

$$\dot{z} = z^2\,(4-z^2)(x^2+y^2),$$

locate all of the stationary points and determine the stability of each point. Explore the nature of the fixed points and the behavior of this system by making suitable 3-dimensional x-y-z plots.

2.3 Numerical Solution of ODEs

In the opening chapter as well as this one, we have encountered nonlinear ODE models for which analytic solutions simply do not exist, in which case use must be made of Maple command structures that solve the given ODE(s) numerically. In this section, we would like to discuss briefly the basis for the Runge–Kutta–Fehlberg 45 (RKF45) method that Maple employs as its default scheme in the `dsolve(ODE,numeric)` command for numerically solving ODEs.

All such numerical schemes make use of the idea of replacing the ODE with a *difference equation* in which each derivative is replaced with a *finite difference approximation* involving a finite step size. The step size may either be fixed or, as in the RKF45 method, variable, taking bigger steps in flatter regions and smaller steps in steeper regions. The difference equation can be iterated using a do loop procedure.

The common finite difference approximations for first and second derivatives are introduced in the first recipe. Our intention in this and subsequent recipes is only to give you the "flavor" of the subject of numerical analysis, referring you to texts such as Burden and Faires [BF89] or *Numerical Recipes* [PFTV89] for more complete treatments.

2.3.1 Finite Difference Approximations

A tool knows exactly how it is meant to be handled, while the user of the tool can only have an approximate idea.
Milan Kundera, Czech author, critic (1929–)

Consider a general function $y(x)$ as schematically depicted by the solid curved line in Figure 2.21. The independent variable has been taken to be x, but it could just as well be the time t. It is desired to find, say, the first and second

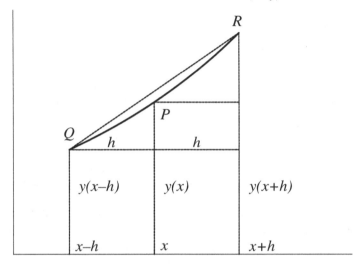

Figure 2.21: Obtaining finite difference approximations to derivatives.

derivatives of y at an arbitrary point P on the curve located at the horizontal position x. Assuming that h is small, two neighboring points R and Q located at $x + h$ and $x - h$, respectively, are considered. At point R, the vertical height is $y(x + h)$ and at Q it is $y(x - h)$.

Since h is assumed to be small, $y(x + h)$ can be Taylor expanded in powers of h about $h = 0$, neglecting terms of, say, order h^4 (i.e., $O(h^4)$) and higher.

```
>  restart:
>  eq1:=y(x+h)=taylor(y(x+h),h=0,4);
```

$$eq1 := y(x + h) =$$
$$y(x) + D(y)(x)\, h + \frac{1}{2}\, (D^{(2)})(y)(x)\, h^2 + \frac{1}{6}\, (D^{(3)})(y)(x)\, h^3 + O(h^4)$$

The $O(h^4)$ term in *eq1* can be removed with the `convert(polynom)` command.

```
>  eq1b:=convert(eq1,polynom);
```

$$eq1b := y(x + h) = y(x) + D(y)(x)\, h + \frac{1}{2}\, (D^{(2)})(y)(x)\, h^2 + \frac{1}{6}\, (D^{(3)})(y)(x)\, h^3$$

Similarly $y(x - h)$ is Taylor expanded and the fourth-order term removed.

> eq2:=y(x-h)=convert(taylor(y(x-h),h=0,4),polynom);

$$eq2 := y(x - h) = y(x) - \mathrm{D}(y)(x)\,h + \frac{1}{2}\,(\mathrm{D}^{(2)})(y)(x)\,h^2 - \frac{1}{6}\,(\mathrm{D}^{(3)})(y)(x)\,h^3$$

On subtracting the second expansion, $eq2$, from the first, $eq1b$,

> eq3:=eq1b-eq2;

$$eq3 := y(x + h) - y(x - h) = 2\,\mathrm{D}(y)(x)\,h + \frac{1}{3}\,(\mathrm{D}^{(3)})(y)(x)\,h^3$$

the h^2 terms cancel, leaving terms involving h and h^3 on the right-hand side of $eq3$. If the latter term is now dropped by substituting $h^3 = 0$ into $eq3$,

> eq3b:=subs(h^3=0,eq3);

$$eq3b := y(x + h) - y(x - h) = 2\,\mathrm{D}(y)(x)\,h$$

the relative error in $eq3b$ will be $\mathrm{O}(h^3)/\mathrm{O}(h) = \mathrm{O}(h^2)$. Solving $eq3b$ for $\mathrm{D}(y)(x)$ yields the *central difference approximation* (CDA) y'_{CDA} to the first derivative. To use the symbol y' in the assignment, y' must be enclosed in left quotes.

> 'y''[CDA]:=solve(eq3b,D(y)(x));

$$y'_{CDA} := \frac{1}{2}\,\frac{y(x + h) - y(x - h)}{h}$$

The quantity y'_{CDA} is just the slope of the chord QR that approaches the slope of the tangent to the $y(x)$ curve at P as $h \to 0$. The reader will recognize this limiting result as the usual definition for the first derivative. The finite difference approximation consists in taking h to be small, but finite.

Now, the above CDA is not the only way of representing y'. The *forward-difference approximation* y'_{FDA} to the first derivative results on dropping the h^2 and h^3 terms in $eq1b$ and solving for $\mathrm{D}(y)(x)$.

> eq4:=subs({h^2=0,h^3=0},eq1b);

$$eq4 := y(x + h) = y(x) + \mathrm{D}(y)(x)\,h$$

> 'y''[FDA]:=solve(eq4,D(y)(x));

$$y'_{FDA} := \frac{y(x + h) - y(x)}{h}$$

The *backward-difference approximation* y'_{BDA} is similarly obtained from $eq2$.

> eq5:=subs({h^2=0,h^3=0},eq2);

$$eq5 := y(x - h) = y(x) - \mathrm{D}(y)(x)\,h$$

> 'y''[BDA]:=solve(eq5,D(y)(x));

$$y'_{BDA} := -\frac{y(x - h) - y(x)}{h}$$

The forward- and backward-difference approximations, both of which have errors of $\mathrm{O}(h)$, correspond to approximating y' at P by the slopes of the chords PR and QP, respectively. For a given (small) value of h, the FDA and BDA are not as accurate as the CDA. Despite this, the FDA is more commonly used to represent first time derivatives in solving initial value problems than the CDA, because, as will be seen, it allows one to explicitly advance forward in time from $t = 0$. Numerical ODE-solving procedures based on the FDA are called *explicit* schemes. If the step size is too large, explicit fixed-step numerical schemes tend to become numerically unstable, so variable-step schemes or *implicit* schemes are also commonly used. Implicit schemes are based on the BDA.

How does one represent the second derivative? As with the first derivative there is more than one way to represent y'' by a finite difference approximation. However, the most commonly used form is obtained by adding *eq1b* and *eq2* and solving for the second derivative.

```
> 'y"':=solve(eq1b+eq2,(D@D)(y)(x));
```

$$y'' := \frac{y(x+h) + y(x-h) - 2\,y(x)}{h^2}$$

This central difference approximation, which has an error $O(h^2)$, is often used in numerically solving diffusion and wave equation models, as shall be illustrated in Chapter 6. A more accurate approximation to y'' is left as a problem.

PROBLEMS:

Problem 2-19: Alternative second-derivative approximation

By also Taylor expanding $y(x+2\,h)$ and $y(x-2\,h)$ show that

$$y''(x) = \frac{[-y(x+2\,h) + 16\,y(x+h) - 30\,y(x) + 16\,y(x-h) - y(x-2\,h)]}{(12\,h^2)} + O(h^4)$$

is an alternative finite difference approximation to the second derivative.

Problem 2-20: Fourth-derivative approximation

Show that

$$y''''(x) = \frac{[y(x+2\,h) - 4\,y(x+h) + 6\,y(x) - 4\,y(x-h) + y(x-2\,h)]}{h^4} + O(h^2)$$

is a finite difference approximation to the fourth derivative.

2.3.2 Rabbits and Foxes: The Sequel

A journey of a thousand miles must begin with a single step.
Lao-tzu, Chinese philosopher (sixth century BC)

Now that we have some idea of how to approximate first and second derivatives by finite differences, numerical ODE-solving algorithms can be created of varying accuracy and CPU (*central processing unit*) time. The ultimate goal, of course, is to produce a highly accurate numerical scheme for a given ODE that takes as little CPU time as possible. Generally, more accurate schemes allow larger step sizes to be used but involve more function evaluations on each step. Larger step sizes reduce the CPU time, while more function evaluations increase the time. Unfortunately, the step size can be increased only so far before the solution becomes quite inaccurate or even displays wild oscillations, an indication of *numerical instability*. How the best compromise between accuracy and time may be reached is the subject matter of numerical analysis courses and beyond the scope of this text. Here, we shall be content to present a few examples that illustrate some of the more important ideas.

The first recipe involves revisiting those natural foes, the sly foxes and pesky jackrabbits of Rainbow County. In this sequel, we will not employ the phase-plane portrait approach but instead replace each time derivative with a forward-difference approximation and convert the ODEs to finite difference equations.

Recall that the predator–prey equations were of the general structure

$$\dot{r} = a\,r - b\,rf \equiv R(r, f), \qquad \dot{f} = -c\,f + d\,rf \equiv F(r, f), \qquad (2.15)$$

where $r(t)$ and $f(t)$ are the rabbit and fox numbers (per unit area) at time t. For the coefficient values we shall take $a = 2$, $b = 0.01$, $c = 1$, and $d = 0.01$, and $r(0) = 300$ rabbits and $f(0) = 150$ foxes as initial conditions.

To solve this set of first-order nonlinear ODEs, we can proceed as follows. If h is the time step size, then the time on step $(n + 1)$ is related to the time on step n by $t_{n+1} = t_n + h$. Introducing the notation $r_n \equiv r(t_n)$, $f_n \equiv f(t_n)$, $r_{n+1} \equiv r(t_{n+1})$, and $f_{n+1} \equiv f(t_{n+1})$, the FDA to the two time derivatives is

$$\dot{r} = \frac{(r_{n+1} - r_n)}{h} \quad \text{and} \quad \dot{f} = \frac{(f_{n+1} - f_n)}{h}.$$

The overall accuracy of the numerical scheme then depends on how $R(r, f)$ and $F(r, f)$ are approximated. Historically, the oldest and simplest approximation is to evaluate these functions on the nth time step, reducing the ODEs to

$$r_{n+1} = r_n + h\,R(r_n, f_n), \qquad f_{n+1} = f_n + h\,F(r_n, f_n). \qquad (2.16)$$

This finite difference approximation, which allows one to advance forward a single step at a time, is referred to as the *forward Euler algorithm*.

The step size h must be taken to be small to obtain an accurate approximation to the solution of the original ODEs. How small? A common approach to answering this question is to cut the step size in two and see what effect it has on the answer. This procedure can be repeated until a sufficiently accurate answer for your purposes has been attained. Some words of caution should be offered, however. One cannot keep cutting the step size in two indefinitely for a fixed number of digits or one will encounter *round-off error*, the answer in fact becoming less accurate. Further, to execute the worksheet out to the same total time will take more and more CPU time as the step size is decreased.

Let's now solve equations (2.16), first loading the plots package and entering the coefficient values.

>   `restart: with(plots):`

>   `a:=2: b:=0.01: c:=1: d:=0.01:`

The initial time, initial rabbit number, and initial fox number are specified and the time step size taken to be $h = 0.02$. We take $N = 1000$ steps, so the total elapsed time for the numerical run will be $0.02 \times 1000 = 20$ time units.

>   `t[0]:=0: r[0]:=300: f[0]:=150: h:=0.02: N:=1000:`

Although the Maple system normally uses 10 digits as its default, the number of digits is specified in case it is desired to increase this number, e.g., to lessen the round-off error mentioned earlier. To compare the speed of different numerical schemes or algorithms, the computer CPU time at which the execution of the algorithm begins is recorded using the `time` command. The CPU time at which the execution is finished will also be recorded, the CPU time for the algorithm then being the difference. The CPU time will depend on the computer being used, the times quoted here being for a 3-GHz PC.

```
>   Digits:=10: begin:=time():
```

Functional operators are formed to calculate R and F for specified x and y.

```
>   R:=(x,y)->a*x-b*x*y: F:=(x,y)->-c*y+d*x*y:
```

A do loop is used to iterate the difference equations (2.16) and the time.

```
>   for n from 0 to N do
>   r[n+1]:=r[n]+h*R(r[n],f[n]);
>   f[n+1]:=f[n]+h*F(r[n],f[n]);
>   t[n+1]:=t[n]+h;
```

A 3-dimensional plotting point is formed for the nth time step.

```
>   pnt[n]:=[t[n],r[n],f[n]];
>   end do:
```

On ending the loop, the temporal evolution of the solution will be observed by creating a three-dimensional plot using a sequence of the plotting points from $n = 0$ to $n = N$ and the `pointplot3d` command.

```
>   pointplot3d([seq(pnt[n],n=0..N)],symbol=cross,color=red,
    axes=normal,labels=["t","r","f"],tickmarks=[2,3,3]);
```

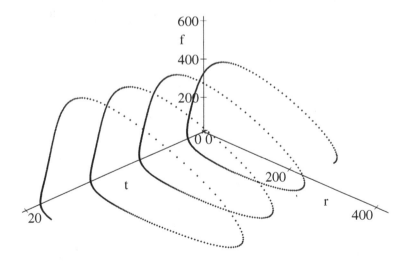

Figure 2.22: Forward Euler solution of the rabbits–foxes equations.

The resulting spiral trajectory, corresponding to a cyclic variation in population numbers, is shown in Figure 2.22. The elapsed CPU time for the run,

```
>   CPUtime:=(time()-begin)*seconds;
```

$$CPUtime := 0.130 \, seconds$$

was only a fraction of a second.

Instead of this "first principles approach" to numerically solving the rabbits–foxes equations, one can alternatively "dial up" the forward Euler approximation as follows. The dependent variables r and f are unassigned,

```
>   unassign('r','f'): begin2:=time():
```

and the ODE system is entered.

```
>   sys:=diff(r(t),t)=R(r(t),f(t)),diff(f(t),t)=F(r(t),f(t)):
```

The initial population numbers are entered as the initial condition.

```
>   ic:=r(0)=300,f(0)=150:
```

The numerical ODE solve command is used with the "classical" forward Euler method included as an option. The step size is 0.02, the same as before.

```
>   sol:=dsolve({sys,ic},{r(t),f(t)},type=numeric,
          method=classical[foreuler],stepsize=0.02):
```

By using the odeplot command and specifying that 1000 time steps are to be taken, exactly the same plot (not displayed here) as in Figure 2.22 is produced.

```
>   odeplot(sol,[t,r(t),f(t)],0..20,numpoints=1000,style=point,
        symbol=cross,axes=normal,labels=["t","r","f"],
        tickmarks=[2,3,3]);
```

If the three-dimensional picture produced by either code is rotated to show the r–f plane, the phase-plane trajectory is as shown in Figure 2.23.

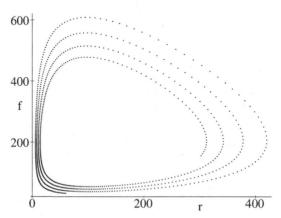

Figure 2.23: Phase-plane trajectory for the rabbits–foxes system.

Although it appears that the trajectory is an outward-growing spiral, this conclusion would not be correct. The trajectory really should be a vortex, i.e., a closed loop. The incorrect behavior is due to *cumulative error* arising from use of the Euler approximation (which has low accuracy) and the finite step size. Keeping in mind the danger of round-off error for a fixed number of digits, the growth of the spiral can in principle be reduced by reducing h. But in practice, the CPU time goes up so rapidly that the Euler method is not used for serious numerical calculations. In the next recipe, the classical *fourth-order Runge–Kutta method* is introduced, a much more accurate numerical scheme than the Euler method for a given step size.

Finally, the CPU time for the forward Euler dial-up method is determined,

```
>  CPUtime2:=(time()-begin2)*seconds;
```

$$CPUtime2 := 0.330 \, seconds$$

and is actually slightly longer than for the first-principles calculation.

PROBLEMS:
Problem 2-21: Van der Pol equation
By setting $\dot{x} = y$, the second-order Van der Pol (VdP) equation

$$\ddot{x} - \epsilon\,(1 - x^2)\,\dot{x} + x = 0$$

may be written as a coupled system of two first-order ODEs. Choosing $\epsilon = 5$ and $x(0) = \dot{x}(0) = 0.1$, solve the VdP system for $h = 0.01$ and $t = 0$ to 15 using the first principles Euler method. Rotate your plot to produce a phase-plane portrait solution and compare the result with what would be obtained using the dial-up Euler option of the `dsolve` command.

Problem 2-22: White dwarf equation
In his theory of white dwarf stars, Chandrasekhar [Cha39] introduced the non-linear equation

$$x\,(d^2y/dx^2) + 2\,(dy/dx) + x\,(y^2 - C)^{3/2} = 0,$$

with the boundary conditions $y(0) = 1$ and $y'(0) = 0$. Write the second-order equation as two first-order equations and solve the system using the first principle's Euler method. Take $h = 0.01$ and 10-digit accuracy, and numerically compute $y(x)$ over the range $0 \le x \le 4$ with $C = 0.1$ and plot the result. (*Hint*: Start at $x = 0.01$ to avoid any problem at the origin.)

Problem 2-23: Baleen whales
May [May80] has discussed the solution of the following normalized equation describing the population of sexually mature adult baleen whales:

$$\dot{x}(t) = -a\,x(t) + b\,x(t - T)(1 - (x(t - T))^N).$$

Here $x(t)$ is the normalized population number at time t, a and b are the mortality and reproduction coefficients, T is the time lag necessary to achieve sexual maturity, and N is a positive parameter. If the term $1 - (x(t - T))^N$ is negative, then this term is to be set equal to zero. Taking $a = 1$, $b = 2$, $T = 2$, step size $h = 0.01$, and 4000 time steps, use the first principles Euler method to solve numerically for $x(t - T)$ versus $x(t)$ and for $x(t)$ for (a) $N = 3.0$, and (b) $N = 3.5$. Plot your results. For (a) you should observe a period-one solution, and for (b) a period-two solution. Discuss how this interpretation can be made from your plots.

Problem 2-24: Artificial example
Consider the nonlinear equation

$$dy/dx = x\,y\,(y-2),$$

with $y(0) = 1$. Taking $h = 0.02$ and 10-digit accuracy, solve for $y(x)$ out to $x = 3$ using the dial-up Euler's method and plot the result.

Problem 2-25: Modified Euler algorithm
If the rabbits–foxes system is written for brevity as $\dot{r} = R(r, f)$ and $\dot{f} = F(r, f)$, the *modified Euler algorithm* for solving the equations is

$$r_{n+1} = r_n + \frac{h}{2}(R_1[n] + R_2[n]), \quad f_{n+1} = f_n + \frac{h}{2}(F_1[n] + F_2[n]),$$

where
$$t_{n+1} = t_n + h, \quad R_1[n] \equiv R(r_n, f_n), \quad F_1[n] \equiv F(r_n, f_n),$$

$$R_2[n] \equiv R(r_n + h\,R_1[n], f_n + h\,F_1[n]), \quad F_2[n] \equiv F(r_n + h\,R_1[n], f_n + h\,F_1[n]).$$

Taking the same parameter values as in the text recipe, but N twice as large, solve the rabbits–foxes equations using the modified Euler algorithm given above. Rotate your plot to show the r–f phase plane and compare the result with that for the Euler method.

Problem 2-26: Onset of numerical instability
Investigate the onset of numerical instability as h is increased in the modified Euler algorithm of the previous problem.

2.3.3   Glycolytic Oscillator

Give me another horse!
William Shakespeare, *King Richard III* (1564–1616)

The Euler algorithm, which is the simplest example of an explicit fixed-step method, is not a very accurate numerical procedure, being of order h accuracy (error $O(h^2)$). A systematic approach to developing more accurate explicit numerical schemes are the fixed-step Runge–Kutta (RK) methods [BF89], which still use the FDA approximation for the first derivatives but create better approximations to the functions on the right-hand side of the ODEs.

Consider an ODE system of the general form

$$\dot{x} = X(t, x, y), \quad \dot{y} = Y(t, x, y), \tag{2.17}$$

where X and Y are known functions of the arguments. Note that any second-order ODE of the general structure $\ddot{x} = Y(t, x, \dot{x})$ can be put into this *standard form* by setting $\dot{x} = y \equiv X$.

In the Euler method, the functions X and Y are evaluated only once on each step. The general RK approach is to increase the accuracy by using more evaluations of the functions X and Y, the extra evaluations being carried out at intermediate points between t_n and $t_n + h$ in such a way that an nth-order RK scheme is equivalent to a Taylor series expansion of the functions truncated at terms of order n. The interested reader is referred to numerical analysis texts such as Burden and Faires [BF89] for the details.

Here it will suffice to illustrate the application of the fourth-order Runge–Kutta (RK4) method, which has been the historical "workhorse" of fixed-step methods. This method, which involves four function evaluations of X and Y on each step, is of order h^4 in accuracy (error $O(h^5)$). Explictly, the RK4 method applied to equation (2.17) takes the following form:

$$x_{n+1} = x_n + \frac{1}{6}(K_1 + 2K_2 + 2K_3 + K_4), \quad y_{n+1} = y_n + \frac{1}{6}(L_1 + 2L_2 + 2L_3 + L_4),$$

$$\text{with} \quad K_1 = h\,X(t_n,\, x_n,\, y_n),$$

$$K_2 = h\,X(t_n + h/2,\, x_n + K_1/2,\, y_n + L_1/2),$$

$$K_3 = h\,X(t_n + h/2,\, x_n + K_2/2,\, y_n + L_2/2),$$

$$K_4 = h\,X(t_n + h,\, x_n + K_3,\, y_n + L_3),$$

and L_1 to L_4 obtained by replacing X with Y in K_1 to K_4.

Since this RK4 scheme is $O(h^4)$ accurate, one can either use the same step size as in the Euler method, producing an answer of greater accuracy, or, alternatively, use a larger step size to speed up the calculation. However, h can be increased only so much before the Taylor expansion on which the RK4 method is based begins to break down and wild numerical oscillations in the output occur. All explicit schemes are prone to this numerical instability if the step size is made too large.

Runge–Kutta methods of even greater accuracy than RK4 can be generated, but these involve even more evaluations of X and Y, which increases the CPU time. The RK4 method is usually deemed to be the optimum fixed-step explicit method that best combines good accuracy with reasonable CPU time.

As an illustration of the RK4 method, let's consider solving the nonlinear ODEs describing the *glycolytic oscillator*. Living cells obtain energy by breaking down sugar, a process known as *glycolysis*. In yeast cells, this process proceeds in an oscillatory way with a period of a few minutes. A simple system of model equations proposed by Sel'kov [Sel68] to describe the oscillations is as follows:

$$\dot{x} = -x + a\,y + x^2\,y, \qquad (2.18)$$

$$\dot{y} = b - a\,y - x^2\,y.$$

Here x and y refer to the concentrations of adenosine diphosphate (ADP) and fructose-6-phosphate (F6P), respectively, and a and b are positive constants. We shall use the nominal values $a = 0.05$ and $b = 0.5$, and take $x_0 = 3$, $y_0 = 3$.

After loading the plots package, specifying the number of digits, and recording the beginning time,

> restart: with(plots): Digits:=10: begin:=time():

the parameter values are entered and $N = 1000$ steps of size $h = 0.05$ are considered.

> t[0]:=0: x[0]:=3: y[0]:=3: a:=0.05: b:=0.5: N:=1000: h:=0.05:

The functional, or arrow, operator is used to produce the right-hand side of each ODE. Whatever forms are inserted for x and y in the entries X(x,y) and Y(x,y) will be operated on as indicated on the right-hand side of the arrow.

> X:=(x,y)-> -x+a*y+x^2*y;

$$X := (x, y) \rightarrow -x + a\,y + x^2\,y$$

> Y:=(x,y)->b-a*y-x^2*y;

$$Y := (x, y) \rightarrow b - a\,y - x^2\,y$$

Similarly, operators are introduced to perform the four function evaluations, where f will be taken to be X and Y. Note that X and Y do not depend explicitly on time here.

> k1:=f->h*f(x[n],y[n]):

> k2:=f->h*f(x[n]+K1[n]/2,y[n]+L1[n]/2):

> k3:=f->h*f(x[n]+K2[n]/2,y[n]+L2[n]/2):

> k4:=f->h*f(x[n]+K3[n],y[n]+L3[n]):

Using the above operators, the fourth-order Runge–Kutta algorithm is iterated from $n = 0$ to $N = 1000$ and a plotting point produced at each step.

> for n from 0 to N do

> K1[n]:=k1(X); L1[n]:=k1(Y);

> K2[n]:=k2(X); L2[n]:=k2(Y);

> K3[n]:=k3(X); L3[n]:=k3(Y);

> K4[n]:=k4(X); L4[n]:=k4(Y);

> x[n+1]:=x[n]+(K1[n]+2*K2[n]+2*K3[n]+K4[n])/6;

> y[n+1]:=y[n]+(L1[n]+2*L2[n]+2*L3[n]+L4[n])/6;

> t[n+1]:=t[n]+h;

> pnt[n]:=[t[n],x[n],y[n]];

> end do:

Using the spacecurve command, the sequence of points is plotted as a solid blue line with normal axes. The tickmarks are controlled, axis labels added, and a particular orientation of the 3-dimensional viewing box chosen.

> spacecurve([seq(pnt[n],n=0..N)],color=blue,axes=normal,
    tickmarks=[2,2,2],labels=["t","x","y"],
    orientation=[23,71]);

The resulting picture is shown in Figure 2.24. It may be rotated on the computer screen by dragging with the mouse.

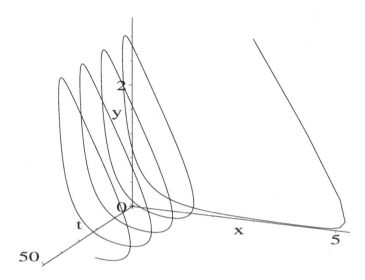

Figure 2.24: Temporal evolution of a glycolytic oscillator.

The spiral behavior characteristic of temporal oscillations is clearly seen. If the three-dimensional figure is rotated on the computer screen to show the x-y phase plane, a closed loop will be observed. One can also view x versus t and y versus t. The CPU time for the first-principles RK4 calculation is

```
>   CPUtime:=(time()-begin)*seconds;
```
$$CPUtime := 0.500\ seconds$$
about one-half a second.

After unassigning the variables x and y,

```
>   unassign('x','y'): begin2:=time():
```
exactly the same 3-dimensional plot is produced (not displayed here) by using the dsolve command with the option method=classical[rk4].

```
>   sys:=diff(x(t),t)=X(x(t),y(t)),diff(y(t),t)=Y(x(t),y(t)):
>   sol:=dsolve({sys,x(0)=3,y(0)=3},{x(t),y(t)},type=numeric,
          method=classical[rk4],stepsize=0.05):
>   odeplot(sol,[t,x(t),y(t)],0..50,axes=normal,numpoints=1000,
          color=blue,style=line,tickmarks=[2,2,2],
          labels=["t","x","y"],orientation=[23,71]);
>   CPUtime2:=(time()-begin2)*seconds;
```
$$CPUtime2 := 0.300\ seconds$$
The CPU time for the dial-up routine is slightly shorter than the first-principles calculation here.

PROBLEMS:

Problem 2-27: Harvesting of fish

In suitably normalized units, the effect of fishing on the normalized population number x of a single species of fish with a limited food supply can be described by the following nonlinear ODE,

$$\dot{x} = x\,(1 - x) - H\,x/(a + x),$$

where H is the harvesting coefficient and a is a parameter. For $H = 0$, the remaining ODE is known as the *logistic equation*. Show that this equation has an analytic solution and plot the result for $a = 0.2$ and $x(0) = 0.1$. Discuss the behavior of the solution.

Then, using the first-principles RK4 method with step size $h = 0.1$, numerically investigate this equation as the harvesting coefficient H is increased from zero. Plot your results and discuss the change in behavior as H increases.

Problem 2-28: Bucky the beaver

Bucky the beaver attempts to swim across a river by steadily aiming at a target point directly across the river. The river is 1 km wide and has a speed of 1 km/hr, while Bucky's speed is 2 km/hr. In Cartesian coordinates, Bucky is initially at $(x = 1, y = 0)$, while the target point is at $(0, 0)$. Bucky is initially swept an infinitesimal distance downstream but recovers almost instantly to continue his swimming motion. His equations of motion are

$$\dot{x} = -\frac{2\,x}{\sqrt{x^2 + y^2}}, \qquad \dot{y} = 1 - \frac{2\,y}{\sqrt{x^2 + y^2}}.$$

(a) Justify the structure of the equations.

(b) Using the first-principles RK4 method with $h = 0.01$, determine how long it takes Bucky to reach the target point.

(c) Determine the analytic solution $y(x)$ for Bucky's path across the river.

(d) Plot the analytic and numerical solutions together for Bucky's path.

Problem 2-29: Chemical reaction

Consider the irreversible chemical reaction

$$2\,K_2Cr_2O_7 + 2\,H_2O + 3\,S \rightarrow 4\,KOH + 2\,Cr_2O_3 + 3\,SO_2,$$

with initially $N_1 = 2000$ molecules of potassium dichromate $(K_2Cr_2O_7)$, $N_2 = 2000$ molecules of water (H_2O), and $N_3 = 3000$ atoms of sulphur (S). The number X of potassium hydroxide (KOH) molecules at time t s is given by the rate equation

$$\dot{X} = k\,(2\,N_1 - X)^2\,(2\,N_2 - X)^2\,(4\,N_3/3 - X)^3,$$

with $k = 1.64 \times 10^{-20}$ s^{-1} and $X(0) = 0$. Determine $X(t)$ by using the first-principles RK4 method with $h = 0.001$. How many KOH molecules are present at $t = 0.1$ s? at $t = 0.2$ s?

2.3.4 Fox Rabies Epidemic

Thought is an infection. In the case of certain thoughts, it becomes an epidemic.
Wallace Stevens, American poet (1879–1955)

As already mentioned, if the step size h is made too large, fixed-step explicit methods not only become increasingly inaccurate, but will become numerically unstable with the solution eventually "blowing up." One approach to avoiding this instability is to make use of the backward-difference approximation to the derivative and create an *implicit* or *semi-implicit* numerical scheme.

In this example, we will illustrate this approach by solving the nonlinear ODEs for the fox rabies epidemic[7] model of Anderson and coworkers [AJMS81] using the *backward Euler algorithm*. The ideas presented here can be extended to higher-order RK methods.

The rabies epidemic model was applied to the fox population in central Europe, the epidemic believed to have originated in Poland in 1939. In this model, the fox population is divided into three categories: susceptibles (population density x foxes/km^2) are foxes that are currently healthy but are susceptible to catching the virus; infected (density y) are foxes that have caught the virus but are not yet capable of passing on the virus; infectious (density z) are foxes that are capable of infecting the susceptibles. The model has no category of recovered immune foxes because very few, if any, survive after acquiring the rabies virus. For other diseases with a lower mortality rate, one would add a recovered category and therefore an additional modeling equation. The relevant equations for the fox rabies epidemic are

$$\dot{x} = a\,x - (b + g\,N)\,x - \beta\,x\,z \equiv X,$$

$$\dot{y} = \beta\,x\,z - (\sigma + b + g\,N)\,y \equiv Y, \qquad (2.19)$$

$$\dot{z} = \sigma\,y - (\alpha + b + g\,N)\,z \equiv Z,$$

with $N = x + y + z$ being the total fox density. The meaning of the various coefficients in (2.19) and their estimated values is as follows:

Symbol	Meaning	Value
a	average per capita birth rate	1/year
b	average per capita intrinsic death rate	0.5/year
β	rabies transmission coefficient	79.67 km^2/year
σ	$1/\sigma$ = average latent period ($\approx$ 28 to 30 days)	13/year
α	death rate of rabid foxes	73/year
g	$g\,N$ represents increased death rate at large N due to depletion of food supply	0.1-5 km^2/year

---

[7]Modeling the spread of diseases is extensively discussed in Murray [Mur89].

Using the BDA for the time derivative to connect time step n to the previous step $n-1$ and approximating the rhs with its value at step n, the backward Euler approximation to equations (2.19) is

$$\frac{x_n - x_{n-1}}{h} = X_n, \quad \frac{y_n - y_{n-1}}{h} = Y_n, \quad \frac{z_n - z_{n-1}}{h} = Z_n,$$

or letting $n \to n+1$ and substituting the forms of X, Y, and Z,

$$x_{n+1} = x_n + h\,(a-b)\,x_{n+1} - h\,g\,N_{n+1}\,x_{n+1} - h\,\beta\,x_{n+1}\,z_{n+1}$$

$$y_{n+1} = y_n + h\,\beta\,x_{n+1}\,z_{n+1} - h\,(\sigma+b)\,y_{n+1} - h\,g\,N_{n+1}\,y_{n+1} \tag{2.20}$$

$$z_{n+1} = z_n + h\,\sigma\,y_{n+1} - h\,(\alpha+b)\,z_{n+1} - h\,g\,N_{n+1}\,z_{n+1}$$

with $N_{n+1} = x_{n+1} + y_{n+1} + z_{n+1}$.

This algorithm is nonlinear in the unknowns x_{n+1}, y_{n+1}, and z_{n+1}, and is an example of an *implicit* algorithm because one cannot explicitly express the unknowns in terms of the known values x_n, y_n, and z_n. The standard approach is to make the scheme *semi-implicit* by linearizing it as follows. For a nonlinear function $f(x_{n+1}, z_{n+1})$, we Taylor expand thus:

$$f(x_{n+1}, z_{n+1}) = f(x_n, z_n) + (x_{n+1} - x_n)\left(\frac{\partial f}{\partial x}\right)_{x_n, z_n} + (z_{n+1} - z_n)\left(\frac{\partial f}{\partial z}\right)_{x_n, z_n} + \cdots$$

For our fox-rabies example, this gives us, for example,

$$f \equiv (x\,z)_{n+1} = x_n\,z_n + z_n(x_{n+1} - x_n) + x_n(z_{n+1} - z_n). \tag{2.21}$$

Applying this procedure to all of the quadratic terms on the rhs of (2.20) and gathering all the "new" values x_{n+1}, etc., on the lhs, we obtain

$$A_{1n}\,x_{n+1} + B_{1n}\,y_{n+1} + C_{1n}\,z_{n+1} = x_n + D_{1n},$$

$$A_{2n}\,x_{n+1} + B_{2n}\,y_{n+1} + C_{2n}\,z_{n+1} = y_n + D_{2n}, \tag{2.22}$$

$$A_{3n}\,x_{n+1} + B_{3n}\,y_{n+1} + C_{3n}\,z_{n+1} = z_n + D_{3n},$$

with

$$A_{1n} = 1 + h\,(b-a) + h\,g\,(N_n + x_n) + h\,\beta\,z_n, \; B_{1n} = h\,g\,x_n, \; C_{1n} = h\,(\beta+g)\,x_n,$$

$$A_{2n} = h\,g\,y_n - h\,\beta\,z_n, \; B_{2n} = 1 + h\,(\sigma+b) + h\,g\,(N_n + y_n), \; C_{2n} = h\,g\,y_n - h\,\beta\,x_n,$$

$$A_{3n} = h\,g\,z_n, \; B_{3n} = h\,g\,z_n - h\,\sigma, \; C_{3n} = 1 + h\,(\alpha+b) + h\,g\,(N_n + z_n),$$

$$D_{1n} = h\,g\,N_n\,x_n + h\,\beta\,x_n\,z_n, \; D_{2n} = h\,g\,N_n\,y_n - h\,\beta\,x_n\,z_n, \; D_{3n} = h\,g\,N_n\,z_n.$$

We now have three linear equations (2.22), with known numerical coefficients determined from the previous step, for the three unknowns x_{n+1}, y_{n+1}, z_{n+1}.

The semi-implicit algorithm is now implemented in the following recipe, taking a total of 5000 time steps of size $h = 0.004$.

```
>   restart: Digits:=10: total:=5000: h:=0.004:
```
The parameter values are entered, with g at the lower end of its range.
```
>   a:=1: b:=0.5: beta:=79.67: g:=0.1: alpha:=73: sigma:=13:
```
Initially, we take $x_0 = 4$, $y_0 = 0.4$, and $z_0 = 0.1$.
```
>   t[0]:=0: x[0]:=4: y[0]:=0.4: z[0]:=0.1:
```
The initial plotting point is formed, and the starting time of the do loop that iterates the algorithm is recorded.
```
>   pnt[0]:=[t[0],x[0],y[0]]: begin:=time():

>   for n from 0 to total do
```
The various coefficients are evaluated on the nth step.
```
>   N[n]:=x[n]+y[n]+z[n];
>   A1[n]:=1+h*(b-a)+h*g*(N[n]+x[n])+h*beta*z[n];
>   B1[n]:=h*g*x[n];
>   C1[n]:=h*(beta+g)*x[n];
>   D1[n]:=h*g*N[n]*x[n]+h*beta*x[n]*z[n];
>   A2[n]:=h*g*y[n]-h*beta*z[n];
>   B2[n]:=1+h*(sigma+b)+h*g*(N[n]+y[n]);
>   C2[n]:=h*g*y[n]-h*beta*x[n];
>   D2[n]:=h*g*N[n]*y[n]-h*beta*x[n]*z[n];
>   A3[n]:=h*g*z[n];
>   B3[n]:=h*g*z[n]-h*sigma;
>   C3[n]:=1+h*(alpha+b)+h*g*(N[n]+z[n]);
>   D3[n]:=h*g*N[n]*z[n];
```
The linear equations for x_{n+1}, y_{n+1}, z_{n+1} are entered,
```
>   E1:=A1[n]*x[n+1]+B1[n]*y[n+1]+C1[n]*z[n+1]=x[n]+D1[n];
>   E2:=A2[n]*x[n+1]+B2[n]*y[n+1]+C2[n]*z[n+1]=y[n]+D2[n];
>   E3:=A3[n]*x[n+1]+B3[n]*y[n+1]+C3[n]*z[n+1]=z[n]+D3[n];
```
and numerically solved.
```
>   sol[n+1]:=(fsolve({E1,E2,E3},{x[n+1],y[n+1],z[n+1]}));
```
The solution is assigned and the values of x_{n+1}, y_{n+1}, and z_{n+1} are recorded at time $t_{n+1} = t_n + h$.
```
>   assign(sol[n+1]);
>   x[n+1],y[n+1],z[n+1];
>   t[n+1]:=t[n]+h;
```
Finally, a plotting point is formed and the loop ended.
```
>   pnt[n+1]:=[t[n+1],x[n+1],y[n+1]];
>   end do:
```

On a 3-GHz PC, the CPU time to execute the loop

```
>  cpu:=time()-begin;
```

$$cpu := 10.565$$

is about 11 seconds.

The points are now plotted, the resulting curve being shown in Figure 2.25. The three-dimensional curve may be rotated on the computer screen by dragging the viewing box with the mouse. What does the model calculation predict?

```
>  plots[spacecurve]([seq(pnt[n],n=0..total)],color=black,
   axes=normal,tickmarks=[3,2,3],labels=["t","x","y"]);
```

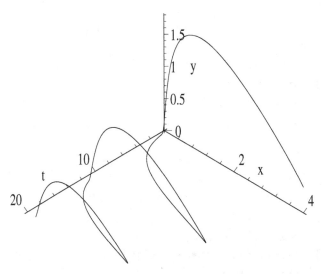

Figure 2.25: Semi-implicit solution of fox rabies epidemic model.

You can confirm that the algorithm remains stable for much larger h, but becomes increasingly inaccurate. In general, semi-implicit schemes are not guaranteed to be stable, but usually are. However, such schemes are costly in terms of computing time if high accuracy is required. For systems of nonlinear ODEs, this is usually the case, so variable- or adaptive-step explicit schemes are preferred by researchers. Such schemes adjust the step size so that larger steps are used when the "terrain" is flatter and smaller steps when it becomes steeper.

Maple's default ODE algorithm is based on the Runge–Kutta–Fehlberg 45 (RKF45) scheme, one of the most widely used adaptive step methods. Without getting into the details, which can be found in standard numerical analysis texts [BF89], the code uses both the fourth-order and fifth-order Runge–Kutta schemes and compares the results of both algorithms on each step, using the difference in answers as a measure of the error. The step size is then adjusted to maintain some maximum tolerance on the error.

PROBLEMS:

Problem 2-30: The arms race

Rapoport [Rap60] has proposed the following system of model equations to describe the arms race between two nations (or two groups of nations),

$$\dot{x} = -m_1 x + a_1 y + b_1 y^2,$$

$$\dot{y} = -m_2 y + a_2 x + b_2 x^2,$$

where x and y are the defense budgets (in suitable units of currency) of the two nations and all constants are positive. Explain the terms in the model.

Apply the semi-implicit backward Euler method to Rapoport's model, taking the nominal values $m_1 = 0.5$, $a_1 = 1$, $b_1 = 0.02$, $m_2 = 0.4$, $a_2 = 0.1$, $b_2 = 0.05$, $x(0) = y(0) = 0.5$. Try different step sizes and plot the results in t vs. x vs. y space. Discuss the results.

Problem 2-31: Van der Pol oscillator

The Van der Pol equation,

$$\ddot{x} - \epsilon (1 - \dot{x}^2) + x = 0,$$

with $\epsilon = 5.03$, $x(0) = 0.1$, and $\dot{x}(0) = 0$ was the subject of recipe **01-1-3**. Choosing a suitable step size, numerically solve this ODE using the semi-implicit backward Euler method and compare with the results obtained in **01-1-3**.

Problem 2-32: Second-order-accurate scheme

Given a system of first-order nonlinear ODEs with a representative equation of the structure $\dot{x} = X(x, \ldots)$, one can create a second-order-accurate semi-implicit numerical scheme by using the backward-difference approximation for the derivative and replacing the Euler approximation X_n on the rhs with the average $(X_n + X_{n+1})/2$.

Derive a second-order-accurate semi-implicit algorithm for the Rössler system of Section 1.2.3. Taking $a = b = 0.2$, $c = 5.0$, $x(0) = 4.0$, $y(0) = z(0) = 0$, and $h = 0.05$, determine the behavior of the system up to $t = 100$ and plot the trajectory in x-y-z space. How do your results compare with those obtained with recipe **01-2-3** for the same parameter values?

Problem 2-33: Stiff ODE system

A *stiff* ODE system is one for which there are two or more very different time or spatial scales of the independent variable. Numerical instability can occur for stiff systems solved with fixed-step explicit schemes unless the step size is taken to be shorter than (about) the shortest time scale for the system. Consider the following set of coupled linear ODEs,

$$\dot{x} = 998\, x + 1998\, y, \quad \dot{y} = -999\, x - 1999\, y,$$

subject to the initial conditions $x(0) = 1$ and $y(0) = 0$.

(a) Analytically solve the system for $x(t)$ and $y(t)$.

(b) Determine the two characteristic times in the solutions and confirm that they are very different from each other.

(c) Using the dial-up RK4 algorithm, confirm the statement on numerical instability and the characteristic time scales.

(d) Using the backward-difference approximation for the time derivatives and the Euler approximation on the right-hand side, show that the numerical instability can be "cured." Choose a step size such that when plotted using a point style, the numerical points lie on the analytic curves for $x(t)$ and $y(t)$.

Problem 2-34: Forced stiff system

Repeat the steps of the previous problem for the forced system

$$\dot{x} = 9\,x + 24\,y + 5\cos t - \frac{1}{3}\sin t, \quad \dot{y} = -24\,x - 51\,y - 9\cos t + \frac{1}{3}\sin t,$$

subject to the initial conditions $x(0) = 4/3$ and $y(0) = 2/3$.

Part II

THE ENTREES

That is the essence of science:
ask an impertinent question,
and you are on the way to a pertinent answer.
Jacob Bronowski, British scientist, author, *The Ascent of Man*, 1973

The whole of science is nothing more
than a refinement of everyday thinking.
Albert Einstein, physics Nobel laureate, *Out of My Later Years*, 1950

The effort to understand the universe is one
of the very few things that lifts human life a little
above the level of farce, and gives it
some of the grace of tragedy.
Steven Weinberg, American physicist, *The First Three Minutes*, 1977

Chapter 3

Linear ODE Models

Among all the mathematical disciplines the theory of differential equations is the most important. It furnishes the explanation of all those elementary manifestations of nature which involve time.
Sophus Lie, Norwegian mathematician (1842–1899)

In the two chapters of the Appetizers, we presented only a few examples of linear ODEs, because the graphical and numerical techniques were more suitable for nonlinear systems where generally exact analytic solutions simply do not exist. Also, modern research often involves nonlinear ODE (and PDE) systems, a subject that is almost completely neglected in undergraduate mathematics training. So, one of the goals of this text is to partially fill in this "hole," showing how a computer algebra system may be used to explore nonlinear models without getting "buried" in messy and complicated mathematical details. However, we would be remiss if we did not provide some coverage of linear differential equation systems, both ordinary and partial. The former are covered in this chapter, while linear PDEs are dealt with in Chapters 5 and 6. In between, we shall explore nonlinear ODEs further in Chapter 4.

For linear ODE systems with constant coefficients, it is always possible to find exact analytic solutions in terms of "elementary" mathematical functions such as sines and cosines, logs, and exponentials. Analytic solutions are also possible for linear ODEs with variable coefficients that are of the so-called *Sturm–Liouville* [MW71] type. Included in this classification are such "famous" ODEs as Bessel's equation and the Legendre and Hermite equations, to mention just a few. The solutions are expressed in terms of "special" functions, defined in terms of either infinite series or finite polynomials.

In this chapter, the emphasis will be on illustrating how the `dsolve` command can be used to easily generate analytic solutions for some physically interesting linear ODE systems, with both constant and variable coefficients. We will not go into mathematical methods here, leaving this to your mathematics education. However, if you would like to see computer algebra recipes that apply mathematical methods to solving ODEs, you are referred to *Computer Algebra Recipes for Mathematical Physics* [Enn05].

3.1   First-Order Models

The most general nth-order linear ODE can be written in the form

$$a_0(t) \frac{d^n x(t)}{dt^n} + a_1(t) \frac{d^{n-1} x(t)}{dt^{n-1}} + \cdots + a_{n-1}(t) \frac{dx(t)}{dt} + a_n(t) x(t) = h(t), \quad (3.1)$$

the equation being labeled as linear because each term on the left-hand side is linear, or of first order, in the dependent variable x. For the sake of definiteness, the independent variable has been taken to be the time t here, but could be a spatial variable. If $h(t) = 0$, the differential equation is said to be *homogeneous*, otherwise it is *inhomogeneous*.

In the following two recipes, we look at two examples of inhomogeneous first-order ODEs, the first involving constant coefficients, the second containing variable coefficients.

3.1.1   How's Your Blood Pressure?

Amid the pressure of great events, a general principle gives no help.
Georg Hegel, German philosopher (1770–1831)

On measuring your blood pressure, your doctor will give you two numbers. A *normal* blood pressure for humans is 120/80, the first number specifying the maximum (*systolic*) pressure (in units of mm of Hg) on the arterial walls, the second number the minimum (*diastolic*) pressure. The blood pressure is generated by the beating of the heart, its variation controlled by the aorta.

The aorta is the large blood vessel into which the arterial blood flows on leaving the heart. During the systolic phase of the heartbeat cycle, blood is pumped under pressure from the heart into one end of the aorta, whose walls then stretch in order to accommodate the blood. The diastolic phase then follows during which there is no flow of blood into the aorta, its walls elastically contracting. The blood is then squeezed out of the aorta and around the body's circulatory system. The following recipe presents a simple model of the variation of blood pressure due to the beating heart and the aorta.

The entry `infolevel[dsolve]` will provide some information about the `dsolve` command, the amount of information generally increasing as the integer (which can vary from 1 to 5) specified on the right of the colon is increased.

```
>  restart: infolevel[dsolve]:=2:
```
Let $V(t)$ be the volume of the aorta and $p(t)$ the pressure within it at time t. Assuming that the aorta expands linearly with increasing p, then $V = V_0 + C\,p$, where V_0 and C are constants. The parameter C, called the *compliance*, is a measure of the stretchability of the aorta. Mentally differentiating this relation produces *ode1*.

```
>  ode1:=diff(V(t),t)=C*diff(p(t),t);
```

$$ode1 := \frac{d}{dt} V(t) = C\left(\frac{d}{dt} p(t)\right)$$

The rate of change of volume (dV/dt) of the aorta is equal to the difference between the rate $f(t)$ at which blood is pumped into the aorta by the heart and the rate $p(t)/R$ at which blood is pumped out of the aorta into the circulatory system. The constant R is referred to as the *systemic resistance*. Thus, the second ODE is is now given by *ode2*.

```
>   ode2:=diff(V(t),t)=f(t)-p(t)/R;
```

$$ode2 := \frac{d}{dt} V(t) = f(t) - \frac{p(t)}{R}$$

An ODE (*ode3*) for pressure alone results on substituting *ode1* into *ode2*.

```
>   ode3:=subs(ode1,ode2);
```

$$ode3 := C\left(\frac{d}{dt} p(t)\right) = f(t) - \frac{p(t)}{R}$$

As a simple model of the forcing function $f(t)$, let's assume that during the systolic (pumping) phase $f(t) = A\sin(\pi t/\tau)$, with A the amplitude and τ the duration of this phase, and $f(t) = 0$ during the diastolic (nonpumping) phase. So $f(t)$ is a piecewise forcing function. If T is the time for one complete cycle (time between heart beats), the piecewise function for two heartbeats can be entered using the following `piecewise` command.

```
>   f(t):=piecewise(t<=tau,A*sin(Pi*t/tau),t<T,0,
              t<=T+tau,A*sin(Pi*(t-T)/tau),t<2*T,0);
```

$$f(t) := \begin{cases} A\sin\left(\dfrac{\pi t}{\tau}\right) & t \leq \tau \\ 0 & t < T \\ A\sin\left(\dfrac{\pi (t-T)}{\tau}\right) & t \leq T + \tau \\ 0 & t < 2T \end{cases}$$

The piecewise function $f(t)$ is automatically substituted into *ode3*, the complete form of the ODE being reproduced below.

```
>   ode3:=ode3;
```

$$ode3 := C\left(\frac{d}{dt} p(t)\right) = \left(\begin{cases} A\sin\left(\dfrac{\pi t}{\tau}\right) & t \leq \tau \\ 0 & t < T \\ A\sin\left(\dfrac{\pi (t-T)}{\tau}\right) & t \leq T + \tau \\ 0 & t < 2T \end{cases}\right) - \frac{p(t)}{R}$$

Using the `dsolve` command, *ode3* is solved for the pressure $p(t)$, subject to the initial condition $p(0) = P$. To simplify the solution, it is assumed that $\tau > 0$, $T > \tau$, $C > 0$, $R > 0$, $A > 0$, and $P > 0$.

```
>   sol:=dsolve({ode3,p(0)=P},p(t))
           assuming (tau>0,T>tau,C>0,R>0,A>0,P>0) ;
```

The `assuming` command applies the assumption only to the command line in which it appears. If you wish to apply assumptions to the entire work sheet, you can use the `assume` command at the beginning of the work sheet. In either case, the assumptions are determined by trial and error, i.e., looking at the form of the answer, which is now given.

Methods for first order ODEs:
  — Trying classification methods —
  trying a quadrature
  trying 1st order linear
  $<-$ *1st order linear successful*

$sol := p(t) =$

$$
\begin{cases}
e^{\left(\frac{-t}{RC}\right)}\left(P + \dfrac{AC\pi\tau R^2}{\tau^2 + \pi^2 R^2 C^2}\right) - \dfrac{AC\pi\tau R^2 \cos\left(\dfrac{\pi t}{\tau}\right)}{\tau^2 + \pi^2 R^2 C^2} + \dfrac{AR\tau^2 \sin\left(\dfrac{\pi t}{\tau}\right)}{\tau^2 + \pi^2 R^2 C^2},\ t \le \tau \\[4ex]
\dfrac{AC\pi\tau R^2\, e^{\left(\frac{\tau-t}{RC}\right)}}{\tau^2 + \pi^2 R^2 C^2} + e^{\left(\frac{-t}{RC}\right)}\left(P + \dfrac{AC\pi\tau R^2}{\tau^2 + \pi^2 R^2 C^2}\right),\ t \le T
\end{cases}
$$

. . . .

In the above solution, we have shown only the analytic form (thus the dots) for the first cycle to save on text space. You can observe the complete solution by executing the recipe on your computer. Because of the inclusion of the `infolevel[dsolve]` command, we are told in the output that Maple recognizes *ode3* as a first-order linear ODE and presumably uses a standard method for solving such an ODE. You may object to this "black box" nature of the `dsolve` command, but obtaining the answer easily in our opinion is more important than the mechanical details leading up to it. However, if the methodology is important to you, we shall mimic a hand calculation in the next recipe.

To make a plot, we shall take the parameter values[1] to be $T = 1$ second (60 heartbeats per minute), $\tau = 0.15$ s, $C = 0.002$ liters/mm Hg, $P = 80$ mm Hg, and $R = 1056$ mm Hg/liter/second.

```
>   tau:=0.15: T:=1: C:=0.002: P:=80: R:=1056:
```

The amplitude parameter A must be such that the right-hand side (rhs) of the solution must be the same at $t = T$ as at $t = 0$, i.e., the heartbeat is periodic.

```
>   eq:=eval(rhs(sol),t=0)=eval(rhs(sol),t=T);
```

$$eq := 80. = \frac{432.0593387\,A\,\pi}{0.0225 + 4.46054400\,\pi^2} + 49.82624166$$

The above equation is numerically solved for A.

```
>   A:=fsolve(eq,A);
```

$$A := 0.9791418590$$

The pressure is then given by the rhs of *sol*, the floating-point evaluation command being used to express the answer to four digits.

```
>   Pressure:=evalf(rhs(sol),4);
```

[1]These approximate values are taken from the Internet.

Pressure:= $\{103.4\,e^{(-0.4735\,t)} - 23.36\cos(20.95\,t) + 0.5280\sin(20.95\,t),\ t \le 0.15$

$\{103.4\,e^{(-0.4735\,t)} + 23.36\,e^{(0.07102-0.4735\,t)},\ t \le 1.$

$\{103.4\,e^{(-0.4735\,t)} + 23.36\,e^{(0.07102-0.4735\,t)} + 23.36\,e^{(0.4735-0.4735\,t)}$

$- 23.36\cos(-20.95 + 20.95\,t) + 0.5280\sin(-20.95 + 20.95\,t),\ t \le 1.15$

$\{103.4\,e^{(-0.4735\,t)} + 23.36\,e^{(0.07102-0.4735\,t)} + 23.36\,e^{(0.4735-0.4735\,t)}$

$+ 23.36\,e^{(0.5445-0.4735\,t)},\ 1.15 < t$

The piecewise forcing function $f(t)$ is plotted over the time interval $t = 0$ to $2T$, axis labels being added to the plot, which is shown in Figure 3.1.

```
>   plot(f(t),t=0..2*T,labels=["t","f(t)"],tickmarks=[3,4]);
```

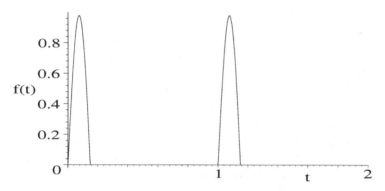

Figure 3.1: Piecewise forcing function $f(t)$.

The blood pressure variation is plotted over the same interval.

```
>   plot(Pressure,t=0..2*T,labels=["t","Pressure"]);
```

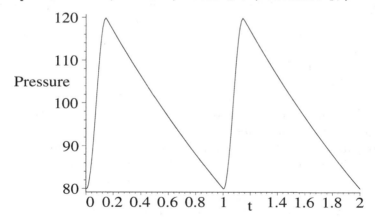

Figure 3.2: Time variation of the blood pressure.

The resulting picture is shown in Figure 3.2, the plot capturing the gross behavior of the variation in blood pressure with time. More detailed behavior can be obtained by using more sophisticated models. You might try an Internet search to learn more about these models.

PROBLEMS:

Problem 3-1: More heartbeats
Modify the forcing function in the text recipe to include four heartbeats. Do an Internet search on blood pressure and discuss how the model could be improved.

Problem 3-2: An RL circuit
A simple electrical circuit consists of a resistor of R ohms (Ω) in series with an inductor of L henries (H) and a battery providing a constant voltage v (in volts (V)). If $i(t)$ is the current in the circuit at time t, the voltage drop across the resistor is $R i$ (*Ohm's law*) and across the inductor is $L (di/dt)$.

(a) Making use of *Kirchhoff's voltage rule*, which states that the sum of the voltage drops is equal to the supplied voltage, derive the first-order ODE for $i(t)$ and solve it, given that the initial current is zero.

(b) Taking $L = 4\,\text{H}$, $R = 12\,\Omega$, and $v = 60\,\text{V}$, plot the solution over a time interval that brings the current to within 1% of its asymptotic value.

(c) The battery is replaced with a generator that produces a variable voltage of $v(t) = 60 \sin(30\,t)$ volts, the resistor and inductor retaining the same values as in part (b). If the initial current is zero, determine the current $i(t)$ for $t \geq 0$. Plot the current over a time interval that brings the current to within 1% of steady state.

Problem 3-3: Learning curves for widget production
Learning curves are used by scientists interested in learning theory. A *learning curve* is a plot of the performance $P(t)$ of someone learning a skill as a function of the training time t. A simple model of learning curve is to assume that the rate dP/dt at which performance improves is proportional to $M - P(t)$, where M is the maximum level of performance. Determine the analytic form of $P(t)$ and discuss what this model implies.

Two new workers (Jimbo and Jumbo) are hired for an assembly line producing widgets. Jimbo (Jumbo) produces 25 (35) widgets during the first hour and 45 (50) widgets the second hour. Estimate the maximum number of widgets per hour that each is capable of producing and plot their learning curves.

Problem 3-4: Population growth
A deer population initially numbers 1000 and has a growth rate of 0.5 when t is measured in months. The population is "harvested" throughout the year at the rate of $h (2 + \cos(\pi t/6))$ per month, where h is a constant. Determine the deer population number as a function of t. What value of h would lead to a zero population growth over the 12-month period from $t = 0$ to $t = 12$? Plot the deer-population curve for this 12-month period.

3.1.2   Greg Arious Nerd's Problem

*It will be found, in fact, that the ingenious are always fanciful,
and the truly imaginative never otherwise than analytic.*
Edgar Allan Poe, American writer, *The Murders in the Rue Morgue*, 1841

Greg Arious Nerd, an eminent mathematician at the Erehwon Institute of Technology (EIT), has assigned the following problem for his undergraduate class to solve using computer algebra. They are not to use `dsolve` to analytically solve the relevant ODE, but instead use command structures found in the DEtools library package that mimic the steps in a hand calculation.

Starting from rest, Evil Knievel weevil experiences a time-dependent drive force $F_{drive} = t^2 e^{-t}$ and a drag force $F_{drag} = -(t^2/(1+t)) v(t)$, per unit mass. Mimicking a hand calculation, determine Evil's velocity $v(t)$ at time t and plot it over the interval $t = 0$ to 10 Erehwonian time units. What distance does Evil travel in this interval?

Here is Professor Nerd's answer key. Loading the DEtools package,

```
>   restart: with(DEtools):
```
the drag and drive forces are entered.

```
>   F[drag]:=-(t^2/(1+t))*v(t); F[drive]:=t^2*exp(-t);
```

$$F_{drag} := -\frac{t^2\, v(t)}{1+t} \qquad F_{drive} := t^2\, e^{(-t)}$$

Applying Newton's second law, the ODE governing Evil Knievel's velocity is

```
>   ode:=diff(v(t),t)=F[drag]+F[drive];
```

$$ode := \frac{d}{dt} v(t) = -\frac{t^2\, v(t)}{1+t} + t^2\, e^{(-t)}$$

Although not requested, Nerd applies the `odeadvisor` command to *ode*. This command will classify the ODE and inclusion of the `help` option causes a relevant Help page with useful hyperlinks to be opened. The Help page should be closed when one is finished reading it.

```
>   odeadvisor(ode,help); #close Help page
```

$$[\_linear]$$

In this case, *ode* is classified as linear, which is obvious by inspection.

Proceeding by hand, one would look for the integrating factor[2] of the first-order inhomogeneous ODE. With Maple, the integrating factor *IF* for *ode* is obtained by entering `intfactor(ode)`, which is then simplified.

```
>   IF:=intfactor(ode);
```

$$IF := e^{\left(\frac{t^2}{2} - t + \ln(1+t)\right)}$$

```
>   IF:=simplify(IF);
```

---

[2]For a general linear first-order ODE, $\dot{x} + f(t)\, x = g(t)$, the integrating factor is $e^{\int f(t)\, dt}$.

$$IF := (1 + t) \, e^{\left(\frac{t \, (t-2)}{2}\right)}$$

With the integrating factor known, the standard mathematical procedure is to multiply *ode* by *IF* and integrate. This is easily accomplished with the first integral (`firint`) command.

> `sol1:=firint(ode*IF);`

$$sol1 := \left(e^{\left(\frac{t \, (t-2)}{2}\right)} + e^{\left(\frac{t \, (t-2)}{2}\right)} t \right) v(t) - 3 \, t \, e^{(1/2 \, t^2 - 2 \, t)} - 4 \, e^{(1/2 \, t^2 - 2 \, t)}$$

$$+ \frac{5}{2} \, I \, \sqrt{\pi} \, e^{(-2)} \, \sqrt{2} \, \mathrm{erf}\left(\frac{1}{2} I \, \sqrt{2} \, t - \sqrt{2} \, I \right) - t^2 \, e^{(1/2 \, t^2 - 2 \, t)} + \_C1 = 0$$

In the solution output, I stands for $\sqrt{-1}$, erf for the *error function,*[3] and $\_C1$ is the integration constant to be determined. Then, *sol1* is solved for $v(t)$, the result (assigned the name V) not being shown here in the text.

> `V:=solve(sol1,v(t));`

Evil Knievel's initial velocity is zero. So the constant $\_C1$ is determined by evaluating the velocity V at $t = 0$, equating the result to 0, and solving for $\_C1$. The constant will be automatically substituted into V.

> `_C1:=solve(eval(V,t=0)=0,_C1);`

$$\_C1 := 4 + \frac{5}{2} I \, \sqrt{\pi} \, e^{(-2)} \, \sqrt{2} \, \mathrm{erf}(\sqrt{2} \, I)$$

On simplifying V, Evil Knievel's velocity is now determined at arbitrary time $t \geq 0$. This is a nontrivial result to derive by hand.

> `Vel:=simplify(V);`

$$Vel := \frac{1}{2} \left(6 \, t \, e^{\left(\frac{t \, (t-4)}{2}\right)} + 8 \, e^{\left(\frac{t \, (t-4)}{2}\right)} - 5 \, I \, \sqrt{\pi} \, e^{(-2)} \, \sqrt{2} \, \mathrm{erf}\left(\frac{1}{2} I \, \sqrt{2} \, t - \sqrt{2} \, I \right) \right.$$

$$\left. + 2 \, t^2 \, e^{\left(\frac{t \, (t-4)}{2}\right)} - 8 - 5 \, I \, \sqrt{\pi} \, e^{(-2)} \, \sqrt{2} \, \mathrm{erf}(\sqrt{2} \, I) \right) e^{\left(-\frac{t \, (t-2)}{2}\right)} / (1 + t)$$

Now, since the original ODE was completely real, the velocity must also be completely real. It can be converted to a real form using the `Re` (real part of) command, assuming that $t > 0$. The new function erfi is the *imaginary error function,*[4] which is a real function here.

> `Vel:=Re(Vel) assuming t>0;`

$$Vel := \frac{1}{2} \left(6 \, t \, e^{\left(\frac{t \, (t-4)}{2}\right)} + 8 \, e^{\left(\frac{t \, (t-4)}{2}\right)} + 5 \, \sqrt{\pi} \, e^{(-2)} \, \sqrt{2} \, \mathrm{erfi}\left(\frac{\sqrt{2} \, t}{2} - \sqrt{2} \right) \right.$$

$$\left. + 2 \, t^2 \, e^{\left(\frac{t \, (t-4)}{2}\right)} - 8 + 5 \, \sqrt{\pi} \, e^{(-2)} \, \sqrt{2} \, \mathrm{erfi}(\sqrt{2}) \right) e^{\left(-\frac{t \, (t-2)}{2}\right)} / (1 + t)$$

---

[3] Defined as $\mathrm{erf}(u) = \frac{2}{\sqrt{\pi}} \int_0^u e^{-t^2} \, dt$.

[4] Defined as $\mathrm{erfi}(u) = -I \, \mathrm{erf}(I \, u) = \frac{2}{\sqrt{\pi}} \int_0^u e^{t^2} \, dt$.

Evil Knievel's velocity is plotted over the time interval $t = 0$ to 10 time units, labels being added to the graph.

```
> plot(Vel,t=0..10,labels=["t","Vel"],tickmarks=[3,3]);
```

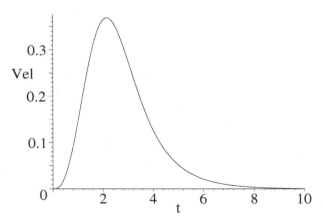

Figure 3.3: Evil Knievel's velocity profile.

The resulting picture is shown in Figure 3.3, the velocity curve increasing from zero to a maximum and then decreasing to zero again as $t \to \infty$.

The distance traveled in the time interval $t = 0$ to 10 is equal to the area under the velocity curve between these limits, this area obtained by performing the integration $\int_0^{10} Vel\, dt$. However, if the integer 10 is used, the integral will not be evaluated. A floating-point answer could be obtained by apply the evalf command. An alternative way used here in Professor Nerd's answer key is to enter the upper limit as 10.0, which forces a floating-point evaluation of the integral. It does not imply any increase in accuracy.

```
> distance:=int(Vel,t=0..10.0); #note floating point number
```

$$distance := 0.9943356328$$

So, Evil Knievel travels just under 1 Erehwonian spatial unit in the time interval.

PROBLEMS:

Problem 3-5: Direct solution
Solve Professor Nerd's problem directly with the dsolve command and check that the answer is identical to that derived in the text recipe.

Problem 3-6: Different drive force
The drive force in the text recipe is replaced with $F_{drive} = \sin(t)\, t^2\, e^{-t}$. Determine the velocity as a function of time and plot the result over the interval $t = 0$ to 10. What is Evil Knievel's maximum velocity and at what time does this occur? What is his maximum displacement from the origin in this interval and at what time does this occur?

3.2 Second-Order Models

3.2.1 Daniel Encounters Resistance

Belief like any other moving body follows the path of least resistance.
Samuel Butler, English writer (1835–1902)

In the introductory recipe, recall that Richard's grandson Daniel threw a small ball with an initial speed of 15 m/s toward a 3.5-meter-high fence located 20 meters from the ball's initial position. The ball left his hand at a height of 2 m above the level ground and just cleared the top of the fence. The gravitational acceleration has the value $g = 9.8$ m/s^2. The ball was regarded as a point particle and air resistance was neglected.

In the recipe, we determined the angle with the horizontal that the ball was thrown, the time for the ball to reach the fence, and animated the motion of the ball with the fence included.

In the following recipe, we shall make the above model calculation more realistic by including the effect of air resistance on the ball. Provided that the ball is smooth and its velocity $\vec{v}$ is not too high, the air resistance is governed by Stokes's resistance law, the drag force taking the form $\vec{F}_{drag} = -m\,k\,\vec{v}$, where m is the mass of the ball and the drag coefficient k is positive.

As previously, the origin is chosen to be on the ground below the initial position of the ball and the x-coordinate is taken to be horizontal and the y-coordinate vertical. The plots package, needed for the animation, is loaded.

```
>  restart: with(plots):
```
With air resistance included, Newton's second law of motion yields the following ODE in the x direction.

```
>  xeq:=diff(x(t),t,t)=-k*diff(x(t),t);
```

$$xeq := \frac{d^2}{dt^2}x(t) = -k\left(\frac{d}{dt}x(t)\right)$$

Although *xeq* is a second-order ODE, it clearly can be reduced to a first-order ODE by integrating both sides once, and then solved by the same "hand" procedure used in the last recipe. However, it's quicker to solve for $x(t)$ using the `dsolve` command, subject to the initial condition $x(0) = xb$ and $\dot{x}(0) = V\cos(\phi)$. Here xb is the initial x-coordinate of the ball, V the initial speed, and ϕ the initial angle of the velocity with the horizontal. For the time being, everything is kept symbolic, the parameter values being substituted later. After applying the `dsolve` command, we take the rhs of the resulting answer, yielding the x-coordinate of the ball at time $t \geq 0$.

```
>  x:=rhs(dsolve({xeq,x(0)=xb,D(x)(0)=V*cos(phi)},x(t)));
```

$$x := \frac{xb\,k + V\cos(\phi)}{k} - \frac{V\cos(\phi)\,e^{(-k\,t)}}{k}$$

Taking *xf* to be the x-coordinate of the fence, we find the time *tf* for the ball to reach the fence by equating x to *xf* and solving for t.

```
>   tf:=solve(x=xf,t);
```

$$tf := -\frac{\ln\left(\dfrac{xb\,k + V\cos(\phi) - xf\,k}{V\cos(\phi)}\right)}{k}$$

In the y direction, the gravitational force on the ball must be included as well as air resistance. Newton's second law yields the ODE given in yeq.

```
>   yeq:=diff(y(t),t,t)=-k*diff(y(t),t)-g;
```

$$yeq := \frac{d^2}{dt^2}y(t) = -k\left(\frac{d}{dt}y(t)\right) - g$$

Solving yeq with the $\texttt{dsolve}$ command, subject to the initial condition $y(0) = yb$, $\dot{y}(0) = V\sin(\phi)$, where yb is the ball's initial y-coordinate, and taking the rhs, yields y.

```
>   y:=rhs(dsolve({yeq,y(0)=yb,D(y)(0)=V*sin(phi)},y(t)));
```

$$y := -\frac{e^{(-k\,t)}\,(V\sin(\phi)\,k + g)}{k^2} - \frac{g\,t}{k} + \frac{yb\,k^2 + V\sin(\phi)\,k + g}{k^2}$$

Evaluating y at $t = tf$, and equating to the fence height yf, yields the following formidable looking transcendental equation eq for the unknown angle ϕ.

```
>   eq:=eval(y,t=tf)=yf;
```

$$eq := -\frac{(xb\,k + V\cos(\phi) - xf\,k)\,(V\sin(\phi)\,k + g)}{k^2\,V\cos(\phi)}$$

$$+\frac{g\ln\left(\dfrac{xb\,k + V\cos(\phi) - xf\,k}{V\cos(\phi)}\right)}{k^2} + \frac{yb\,k^2 + V\sin(\phi)\,k + g}{k^2} = yf$$

To solve the transcendental equation for ϕ, the given values $xb = 0\,\mathrm{m}$, $yb = 2\,\mathrm{m}$, $xf = 20\,\mathrm{m}$, $yf = 3.5\,\mathrm{m}$, $V = 15\,\mathrm{m/s}$, and $g = 9.8\,\mathrm{m/s^2}$, are entered. The drag coefficient is taken to be $k = 0.01\,\mathrm{s^{-1}}$.

```
>   xb:=0: yb:=2: xf:=20: yf:=3.5: V:=15: g:=9.8: k:=0.01:
```

The transcendental equation must be solved numerically. This is done in the following two command lines, two different search ranges (in radians) being specified for ϕ to determine the two possible angles, Φ_1 and Φ_2, at which the ball can be thrown to just clear the fence. Making use of the ditto operator, we convert the first angle to degrees.

```
>   Phi[1]:=fsolve(eq,phi,0..0.8); evalf(convert(%,degrees));
```

$$\Phi_1 := 0.6696475750 \qquad 38.36797980\ degrees$$

```
>   Phi[2]:=fsolve(eq,phi,0.8..Pi/2);
```

$$\Phi_2 := 0.9693759172$$

Without air resistance, it was previously found that the smaller of the two possible angles was 37.47 degrees, about 1 degree less than the value found above. It is left as an exercise for you to compare the values for the upper angle, with and without air resistance.

To animate the ball, the first angle Φ_1 will be selected.

```
>  phi:=Phi[1]: #select angle
```
In this case, the time *tf* to reach the fence is calculated
```
>  tf:=evalf(tf);
```
$$tf := 1.715218642$$
to be about 1.72 s, slightly more than the 1.68 s without air resistance.

Setting $y = 0$, we determine the time T for the ball to hit the ground,
```
>  T:=solve(y=0,t);
```
$$T := 2.090175405, \ -0.1946627301$$
the positive angle (first solution here) being selected.
```
>  T2:=T[1]; #choose positive time
```
$$T2 := 2.090175405$$
The ball hits the ground after 2.09 seconds, compared to 2.06 seconds without air resistance.

The fence is plotted as a quite thick (default color red) line.
```
>  fence:=plot([[xf,0],[xf,yf]],style=line,thickness=3):
```
Using exactly the same syntax as in the introductory recipe, the motion of the ball is animated with the fence as background.
```
>  animate(pointplot,[[[x,y]],symbol=circle,symbolsize=14],
       t=0..T2,frames=200,background=fence,scaling=constrained);
```
To see this animation, execute the recipe on your computer, click on the opening frame, and then on the start arrow in the Maple tool bar.

PROBLEMS:

Problem 3-7: Maximum drag coefficient
With all other parameters the same as in the text recipe, what is the maximum value of the drag coefficient k such that the ball just clears the fence? What are the two corresponding initial angles?

Problem 3-8: How high?
If the drag coefficient k is equal to $0.1\,\mathrm{s}^{-1}$, how high can the fence be for the ball to just clear it, all other parameters the same as in the text recipe? What are the two initial angles in this case?

Problem 3-9: A falling raindrop
For a sphere of diameter d meters moving in air, the approximate value of the constant k in Stokes's linear resistance law, $\vec{F}_{drag} = -k\,\vec{v}$ newtons, is given by $k = 1.55 \times 10^{-4}\,d$. For a small spherical raindrop (density $\rho = 10^3$ kg/m^3, $d = 10^{-4}$ m) falling from rest, determine the distance through which it falls in t seconds and its velocity then. Plot the distance and velocity separately over a time interval $t = 0$ to the time at which the velocity reaches 99 percent of its *terminal* (asymptotic) value.

3.2.2   Meet Mr. Laplace

The weight of evidence for an extraordinary claim must be
proportioned to its strangeness (known as the principle of Laplace).
Pierre-Simon Laplace, French mathematician (1749–1827)

Pierre-Simon Laplace was one of the most influential scientists in history. Indeed, he has been referred to as the Newton of France. Not only did he make outstanding contributions to astronomy, for example, mathematically investigating the stability of the solar system, he developed probability theory as a coherent body of knowledge from a set of miscellaneous problems, and played a leading role in forming the modern discipline of mathematical physics. One of his most important contributions to the latter is the Laplace transform, which can be used to solve linear ODEs.

The Laplace transform of a function $f(t)$ is defined as

$$L(f(t)) \equiv F(s) = \int_0^\infty f(t)e^{-st}dt.$$

By integrating by parts and assuming that $e^{-st}f(t) \to 0$ as $t \to \infty$, it is easy to show that

$$L\left(\dot{f}(t)\right) = s\,F(s) - f(0) \quad \text{and} \quad L\left(\ddot{f}(t)\right) = s^2\,F(s) - s\,f(0) - \dot{f}(0).$$

The Laplace transform method of solving a linear ODE (system) with constant coefficients is to Laplace transform the ODE, solve the resulting algebraic equation for $F(s)$, and then perform the inverse Laplace transform to obtain the solution $f(t)$. In the era before computer algebra existed, tables of Laplace transforms and their inverses were almost as common as integral tables, and one of their main uses was in solving linear ODEs. With the Maple computer algebra system, these tables are obsolete, since Maple has an integral transform library package that includes the Laplace transform and its inverse.

The use of this package is now demonstrated in the first of the two recipes that follow. It is desired to solve the following forced oscillator ODE for $x(t)$:

$$\ddot{x} + 4\,\dot{x} + 13\,x = \sin(t), \quad \text{with } x(0) = 1 \text{ and } \dot{x}(0) = -5.$$

Entering the integral transform package and using a semicolon to display its contents, we see that it contains the Laplace transform (`laplace`) command and its inverse (`invlaplace`).

```
>  restart: with(inttrans);
```

   [*addtable, fourier, fouriercos, fouriersin, hankel, hilbert, invfourier,*
      invhilbert, invlaplace, invmellin, laplace, mellin, savetable]

The forced oscillator *ode* is entered,

```
>  ode:=diff(x(t),t,t)+4*diff(x(t),t)+13*x(t)=sin(t);
```

$$ode := \left(\frac{d^2}{dt^2}x(t)\right) + 4\left(\frac{d}{dt}x(t)\right) + 13\,x(t) = \sin(t)$$

along with the initial conditions.

```
>  x(0):=1: D(x)(0):=-5:
```

The `alias` command is used to replace $laplace(x(t), t, s)$, which would otherwise appear in the output, with the symbol F. That is to say F will be the Laplace transform of $x(t)$, the transformed independent variable being s.

```
>  alias(F=laplace(x(t),t,s)):
```

The Laplace transform of *ode* is performed,

```
>  eq:=laplace(ode,t,s);
```

$$eq := s^2\, F + 1 - s + 4\, s\, F + 13\, F = \frac{1}{s^2 + 1}$$

and the algebraic equation *eq* solved for F.

```
>  F:=solve(eq,F);
```

$$F := \frac{s\,(-s + s^2 + 1)}{s^4 + 14\, s^2 + 4\, s^3 + 4\, s + 13}$$

The inverse Laplace transform of F is calculated, yielding the analytic solution (labeled X) of the forced oscillator ODE.

```
>  X:=invlaplace(F,s,t);
```

$$X := -\frac{1}{40} \cos(t) + \frac{3}{40} \sin(t) + \frac{1}{120}\,(123 \cos(3\,t) - 121 \sin(3\,t))\,e^{(-2\,t)}$$

The steps carried out above mimic a hand calculation. An easier way to derive exactly the same form of $x(t)$ is to use the `dsolve` command with the Laplace transform option, i.e., include `method=laplace`.

```
>  dsolve({ode,x(0)=1,D(x)(0)=-5},x(t),method=laplace);
```

$$x(t) = -\frac{1}{40} \cos(t) + \frac{3}{40} \sin(t) + \frac{1}{120}\,(123 \cos(3\,t) - 121 \sin(3\,t))\,e^{(-2\,t)}$$

The implementation of the Laplace transform method with Maple can be extended to systems of linear ODEs, as is demonstrated in the following problem, a problem that is quite challenging to do by hand.

Masses m_1 and m_2, with equilibrium positions at $x=2$ and $x=5$, are free to move horizontally on a smooth surface (the x-axis). The mass m_1 is connected to a fixed wall on its left by a linear spring (spring constant k) and on its right to m_2 with an identical spring. A driving force $F = f\, \sin(\omega\, t)$ acts to the right on m_2. A fluid resistance is present, given by Stokes's drag law, $F_{drag} = -a\, v$.

(a) Derive the governing ODEs for the displacements $x_1(t)$ and $x_2(t)$ of m_1 and m_2 from equilibrium. Taking $m_1 = 2$, $m_2 = 1$, $k = 1$, $a = 1$, $\omega = 2$, $f = 2$, $x_1(0) = 1$, $x_2(0) = 0$, $\dot{x}_1(0) = 0$, and $\dot{x}_2(0) = 0$, solve the ODEs using the Laplace transform option in the `dsolve` command.

(b) Extract the steady-state and transient parts of x_1 and x_2 separately and plot them. Discuss the results.

(c) Animate the motion of m_1 and m_2 about their equilibrium positions over a time interval for which the transients become small.

Loading the plots package, we use the `alias` command. Entering `x1`, `x2`, `m1`, `m2` will produce the subscripted quantities x_1, x_2, m_1, and m_2 in the output.

```
> restart: with(plots):
```
```
> alias(x[1]=x1,x[2]=x2,m[1]=m1,m[2]=m2):
```

When m_2 is displaced from equilibrium by amount x_2 at time t in, say, the positive x-direction, it exerts a force on m_1 through the connecting spring. If m_1 is displaced from equilibrium by amount $x_1(t)$, the force exerted on m_1 by m_2 is $k\,(x_2(t) - x_1(t))$. The spring connected to the wall will exert a force $-k\,x_1(t)$ in the opposite direction. Including damping, Newton's second law yields the ODE *ode1* governing the displacement $x_1(t)$ of m_1.

```
> ode1:=m1*diff(x1(t),t,t)
           =k*(x2(t)-x1(t))-k*x1(t)-a*diff(x1(t),t);
```

$$ode1 := m_1 \left(\frac{d^2}{dt^2} x_1(t) \right) = k\,(x_2(t) - x_1(t)) - k\,x_1(t) - a \left(\frac{d}{dt} x_1(t) \right)$$

The mass m_1 will exert an equal and opposite force on m_2 through the connecting spring, i.e., $-k\,(x_2(t) - x_1(t))$. Including the damping and driving forces, the displacement $x_2(t)$ of m_2 is given by the ODE *ode2*.

```
> ode2:=m2*diff(x2(t),t,t)
           =-k*(x2(t)-x1(t))-a*diff(x2(t),t)+f*sin(omega*t);
```

$$ode2 := m_2 \left(\frac{d^2}{dt^2} x_2(t) \right) = -k\,(x_2(t) - x_1(t)) - a \left(\frac{d}{dt} x_2(t) \right) + f \sin(\omega t)$$

One has a system of two coupled, linear, second-order, inhomogeneous ODEs. To solve them, let's first enter the given parameter values,

```
> m1:=2: m2:=1: k:=1: a:=1: omega:=2: f:=2:
```

and the initial conditions.

```
> ic:=x1(0)=1,x2(0)=0,D(x1)(0)=0,D(x2)(0)=0:
```

The set of ODEs is analytically solved for $x_1(t)$ and $x_2(t)$, subject to the initial conditions, using the `dsolve` command with the Laplace transform option. The same answer could be obtained by mimicking the hand calculation in the previous example. This is left as a problem.

```
> sol:=dsolve({ode1,ode2,ic},{x1(t),x2(t)},method=laplace);
```

$$sol := \{ x_1(t) = \frac{36}{493} \cos(2t) + \frac{26}{493} \sin(2t)$$

$$- \frac{1}{194242} \left(\sum_{\_\alpha = \%1} (56380 + 339955\_\alpha + 193087\_\alpha^2 + 188250\_\alpha^3)\, e^{(-\alpha t)} \right) \},$$

$$\{ x_2(t) = -\frac{164}{493} \cos(2t) - \frac{228}{493} \sin(2t)$$

$$- \frac{1}{194242} \left(\sum_{\_\alpha = \%1} (107923 + 533529\_\alpha + 249954\_\alpha^2 + 293736\_\alpha^3)\, e^{(-\alpha t)} \right) \}$$

$$\%1 := \text{RootOf}(2\_Z^4 + 3\_Z^3 + 5\_Z^2 + 3\_Z + 1)$$

The sine and cosine terms in the solution survive as $t \to \infty$ and represent the steady-state parts, while the exponential terms decay to zero and thus correspond to the transient parts of the solution. The summation is over the four roots of the quartic polynomial given in the subexpression %1. To manipulate the solution further, it is now assigned.

```
>  assign(sol):
```

Using the **remove** command, the exponential terms are removed from $x_1(t)$ and $x_2(t)$ to yield the steady-state parts, *x1ss* and *x2ss*, separately.

```
>  x1ss:=remove(has,x1(t),exp); x2ss:=remove(has,x2(t),exp);
```

$$x1ss := \frac{36}{493}\cos(2\,t) + \frac{26}{493}\sin(2\,t) \quad x2ss := -\frac{164}{493}\cos(2\,t) - \frac{228}{493}\sin(2\,t)$$

The **select** command is used to extract the exponential terms from $x_1(t)$ and $x_2(t)$, thus producing the transient parts separately.

```
>  x1tr:=select(has,x1(t),exp): x2tr:=select(has,x2(t),exp):
```

Because the roots of a quartic polynomial must be found, the numerical floating-point evaluation command is applied to each of the transient parts. The complex evaluation command is then used to simplify the outputs, which are given by *x1tr* and *x2tr*.

```
>  x1tr:=evalc(evalf(x1tr)); x2tr:=evalc(evalf(x2tr));
```

$$\begin{aligned}
x1tr := \ & 0.5591296681\, e^{(-0.3923340189\,t)} \cos(0.3903466128\,t) \\
& + 0.9914981558\, e^{(-0.3923340189\,t)} \sin(0.3903466128\,t) \\
& + 0.3678480196\, e^{(-0.3576659811\,t)} \cos(1.226572647\,t) \\
& - 0.1154210539\, e^{(-0.3576659811\,t)} \sin(1.226572647\,t)
\end{aligned}$$

$$\begin{aligned}
x2tr := \ & 0.6802842310\, e^{(-0.3923340189\,t)} \cos(0.3903466128\,t) \\
& + 1.721342502\, e^{(-0.3923340189\,t)} \sin(0.3903466128\,t) \\
& - 0.3476270302\, e^{(-0.3576659811\,t)} \cos(1.226572647\,t) \\
& + 0.3225193055\, e^{(-0.3576659811\,t)} \sin(1.226572647\,t)
\end{aligned}$$

You might have thought our earlier identification of the exponential terms as the transient contribution a bit premature, but now one can clearly see that they decay to zero as t goes to infinity.

The steady-state and transient contributions are now plotted separately. Lists are used here so that the steady-state and transient parts of $x_1(t)$ are colored red on the computer screen, while the corresponding parts for $x_2(t)$ are colored blue.

```
>  plot([x1ss,x2ss],t=0..20,color=[red,blue],tickmarks=[3,4]);
```

```
>  plot([x1tr,x2tr],t=0..20,color=[red,blue],tickmarks=[3,4]);
```

The black-and-white versions of the two pictures are shown in Figure 3.4, the steady-state contributions on the left, the transient contributions on the right.

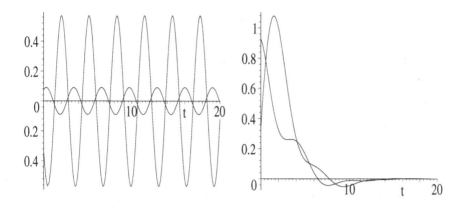

Figure 3.4: Steady-state (left) and transient (right) contributions.

In the steady-state picture, the shorter of the two oscillatory curves corresponds to $x_1(t)$, the taller one to $x_2(t)$. The motion of mass m_1 is approximately 180 degrees out of phase with m_2. The period of oscillation of both masses is identical with that of the driving force.

In the transient picture, the curve that rises above 1 is $x_2(t)$. Both curves decay to essentially zero in less than 20 time units.

To animate the solution, the complete (steady-state plus transient) time-dependent displacements are added to the given equilibrium coordinates.

```
>  X1:=2+x1ss+x1tr:  X2:=5+x2ss+x2tr:
```

The motion of the two masses is now animated, each mass being represented by a size-10 (default color red) circle. To produce motion along the horizontal axis, the horizontal coordinate of each mass is entered in a list with the vertical coordinate set equal to zero. The time range is such that the transient parts vanish, revealing the steady-state motion. You can take a larger time interval if you desire. To obtain a smooth animation, 200 frames are used.

```
>  animate({[X1,0],[X2,0]},t=0..20,frames=200,style=point,
    symbol=circle,symbolsize=20,tickmarks=[3,2]);
```

Execute the command on your computer to see the animated motion.

PROBLEMS:
Problem 3-10: Hand calculation
Making use of the integral transform package, mimic a hand calculation to derive the solution for the second example.

Problem 3-11: Transform package approach
Use the integral transform package and the Laplace transform method to solve the following ODEs. Confirm the solutions using the `method= laplace` option in the `dsolve` command. Plot each solution over a suitable time range that includes the steady-state regime. Extract the transient part of the solution and

estimate the time interval over which it lasts.

(a) $\dot{y} + 5\,y = \cos(t) + e^{-t}, \qquad y(0) = 1;$

(b) $\ddot{y} - 5\,\dot{y} + 6\,y = 0, \qquad y(0) = 2,\ \dot{y}(0) = 5;$

(c) $9\,\ddot{y} + 6\,\dot{y} + y = 5, \qquad y(0) = 6,\ \dot{y}(0) = 1;$

(d) $5\,\ddot{y} + 2\,\dot{y} + y = \sin^2(t), \qquad y(0) = -3,\ \dot{y}(0) = 1;$

(e) $\dot{x} = -4\,x + y + 3, \quad \dot{y} = -4\,x - 4\,y + 5, \qquad x(0) = 1,\ y(0) = -1;$

(f) $\ddot{y} + \dot{y} = 2\,\cos^4(t)\,e^{-t}, \qquad y(0) = 0,\ \dot{y}(0) = -2;$

(g) $\ddot{y} + 2\,\dot{y} = t\,\cos(t)\,e^{-t} + t^2\,\cos(2\,t)\,e^{-2\,t}, \qquad y(0) = 0,\ \dot{y}(0) = 0.$

3.2.3   Jennifer's Formidable Series

Isn't life a series of images that change as they repeat themselves?
Andy Warhol, American pop artist (1928–1987)

Jennifer is currently teaching a course on ordinary differential equations at MIT. As part of an assignment, she has assigned the following problem to be done with the Maple computer algebra system.

Consider the following second-order linear ODE with variable coefficients,

$$2\,x\,y''(x) + (1 - 2\,x)\,y'(x) - x^2\,y(x) = 0, \quad y(0) = 1, y'(0) = 0.$$

(a) Show that Maple is unable to produce a closed-form analytic solution.

(b) Derive a procedure for numerically determining $y(x)$ and $y'(x)$ at an arbitrary x value. Evaluate y and y' at $X = 4$.

(c) Derive a series solution for $y(x)$, keeping sufficient terms to achieve 1% agreement with the numerical answer at X.

(d) Plot the series derived in part (c) and the numerical solution together in the same figure over the range $x = 0$ to X.

Vectoria, who is in Jennifer's class, has submitted the following recipe to solve the problem. In order to have the ODE appear on the computer screen in the prime notation used in the problem wording, she loads the PDEtools package, which contains the declare command for achieving this notation.

```
>   restart: with(plots): with(PDEtools):
```

As indicated by the output, the following declare command will cause $y(x)$ to be displayed as y and introduce primes for the derivatives with respect to x.

```
>   declare(y(x),prime=x):
```

$y(x)$ *will now be displayed as* y

derivatives with respect to x of functions of one variable

will now be displayed with $'$

On entering the given differential equation, Vectoria observes that the output in *de* is expressed in terms of the prime notation.

```
> de:=2*x*diff(y(x),x,x)+(1-2*x)*diff(y(x),x)-x^2*y(x)=0;
```

$$de := 2\,x\,y'' + (1 - 2\,x)\,y' - x^2\,y = 0$$

The initial condition is specified, as well as the value of X and a parameter d that will control the spacing of the numerical points in the graph.

```
> ic:=y(0)=1,D(y)(0)=0: X:=4: d:=8:
```

Using `dsolve`, an attempt is made to analytically solve *de* for $y(x)$.

```
> dsolve({de,ic},y(x)); #no analytic solution
```

No output is generated, Maple being unable to provide an analytic answer. Although she has now omitted it, Vectoria had included `infolevel[dsolve]:=5;` prior to the `dsolve` command. On doing so she was able to view a very lengthy list of unsuccessful ODE solving methods tried by Maple. If you wish to see these attempts, include the above `infolevel` command.

To generate a procedure for numerically evaluating y and y' at arbitrary x, Vectoria includes the `numeric` option in `dsolve` and requests that the output be given as a "list procedure."

```
> numsol:=dsolve({de,ic},y(x),numeric,output=listprocedure);
```

$$numsol := [x = (\mathbf{proc}(x) \ldots \mathbf{end\ proc}), y = (\mathbf{proc}(x) \ldots \mathbf{end\ proc}),$$
$$y' = (\mathbf{proc}(x) \ldots \mathbf{end\ proc})]$$

Evaluating $y(x)$ with the *numsol* procedure will generate a numerical answer for y at x when the value of x is supplied as the argument of `Ynum` defined below. Then, entering `Ynum(X)` gives the numerical value of y at X (4, here), the answer being given to 18 digits, more than the "normal" 10-digit accuracy.

```
> Ynum:=eval(y(x),numsol): Ynum(X);
```

$$41.1065491999371986$$

Thus, $y \approx 41$ at $x = X = 4$. A similar command structure is used to numerically evaluate the derivative of y, yielding $y' \approx 76$ at X.

```
> Ynumder:=eval(diff(y(x),x),numsol): Ynumder(X);
```

$$75.6306643796735756$$

Vectoria will use a do loop to systematically calculate series solutions of *de* as a function of the order n, terms of order x^n and larger being neglected. The do loop will include a conditional statement that will terminate the loop when the absolute percentage difference $|100\,(y_{num}(X) - y_{series}(X))/y_{num}(X)|$ drops below one percent. As a "seed number" to implement the conditional statement, she calculates the absolute percentage difference between the starting value of $y(0) = 1$ and the numerical value of y at $x = X = 4$.

```
> percent[0]:=abs(evalf(100*(Ynum(X)-1)/Ynum(X))); #seed
```

$$percent_0 := 97.56729762$$

The do loop runs from 1 to 50, the number 50 being chosen to be large enough to achieve a percentage error below one percent. The loop will calculate the series solution to order n as long as the percentage error in the previous order is above 1 percent.

```
>  for n from 1 to 50 while percent[n-1]>1 do
```

The order of the series is taken to be n. The error in the series is of order x^n, terms of this order and larger being neglected. If the order is not specified, the default order is 6.

```
>  Order:=n:
```

The differential equation is solved for $y(x)$, subject to the initial condition, with the **series** option included. The **dsolve** command uses several methods when trying to find a series solution to an ODE or ODE system. When initial conditions are given, such as in this problem, the series is calculated at the given point. Otherwise, the series is calculated at the origin. The first method used is a Newton iteration, the second involves a direct substitution to generate a system of equations which must be solved, while the third is the method of Frobenius discussed in standard ODE texts. Useful references may be found by entering the topic **dsolve,series** in Maple's Topic Search.

```
>  sol[n]:=dsolve({de,ic},y(x),'series');
```

To remove the "order of" term that appears in the output of the last command line, the **convert(,polynom)** command is applied to the right-hand side of the nth order solution. The largest term in the series for Y_n is x^{n-1}.

```
>  Y[n]:=convert(rhs(sol[n]),polynom);
```

The absolute percentage error at $x = X$ is calculated for each order,

```
> percent[n]:=abs(evalf(100*(Ynum(X)-eval(Y[n],x=X))/Ynum(X)));
```

and the do loop ended with a colon to suppress the lengthy output.

```
>  end do:
```

The loop stops at $N = n - 1$, in this case order 17, for which the percentage error (to 3 digits) is 0.696 percent. As you can check by either replacing the colon with a semicolon in the do loop or taking $N = n - 2$, the percentage error in order 16 is 1.31 percent. Order 17 is the first order for which the percentage error drops below one percent. The corresponding polynomial representation of the solution is given by Y_{17}.

```
>  N:=n-1; evalf(percent[N],3)*percent; Y[N]:=Y[N];
```

$$N := 17 \qquad 0.696 \, percent$$

$$Y_{17} := 1 + \frac{1}{15} x^3 + \frac{1}{70} x^4 + \frac{4}{1575} x^5 + \frac{29}{20790} x^6 + \frac{43}{126126} x^7 + \frac{61}{1001000} x^8$$

$$+ \frac{16013}{1033782750} x^9 + \frac{498307}{152770117500} x^{10} + \frac{7710323}{14115958857000} x^{11}$$

$$+ \frac{2117887}{21250934424720} x^{12} + \frac{96270661}{5534097506437500} x^{13} + \frac{121568353967}{46021554863534250000} x^{14}$$

$$+ \frac{1141494706493}{2859910909376771250000} x^{15} + \frac{2799853112879}{47283860368362618000000} x^{16}$$

Using the sequence command, we graph the numerical solution from $x = 0$ to X in steps of size $1/d$. Each numerical point is plotted as a size-16 blue circle.

```
>  gr1:=plot([seq([i/d,Ynum(i/d)],i=0..d*X)],style=point,
            symbol=circle,symbolsize=16,color=blue):
```
The series solution is plotted as a solid thick curve over the same range of x,

```
>  gr2:=plot(Y[N],x=0..X,thickness=2):
```
and the two graphs superimposed to produce Figure 3.5.

```
>  display({gr1,gr2},labels=["x","y"]);
```

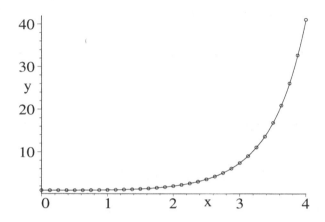

Figure 3.5: Comparison of numerical (circles) and series (curve) solutions.

Of course, because of the imposition of the conditional statement, the fit of the series solution to the numerical points is very good.

Vectoria hopes that Jennifer will like her solution to this problem. By modifying Vectoria's recipe, you should be able to solve the following problems, which also appeared on Jennifer's assignment.

PROBLEMS:
Problem 3-12: What's special about this ODE?
Consider the following linear second-order ODE:

$$x\,y''(x) + y'(x) + x\,y(x) = 0, \quad y(0) = 1, y'(0) = 0.$$

(a) Show that Maple is able to produce a closed-form analytic solution. Identify the ODE and the "special" function that appears in the solution.

(b) Derive a series solution for $y(x)$, keeping sufficient terms to achieve 1% agreement with the analytic answer at $x = X = 10$.

(c) Plot the series derived in part (b) and the analytic solution together in the same figure over the range $x = 0$ to X, representing the analytic answer by appropriately sized and spaced circles.

Problem 3-13: Another ODE
Consider the following linear second-order ODE:

$$x\,y''(x) + 2\,y'(x) + x\,y(x) = 0, \quad y(0) = 1, y'(0) = 0.$$

(a) Show that Maple is able to produce a closed-form analytic solution.

(b) Derive a series solution for $y(x)$, keeping sufficient terms to achieve 1% agreement with the analytic answer at $x = X = 20$.

(c) Plot the series derived in part (b) and the analytic solution together in the same figure over the range $x = 0$ to X, representing the analytic answer by appropriately sized and spaced circles.

Problem 3-14: Variation on the text problem
Carry out the same steps as in Jennifer's text problem for the ODE

$$y''(x) - x\,y'(x) - 1 = 0, \quad y(0) = 1, y'(0) = 0.$$

3.3   Special Function Models

Many boundary-value problems of mathematical physics lead to variable-coefficient ODEs of the *Sturm–Liouville* (S–L) form,

$$\frac{d}{dx}\left[p(x)\,\frac{dy(x)}{dx}\right] - q(x)\,y(x) = -\lambda\,w(x)\,y(x), \qquad (3.2)$$

where $p(x)$, $q(x)$, and $w(x)$ are real functions of the spatial variable x, λ is a real constant, and certain conditions on y and its first derivative are specified at the boundaries. Often the only solution to such a problem is the trivial solution $y(x) = 0$, but for special values (called the *eigenvalues*) λ_1, λ_2, ... of λ, nontrivial solutions (called the *eigenfunctions*) y_1, y_2, ... can occur.

For certain choices of the functions $p(x)$, $q(x)$, $w(x)$, and λ, the well-known *Bessel, Legendre, Hermite, Chebyshev, Mathieu*, etc., ODEs result. The solutions of these ODEs, which take the form of infinite series or finite polynomials in x, have been historically labeled as *special* functions, to distinguish them from such *ordinary* functions as the sine, cosine, log, and exponential functions.

Special-function solutions can also occur for initial-value problems with x replaced with the time t. The systematic study of the S–L ODE and special functions and their properties is beyond the scope of this text. The interested reader is referred to standard mathematical texts (e.g., [MW71], [Boa83]) for the theoretical aspects and to [Enn05] for recipes on applications. Here, we shall be content to introduce two of the more famous members of the S–L family and then look at two special-function models.

3.3.1 Jennifer Introduces a Special Family

"That's a great deal to make one word mean," Alice said in a thoughtful tone. "When I make a word do a lot of work like that," said Humpty Dumpty, "I always pay it extra."
Lewis Carroll, *Through the Looking Glass*, 1872

Although we shall not emulate Humpty Dumpty and pay the word "special" extra for making it work so hard in this section, we would tell Alice, if she were here, that the phrase "special" function does encompass a great many functions with their associated mathematical properties. *The Handbook of Mathematical Functions* by Abramowitz and Stegun [AS72], for example, contains over 1000 pages dealing with the properties of special functions. This handbook, produced by the U.S. National Bureau of Standards, is one of the standard reference books for these functions and has been reprinted many times since it was first issued.

We have asked our MIT mathematician friend Jennifer to introduce two of the more prominent members of this special family of functions.

Letting the letter L stand for λ, Jennifer forms a functional operator SL for generating specific cases of the Sturm–Liouville ODE when the forms of p, q, w, and L are supplied as arguments.

```
> restart: with(plots):
> SL:=(p,q,w,L)->diff(p*diff(y(x),x),x)-q*y(x)=-L*w*y(x);
```

$$SL := (p,\, q,\, w,\, L) \rightarrow \left(\frac{d}{dx} \left(p \left(\frac{d}{dx} y(x) \right) \right) \right) - q\, y(x) = -L\, w\, y(x)$$

A second functional operator *sol* is introduced to provide the general analytic solution $y(x)$ of a specified ODE.

```
> sol:=ode->dsolve(ode,y(x)):
```

Now Jennifer will make two choices for p, q, w, and L that lead to probably the best-known two members of the family of special functions. Taking $p = x$, $q = -x$, $w = 1/x$, and $L = -m^2$ as arguments in the Sturm–Liouville operator produces *ode1*.

```
> ode1:=SL(x,-x,1/x,-m^2);
```

$$ode1 := \left(\frac{d}{dx} y(x) \right) + x \left(\frac{d^2}{dx^2} y(x) \right) + x\, y(x) = \frac{m^2\, y(x)}{x}$$

To illustrate that Maple is able to identify this ODE, Jennifer loads the DEtools package and applies the `odeadvisor` command to *ode1*.

```
> with(DEtools): odeadvisor(ode1);
```

$$[\_Bessel]$$

So, *ode1* is Bessel's differential equation, whose general solution $y(x)$

```
> sol(ode1);
```

$$y(x) = \_C1\, \mathrm{BesselJ}(m,\, x) + \_C2\, \mathrm{BesselY}(m,\, x)$$

is a linear combination of a *Bessel function of the first kind* (BesselJ(m, x)) and of the *second kind* (BesselY(m, x)), with _C1 and _C2 arbitrary constants. In traditional mathematical notation, the Bessel functions are written as $J_m(x)$ and $Y_m(x)$, m being referred to as the *order* of the Bessel function. Highlighting BesselJ(m, x) or BesselY(m, x) on the computer screen will open a Help page with some (but not much) information about these special functions.

If you wish to know what other special functions are known to Maple, execute the following `inifcns` (initially known mathematical functions) command line and use the hyperlinks provided in the lengthy list of functions. Remember to close the Help page when you're done.

```
>   ?inifcns;
```

The Bessel functions are actually infinite *Frobenius power series* solutions of the Bessel ODE, i.e., a series expansion about $x = 0$ is sought of the form $y(x) = \sum_{m=0}^{\infty} c_m x^{m+s}$. The allowed values of the *index* s and the forms of the coefficients c_m are determined[5] by substituting $y(x)$ into the Bessel ODE.

The `series` command can be used to obtain the form of the Frobenius series. For example, since the most commonly occurring order in physical problems involves positive-integer values of m, Jennifer calculates the Frobenius series of $J_m(x)$ for $m = 0$ to 2, terms of order x^8 and higher being omitted here.

```
>   seq(J[m]=series(BesselJ(m,x),x=0,8),m=0..2);
```

$$J_0 = 1 - \frac{1}{4} x^2 + \frac{1}{64} x^4 - \frac{1}{2304} x^6 + O(x^8),$$

$$J_1 = \frac{1}{2} x - \frac{1}{16} x^3 + \frac{1}{384} x^5 - \frac{1}{18432} x^7 + O(x^8),$$

$$J_2 = \frac{1}{8} x^2 - \frac{1}{96} x^4 + \frac{1}{3072} x^6 + O(x^8)$$

Note that at $x = 0$, we have $J_0 = 1$ and $J_1 = J_2 = 0$. As is easily confirmed by increasing the range of m in the `seq` command, $J_m(0) = 0$ for $m = 3, 4 \ldots$

Looking at the above truncated series does not convey much of an idea of the shapes of $J_0(x)$, $J_1(x)$, and $J_2(x)$. Jennifer will now plot these functions and attach appropriate identifying labels to each curve. First, in the graph `gr1` she plots $J_m(x)$ over the range $x = 0$ to 20 for $m = 0$ to 2.

```
>   gr1:=plot({seq(BesselJ(m,x),m=0..2)},x=0..20,thickness=2):
```

She uses the `textplot` command in `gr2` to place appropriate labels on the Bessel function curves. The horizontal and vertical coordinates of the names (entered as strings) were determined by observing the picture generated by `gr1`.

```
>   gr2:=textplot([[1.5,0.9,"J0"],[3,0.6,"J1"],[5,0.4,"J2"]]):
```

The two graphs are superimposed to yield Figure 3.6.

```
>   display({gr1,gr2},tickmarks=[2,2]);
```

The $J_m(x)$ are oscillatory with amplitudes that decrease with increasing x. Unlike $\sin(x)$ or $\cos(x)$, the $J_m(x)$ do not cross the horizontal axis at equal

---

[5]A recipe is given in *Computer Algebra Recipes for Mathematical Physics* [Enn05].

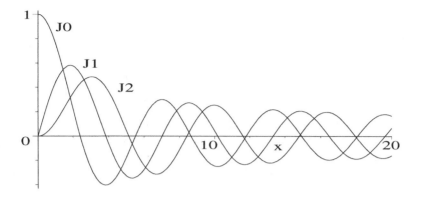

Figure 3.6: Bessel functions of the first kind of order 0, 1, and 2.

intervals of x. The zeros of a particular Bessel function, for example J_2, can be approximately determined by clicking the mouse on the computer screen plot with the cursor placed at the location of a zero. The coordinates of the cursor are displayed in a small window at the top left of the computer screen. More-accurate values of the Bessel J_2 zeros can be determined with the `BesselJZeros` command. The first six zeros are now determined to 4-digit accuracy.

```
>  Zeros:=evalf(BesselJZeros(2,0..5),4);
```
$$Zeros := 0,\ 5.136,\ 8.417,\ 11.62,\ 14.80,\ 17.96$$

What do the integer-order Bessel functions of the second look like? Jennifer first examines the analytic forms of the Frobenius series for $Y_0(x)$, $Y_1(x)$, and $Y_2(x)$, terms of order x^8 and higher again being omitted. Because the expressions are lengthy, only Y_0 is shown here in the text.

```
>  seq(Y[m]=series(BesselY(m,x),x=0,8),m=0..2);
```

$$Y_0 = \left(\frac{2\,(-\ln(2) + \ln(x))}{\pi} + \frac{2\gamma}{\pi}\right) + \left(-\frac{1}{2}\,\frac{-\ln(2) + \ln(x)}{\pi} - \frac{-\frac{1}{2} + \frac{\gamma}{2}}{\pi}\right) x^2$$

$$+ \left(-\frac{\frac{3}{64} - \frac{\gamma}{32}}{\pi} + \frac{1}{32}\,\frac{-\ln(2) + \ln(x)}{\pi}\right) x^4$$

$$+ \left(-\frac{-\frac{11}{6912} + \frac{\gamma}{1152}}{\pi} - \frac{1}{1152}\,\frac{-\ln(2) + \ln(x)}{\pi}\right) x^6 + O(x^8)$$

The constant γ in the above output is the *Euler–Mascheroni constant*. It is defined as

$$\gamma = \lim_{n\to\infty} \left(\sum_{i=1}^{n} \frac{1}{i} - \ln(n) \right) \approx 0.5772157.$$

Because of the $\ln(x)$ terms, $Y_0(x)$ diverges to $-\infty$ at $x = 0$. A similar behavior occurs for $Y_1(x)$, $Y_2(x)$, etc. Because of this divergence, positive integer-order Bessel functions of the second kind are rejected in physical problems (where they often occur) in regions that include the origin.

Again the truncated series do not reveal the shapes of $Y_0(x)$, $Y_1(x)$, etc. In the next three command lines, Jennifer plots $Y_m(x)$ over the range $x = 0$ to 20 for $m = 0$, 1, 2 and adds identifying labels to the curves. Note that in the `display` command, she limits the vertical range of the final figure to be between -1 and $+1$ so the oscillations are clearly seen.

```
>   gr3:=plot({seq(BesselY(m,x),m=0..2)},x=0..20,thickness=2):
>   gr4:=textplot([[2,.65,"Y0"],[3.7,.55,"Y1"],[6,.4,"Y2"]]):
>   display({gr3,gr4},view=[0..20,-1..1],tickmarks=[2,2]);
```

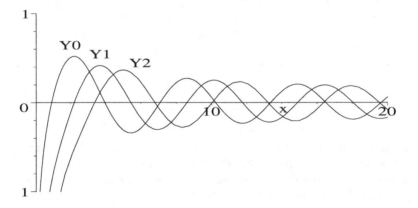

Figure 3.7: Bessel functions of the second kind of order 0, 1, and 2.

The Bessel functions $Y_0(x)$, $Y_1(x)$, and $Y_2(x)$ are shown in Figure 3.7. As with the Bessel functions of the first kind, the zeros are not evenly spaced along the x-axis. Their locations can be determined with the `fsolve` command.

As a second example of a special-function solution to a S–L ODE, Jennifer takes $p = 1 - x^2$, $q = 0$, $w = 1$, and $L = n(n+1)$ as arguments in the Sturm–Liouville functional operator and then derives the general solution.

```
>   ode2:=SL(1-x^2,0,1,n*(n+1)); sol(ode2);
```

$$ode2 := -2x \left(\frac{d}{dx} y(x) \right) + (1 - x^2) \left(\frac{d^2}{dx^2} y(x) \right) = -n(n+1)y(x)$$

$$y(x) = \_C1 \, \mathrm{LegendreP}(n, x) + \_C2 \, \mathrm{LegendreQ}(n, x)$$

The general solution $y(x)$ is a linear combination of a *Legendre function of the first kind* (LegendreP(n, x)) and of the *second kind* (LegendreQ(n, x)), n indicating the order. In problems of physical interest, n often takes on positive-integer values, the Legendre functions of the first kind then being finite polynomials, denoted by the symbol $P_n(x)$, with x ranging[6] from $x = -1$ to $+1$. Jennifer now derives the first few *Legendre polynomials*,

```
>  polynomials:=seq(P[n]=simplify(LegendreP(n,x)),n=0..4);
```

$$polynomials := P_0 = 1,\ P_1 = x,\ P_2 = \frac{3\,x^2}{2} - \frac{1}{2},\ P_3 = \frac{5}{2}x^3 - \frac{3}{2}x,$$

$$P_4 = \frac{35}{8}x^4 - \frac{15}{4}x^2 + \frac{3}{8}$$

and plots them with identifying labels added to the curves.

```
>  gr5:=plot({seq(LegendreP(n,x),n=0..4)},x=-1..1,thickness=2):
>  gr6:=textplot([[.3,.9,"P0"],[.5,.65,"P1"],[.2,-.55,"P2"],
        [-.5,.55,"P3"],[.15,.45,"P4"]]):
>  display({gr5,gr6},tickmarks=[2,3]);
```

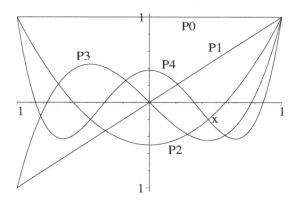

Figure 3.8: The Legendre polynomials $P_m(x)$ for $m = 0$ to 4.

The Legendre polynomials are shown in Figure 3.8. They take on the values -1 and $+1$ at the ends of the range.

What do the Legendre functions of the second kind look like? For physical problems where x varies from -1 to $+1$, the following environment command line must be entered to pick out the correct solution branch.

```
>  _EnvLegendreCut:=1..infinity:
```

The analytic forms of $Q_m(x)$ are then calculated for $m = 0$, 1, and 2.

```
>  seq(Q[m]=simplify(LegendreQ(m,x)),m=0..2);
```

---

[6]The variable x can be the cosine of a polar angle, whose range is from 0 to π radians. So $x = \cos\theta$ varies from -1 to $+1$.

$$Q_0 = \frac{1}{2}\ln(x+1) - \frac{1}{2}\ln(1-x), \quad Q_1 = \frac{1}{2}x\ln(x+1) - \frac{1}{2}x\ln(1-x) - 1,$$

$$Q_2 = -\frac{1}{4}\ln(x+1) + \frac{1}{4}\ln(1-x) + \frac{3}{4}x^2\ln(x+1) - \frac{3}{4}x^2\ln(1-x) - \frac{3x}{2}$$

The $Q_m(x)$ for m equal to zero or a positive integer diverge to infinity at $x = -1$ and $+1$ and must therefore be rejected as being unphysical.

PROBLEMS:

Problem 3-15: A Sturm–Liouville equation

Show that the following ODE is of the Sturm–Liouville form:

$$y''(x) - 2\,x\,y'(x) + 2\,n\,y(x) = 0.$$

Determine the general solution of this ODE and plot the included special functions over a suitable range of the independent variable x.

Problem 3-16: Recurrence Formula

Bessel functions of different orders can be related through *recurrence relations*. Use Maple to prove the following Bessel function recurrence relations.

- $J_{m-1}(x) + J_{m+1}(x) = (2\,m/x)\,J_m(x)$;

- $4\dfrac{d^2 J_m(x)}{dx^2} = J_{m+2}(x) - 2\,J_m(x) + J_{m-2}(x)$.

Problem 3-17: Bessel function solutions

Find the general solution of each of the following ODEs in terms of Bessel functions and identify the order. Identify any other new functions that occur.

(a) $y''(x) + y(x)/\sqrt{x} = 0$;

(b) $y''(x) + x\,y(x) = 0$, Hint: Use `convert( ,Bessel)`;

(c) $\dfrac{d^2}{dx^2}\left(x^{\frac{16}{5}}\dfrac{d^2 y}{dx^2}\right) - x^{\frac{8}{5}}\,y(x) = 0$;

Problem 3-18: Orthogonality

An important general property that all solutions $y_n(x)$ of the Sturm–Liouville equation corresponding to a given λ_n possess is *orthogonality*. Provided that $y(x)$ or $y'(x)$ or $p(x)$ vanishes at the endpoints a and b of the range (referred to as *Sturm–Liouville boundary conditions*), then

$$\int_a^b w(x)\,y_m(x)\,y_n(x)\,dx = 0, \quad \text{for } m \neq n, \tag{3.3}$$

where $w(x)$ is referred to as the *weight function*. Confirm the orthogonality property for Legendre functions of the second kind of orders 2 and 3 over the range $x = -1$ to $+1$. Which of the possible Sturm–Liouville boundary conditions is satisfied?

3.3.2 The Vibrating Bungee Cord

Wisdom consists in being able to distinguish among dangers and make a choice of the least harmful.
Niccolò Machiavelli, from *The Prince* (1469–1527)

While on vacation in Rainbow County, Jennifer is nervously watching her sister Heather getting ready to bungee jump from an abandoned railway bridge spanning a deep gorge. At present, the uniform elastic bungee cord has no one attached to its lower end, but is simply hanging vertically downward and displaying small vibrations transverse to its length. This reminds Jennifer that the famous mathematician Daniel Bernoulli first studied this vibrational problem nearly 300 years ago, and was able to solve it. To put a modern spin on an old problem, Jennifer recently showed her mathematical physics class a computer algebra derivation of the solution.

With Jennifer's permission, we shall now reproduce her treatment. To aid in understanding the physics of the problem, a *free-body diagram* is shown in Figure 3.9, which shows the relevant forces on the bungee cord and introduces the notation that will be used.

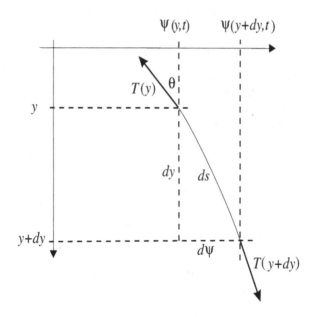

Figure 3.9: Free-body diagram for a segment of vibrating bungee cord.

Jennifer begins her recipe by loading the plots and plottools library packages, which will be needed for the animation of the transverse vibrations of the vertical cord.

```
>  restart: with(plots): with(plottools):
```
Measuring y vertically downward, the origin $y = 0$ is chosen to be at the top end of the bungee cord (length L) where it is attached to the bridge. If the cord has a mass density ϵ per unit length, and g is the acceleration due to gravity, the tension T in the cord is given by $T = \epsilon\, g\,(L - y)$, which is entered.

```
>  T:=epsilon*g*(L-y);
```

$$T := \epsilon\, g\,(L - y)$$

At the top end ($y = 0$), the cord has a tension $T = \epsilon\, g\, L$, because the cord must support its entire weight. At the bottom end, $y = L$ and the tension is zero.

The cord is now allowed to undergo a small transverse displacement $d\psi(y, t)$ at time t from the equilibrium position. The above expression for T will still be valid. To understand this, consider the cord segment of arc length

$$ds = \sqrt{(dy)^2 + (d\psi)^2} = \sqrt{1 + (d\psi/dy)^2}\, dy$$

shown in Figure 3.9. Letting θ be the angle with the vertical, and noting that the forces still balance in the y-direction, then

$$(T\cos\theta)_y - (T\cos\theta)_{y+dy} = (\epsilon\, ds)\, g, \tag{3.4}$$

with $\cos\theta = dy/ds$. Assuming that $d\psi/dy \ll 1$, one has $ds \approx dy$ and $\cos\theta \approx 1$, so the vertical force equation (3.4) reduces to

$$T(y) - T(y + dy) = (\epsilon\, dy)\, g. \tag{3.5}$$

Taylor expanding the left-hand side of (3.5) to the first power in dy, and dividing both sides by dy, yields $\partial T/\partial y = -\epsilon\, g$. The expression $T = \epsilon\, g\,(L - y)$ follows on integrating and evaluating the constant at $y = 0$.

In the ψ-direction, Newton's second law yields

$$(T\sin\theta)_{y+dy} - (T\sin\theta)_y = (\epsilon\, ds)\, \frac{\partial^2\psi}{\partial t^2} \tag{3.6}$$

with $\sin\theta = d\psi/ds$. For $d\psi/dy \ll 1$, $\sin\theta \approx \partial\psi/\partial y$, partial derivatives being used because one also has time as an independent variable. So (3.6) becomes

$$\left(T(y)\frac{\partial\psi}{\partial y}\right)_{y+dy} - \left(T(y)\frac{\partial\psi}{\partial y}\right)_y = (\epsilon\, dy)\, \frac{\partial^2\psi}{\partial t^2}, \tag{3.7}$$

or, on Taylor expanding the left-hand side for small dy, dividing by dy, and taking the limit $dy \to 0$,

$$\frac{\partial}{\partial y}\left(T(y)\frac{\partial\psi}{\partial y}\right) = \epsilon\, \frac{\partial^2\psi}{\partial t^2}. \tag{3.8}$$

This equation of motion is now entered,

```
>  eq:=diff(T*diff(psi(y,t),y),y)=epsilon*diff(psi(y,t),t,t);
```

$$eq := -\epsilon\, g\left(\frac{\partial}{\partial y}\, \psi(y, t)\right) + \epsilon\, g\,(L - y)\left(\frac{\partial^2}{\partial y^2}\, \psi(y, t)\right) = \epsilon\left(\frac{\partial^2}{\partial t^2}\, \psi(y, t)\right)$$

the expression for the tension being automatically substituted.

The equation of motion is a linear partial differential equation. Solving linear PDEs is the subject matter of Chapters 5 and 6. The PDE may be converted into an ODE by assuming a solution of the form $\psi(y,t) = X(y)\cos(\omega t)$, with ω taken to be a positive angular frequency.

```
>   psi(y,t):=X(y)*cos(omega*t);
```

$$\psi(y,\ t) := X(y)\cos(\omega t)$$

With the assumed solution automatically substituted, the resulting output of *eq* is divided by $\cos(\omega t)$ and expanded to produce the ODE *eq2*.

```
>   eq2:=expand(eq/cos(omega*t));
```

$$eq2 := -\epsilon g\left(\frac{d}{dy}X(y)\right) + \epsilon g\left(\frac{d^2}{dy^2}X(y)\right)L - \epsilon g\left(\frac{d^2}{dy^2}X(y)\right)y = -\epsilon X(y)\omega^2$$

The second derivative terms are collected in *eq2*,

```
>   eq3:=collect(eq2,diff(X(y),y,y));
```

$$eq3 := (\epsilon g L - \epsilon g y)\left(\frac{d^2}{dy^2}X(y)\right) - \epsilon g\left(\frac{d}{dy}X(y)\right) = -\epsilon X(y)\omega^2$$

and a general analytic solution to *eq3* obtained.

```
>   sol:=dsolve(eq3,X(y));
```

$$sol := X(y) = \_C1\,\text{BesselJ}\left(0,\ 2\sqrt{\frac{L-y}{g}}\,\omega\right) + \_C2\,\text{BesselY}\left(0,\ 2\sqrt{\frac{L-y}{g}}\,\omega\right)$$

The general solution is a linear combination of zeroth-order Bessel functions of the first and second kinds with arbitrary constants $\_C1$ and $\_C2$. The Bessel Y_0 function diverges to $-\infty$ at $y = L$ so must be removed on physical grounds.

```
>   X:=remove(has,rhs(sol),BesselY);
```

$$X := \_C1\,\text{BesselJ}\left(0,\ 2\sqrt{\frac{L-y}{g}}\,\omega\right)$$

To remove the arbitrary constant $\_C1$, the op command is used to select the second operand in X.

```
>   X:=op(2,X);
```

$$X := \text{BesselJ}\left(0,\ 2\sqrt{\frac{L-y}{g}}\,\omega\right)$$

Taking the bungee cord length to be $L = 30$ meters, its density $\epsilon = \frac{1}{2}$ kilogram/meter, and $g = 9.8$ meter/second^2, the form of X is as follows:

```
>   L:=30: g:=9.8: epsilon:=1/2: X:=X;
```

$$X := \text{BesselJ}(0,\ 2\sqrt{3.061224489 - 0.1020408163\,y}\,\omega)$$

The transverse displacement X of the cord at $y = 0$ is zero, which is entered as a boundary condition, *bc*.

```
>   bc:=eval(X,y=0)=0;
```

$$bc := \text{BesselJ}(0,\ 3.499271060\,\omega) = 0$$

The boundary condition implies that there are certain allowed frequencies (*eigen-frequencies*) ω of vibration, each frequency corresponding to a possible *normal mode* solution. A general transverse motion of the cord will involve a linear combination of normal modes, the combination depending on the initial conditions imposed on the cord.

The eigenfrequencies will now be determined. First, the op command is used to extract the first element of the second argument on the lhs of *bc*.

```
>   c:=op([2,1],lhs(bc));
```

$$c := 3.499271060$$

Using the BesselJZeros command, the first 10 eigenfrequencies are numerically evaluated, and assigned.

```
>   freq:=seq(omega[n]=evalf(BesselJZeros(0,n)/c),n=1..10);
```

$$freq := \omega_1 = 0.6872361462, \; \omega_2 = 1.577493717, \; \omega_3 = 2.473008739,$$
$$\omega_4 = 3.369711645, \; \omega_5 = 4.266865142, \; \omega_6 = 5.164236682,$$
$$\omega_7 = 6.061730076, \; \omega_8 = 6.959298411, \; \omega_9 = 7.856916100,$$
$$\omega_{10} = 8.754568007$$

```
>   assign(freq):
```

A functional operator for creating the nth normal mode, with the time dependence included, is formed.

```
>   mode:=n->eval(X,omega=omega[n])*cos(omega[n]*t):
```

As an explicit example, choosing $n = N = 3$ produces the normal mode X_3 corresponding to the third eigenfrequency.

```
>   N:=3: X[N]:=mode(N);
```

$$X_3 := \text{BesselJ}(0, 4.946017478\sqrt{3.061224489 - .1020408163\,y})\cos(2.473008739\,t)$$

The normal mode is now animated, the plot being rotated by $-\frac{\pi}{2}$ radians using the rotate command, so that the animated string is hanging vertically in equilibrium, rather than being pictured as horizontal. The scaling is constrained, so that the transverse displacement is small compared to the length of the cord. The axes are also removed so that the small vibrations are better viewed.

```
>   rotate(animate(plot,[X[N],y=0..L],t=0..10,color=red,
      frames=100,scaling=constrained,axes=none),-Pi/2);
```

The animation can be observed on your computer by executing the above command line, clicking on the computer plot, and on the start arrow in the tool bar. You should also look at the other transverse vibrational modes as well, by changing the value of N from 3 to some other integer between 1 and 10. Or, if you want, you could generate even higher-integer normal modes.

While we have been looking at Jennifer's computer algebra treatment of the transverse vibrations of the cord, Heather has completed her bungee jump, the largest of the vertical oscillations bringing her head to within a meter of the river far below the bridge. Looking rather pale, Heather attempts to persuade Jennifer to also do a bungee jump, but Jennifer sensibly declines.

PROBLEMS:

Problem 3-19: Vibrations of a weighted bungee cord

Determine the transverse normal modes of vibration of the bungee cord if a mass $M = 60$ kg hangs from the lower end. Animate one of the normal modes.

Problem 3-20: A stiffening spring

A vibrating spring is governed by the ODE

$$\ddot{x}(t) + a\,\dot{x}(t) + k\,x(t) = 0,$$

where the spring coefficient is $k = b + c\,e^{dt}$, and initially $x(0) = A$, $\dot{x}(0) = 0$. Determine the analytic form of $x(t)$. Taking $a = \sqrt{5}$, $b = \frac{1}{4}$, $c = 1$, $d = 1$, and $A = 1$, plot $x(t)$ over a time interval for which the oscillations effectively vanish. Determine the threshold on a for critical damping.

Problem 3-21: The growing pendulum

Consider a pendulum that consists of a point mass m at the bottom end of a light supporting rod of length L that is allowed to move in a vertical plane about a pivot point at its top end. Suppose that L increases at a steady rate, i.e., $L = L_0 + v\,t$, where $v > 0$ is a constant speed and t the time. Letting the rod make an angle $\theta(t)$ with the vertical and neglecting drag, use Newton's second law to derive the ODE for small θ. Solve the ODE for $L_0 = 1$ meter, $g = 9.8$ m/s^2, $\theta(0) = \pi/6$ radians, $\dot{\theta}(0) = 0$ radians/s, and $v = 0.5$ m/s. Animate the motion of the pendulum arm (representing it as a thick line) over the time interval $t = 0$ to 100 seconds, taking 100 frames and using constrained scaling.

Problem 3-22: Onset of bending

A thin, vertical steel wire of length L and circular cross section of radius a is clamped at its bottom and is free at its top. Let θ be the angular deflection of the wire from the vertical at a distance y from the top. If L is small, the wire is *stable* in the vertical position, i.e., $\theta = 0$ for all values of y. As L increases, there is a critical value L_{cr} beyond which the wire is unstable and will bend from the vertical.

The relevant ODE for small angular displacements θ is

$$\frac{d^2\theta}{dy^2} = -c^2\,y\,\theta, \qquad \text{where } c = \frac{2}{a}\sqrt{\frac{\rho g}{Y}}. \tag{3.9}$$

Here ρ is the mass density, g is the acceleration due to gravity, and Y is Young's modulus.

(a) Determine the solution of the ODE, subject to the boundary conditions $\theta = 0$ at $y = L$ (wire clamped at bottom) and $d\theta/dy = 0$ at $y = 0$ (wire is free at top).

(b) Show that the onset of bending occurs at $L_{cr} \approx (2.8/c)^{2/3}$.

(c) Determine L_{cr} for a steel $(Y = 2.1 \times 10^{11}$ N/m^2 and $\rho = 7800$ kg/m$^3)$ wire of radius 1 mm. Take $g = 9.8$ m/s^2.

3.3.3 Mathieu's Spring

In every tyrant's heart there springs in the end
This poison, that he cannot trust a friend.
Aeschylus, Greek dramatist, Prometheus, in *Prometheus Bound* (525–456 BC)

Emile Mathieu (1835–1890), a French mathematician, introduced the special functions that bear his name as solutions to the problem of determining the transverse vibrations of an elliptically shaped elastic membrane. As illustrated in the following recipe, *Mathieu functions* also occur for the vibrations of a unit mass attached to the end of a linear spring having a time-dependent spring coefficient $k = a + b \cos(ct)$, where a, b, and c are real constants. The mass is allowed to slide on a smooth horizontal surface, but experiences a viscous drag force given by Stokes's law, $F_{drag} = -\Gamma(dx/dt)$, where $x(t)$ is the displacement of the mass from equilibrium at time t and Γ is the damping coefficient.

To see what mathematical approach Maple uses in solving the relevant ODE, the `infolevel[dsolve]` command is set to 2.

>   `restart: infolevel[dsolve]:=2:`

The mathematical form of the spring coefficient is entered.

>   `k:=a+b*cos(c*t);`

$$k := a + b \cos(ct)$$

A functional operator *ode* is introduced to generate the ODE governing the motion of the unit mass for a given value of the damping coefficient Γ.

>   `ode:=Gamma->diff(x(t),t,t)+Gamma*diff(x(t),t)+k*x(t)=0;`

$$ode := \Gamma \rightarrow \left(\frac{d^2}{dt^2}x(t)\right) + \Gamma\left(\frac{d}{dt}x(t)\right) + k\,x(t) = 0$$

An operator X is formed to analytically solve *ode* for $x(t)$ for a specified Γ value and the initial condition $x(0) = A$, $\dot{x}(0) = 0$.

>   `X:=Gamma->rhs(dsolve({ode(Gamma),x(0)=A,D(x)(0)=0},x(t))):`

As specific parameter values, let's take $a = 10$, $b = 2$, $c = 2$, and $A = 1$.

>   `a:=10: b:=2: c:=2: A:=1:`

The analytic solution for, say, $\Gamma = 2$ is generated.

>   `sol:=simplify(X(2)); #example`

   Methods for second order ODEs:
   — Trying classification methods —
   ....

   −> Trying a Liouvillian solution using Kovacic's algorithm
   <− No Liouvillian solutions exists
   −> Trying a solution in terms of special functions:
   −> Bessel
   −> elliptic
   −> Legendre
   −> Whittaker

-> *hypergeometric*

. . . .

-> *Mathieu*

<- *Mathieu successful*

<- *special function solution successful*

. . . .

Change of variables used:

[t = 1/2*arccos(t)]

Linear ODE actually solved:

(5+t)*u(t)+(-2*(1-t^2)^(1/2)-2*t)*diff(u(t),t)

+(2-2*t^2)*diff(diff(u(t),t),t) = 0

<- change of variables successful

$$sol := e^{(-t)} \left(\text{MathieuC}(9, -1, t) + \text{MathieuS}(9, -1, t) \right)$$

The solution is expressed in terms of the even (MathieuC) and odd (MathieuS) general Mathieu functions. Information about these functions may be obtained by highlighting, for example, **MathieuC**, opening the **Help** window, and clicking on **Help on "MathieuC"**. From the partial output shown here in the text, one can see that the answer was arrived at by replacing the independent variable t with $\frac{1}{2} \arccos(t)$ and solving the resultant ODE.

The information level is now turned off by setting the integer to zero, and the plots library package is loaded so that the motion of the unit mass can be animated.

```
>  infolevel[dsolve]:=0: with(plots):
```

The horizontal displacement of the unit mass from equilibrium is animated over the time interval $t = 0$ to 6 for $\Gamma = 2$ and 0, 150 frames being used. The mass is represented by a size-20 red box for $\Gamma = 2$ and a blue box for zero damping.

```
>  animate(pointplot,[[[X(2),0],[X(0),0]],symbol=box,
   color=[red,blue],symbolsize=20],t=0..6,frames=150,
   tickmarks=[3,0]);
```

Execute the recipe to see the motion, and feel free to experiment with different parameter values and initial conditions.

PROBLEMS:

Problem 3-23: Critical damping

In the text recipe, at what critical value of Γ does the oscillatory motion cease and the unit mass approach the equilibrium position monotonically?

Problem 3-24: Other parameter values

Explore the text recipe for other parameter values and discuss any interesting behavior.

3.3.4   Quantum-Mechanical Tunneling

If we see light at the end of the tunnel,
It's the light of the oncoming train.
Robert Lowell, American poet commenting on pessimism (1917–1977)

The *Schrödinger equation* [Gri95] describing the one-dimensional motion of a particle of mass m moving in a potential $V(x)$ at time t is

$$-\frac{\hbar^2}{2\,m}\frac{\partial^2 \Psi(x,t)}{\partial x^2} + V(x)\,\Psi(x,t) = I\,\hbar\,\frac{\partial \Psi(x,t)}{\partial t}, \tag{3.10}$$

where $\Psi(x,t)$ is the *wave function*, $I = \sqrt{-1}$, and $\hbar = h/(2\,\pi)$, where h is Planck's constant. Assuming a *stationary-state* solution, $\Psi(x,t) = \psi(x)\,e^{-I\,E\,t/\hbar}$, reduces the PDE (3.10) to the *time-independent* Schrödinger ODE,

$$\frac{d^2\psi(x)}{dx^2} + \frac{2\,m}{\hbar^2}(E - V(x))\,\psi(x) = 0 \tag{3.11}$$

The *probability* of finding the particle between x and $x + dx$ at time t is $|\Psi(x,t)|^2\,dx = |\psi(x)|^2\,dx \equiv P(x)\,dx$. The total probability of finding the particle *somewhere* in the range $x = -\infty$ to $+\infty$ is $\int_{-\infty}^{+\infty} P(x)\,dx = 1$.

Although classically impossible, a particle with energy E incident on a finite potential barrier of maximum height $V_{max} > E$ has a nonzero probability of quantum-mechanically *tunneling* through the barrier to the opposite side. This is the basis of *α-emission*, in which α particles are able to escape from the nucleus through the nuclear potential barrier. For a particle incident on a barrier with incident amplitude ψ_{inc}, reflected amplitude ψ_{refl}, and transmitted amplitude ψ_{trans}, one can define the *reflection coefficient* $R = |\psi_{refl}|^2/|\psi_{inc}|^2$ and the *transmission coefficient* $T = \sqrt{(E - V_{trans})/(E - V_{inc})}\,|\psi_{trans}|^2/|\psi_{inc}|^2$.

The factor $2\,m/\hbar^2$ can be removed from equation (3.11) by rescaling the spatial variable, letting x now stand for $(\sqrt{2\,m}/\hbar)\,x$. Using the "scaled" ODE, our goal in this recipe is to determine the reflection and transmission coefficients for a particle of energy $E > 0$ incident on a potential barrier $V(x) = a\,x^2$, with $a = 1$, located between $x = 0$ and $x = L = 1$. Outside the barrier, $V(x) = 0$, i.e., $V_{trans} = V_{inc} = 0$ so $T = |\psi_{trans}|^2/|\psi_{inc}|^2$. Then we will plot both coefficients as a function of the energy E.

The DEtools library package is loaded, because it contains the **expsols** command, which will enable us to generate exponential solutions to the Schrödinger ODE rather than the default sine and cosine solutions that would otherwise occur in the regions $x < 0$ and $x > L$, where $V(x) = 0$.

```
>  restart: with(DEtools):
```
The parameter values $a = 1$ and $L = 1$ are entered, along with the assumed condition $E > 0$.

```
>  a:=1: L:=1: assume(E>0):
```
The scaled Schrödinger ODE is entered for $V(x) = 0$,

```
>  de1:=diff(psi(x),x,x)+E*psi(x);
```

$$de1 := \left(\frac{d^2}{dx^2}\psi(x)\right) + E\,\psi(x)$$

and two independent exponential solutions obtained.

> soll:=expsols(de1,psi(x));

$$soll := \left[e^{(\sqrt{E}\,x\,I)},\ e^{(-I\,\sqrt{E}\,x)}\right]$$

Remembering that the spatial forms are to be mentally multiplied by the time factor $e^{-I\,E\,t/\hbar}$, the first term in *soll* corresponds to a plane wave traveling to the right, the second term to a plane wave traveling to the left. In region 1 $(x < 0)$, the total amplitude can be taken to be of the form $\psi1 = e^{(\sqrt{E}\,x\,I)} + A\,e^{(-I\,\sqrt{E}\,x)}$, the first term representing the incident wave (particle), the second term the reflected wave. Without loss of generality, the coefficient of the incident wave has been taken to be 1, because R and T involve ratios of amplitudes. The reflection coefficient then is $R = |A|^2$. Making use of *soll*, the wave form $\psi1$ is entered.

> psi1:=soll[1]+A*soll[2];

$$\psi1 := e^{(\sqrt{E}\,x\,I)} + A\,e^{(-I\,\sqrt{E}\,x)}$$

In region 3 $(x > L)$, there is only a transmitted wave traveling to the right, which is now entered. The transmission coefficient will be $T = |B|^2$.

> psi3:=B*soll[1];

$$\psi3 := B\,e^{(\sqrt{E}\,x\,I)}$$

In the barrier $(0 < x < L)$ region 2, the potential is $V(x) = a\,x^2$. The scaled Schrödinger ODE is entered for this region,

> de2:=diff(psi(x),x,x)+(E-a*x^2)*psi(x);

$$de2 := \left(\frac{d^2}{dx^2}\psi(x)\right) + (E - x^2)\,\psi(x)$$

and the general solution $\psi2$ obtained in terms of two undetermined coefficients _C1 and _C2 and the *Whittaker* functions of the first and second kinds. To learn a little more about these special functions, highlight either one in the ouput on the computer screen and consult Maple's Help.

> psi2:=rhs(dsolve(de2,psi(x)));

$$\psi2 := \frac{\_C1\ \text{WhittakerM}\left(\dfrac{E}{4},\dfrac{1}{4},x^2\right)}{\sqrt{x}} + \frac{\_C2\ \text{WhittakerW}\left(\dfrac{E}{4},\dfrac{1}{4},x^2\right)}{\sqrt{x}}$$

Assuming that $x > 0$, we try to convert the Whittaker functions in $\psi2$ to the more "standard" *Hermite* functions.

> psi2:=convert(psi2,Hermite) assuming x>0;

$$\psi2 := \frac{\_C1\ \text{WhittakerM}\left(\dfrac{E}{4},\dfrac{1}{4},x^2\right)}{\sqrt{x}} + \frac{\_C2\ \text{HermiteH}\left(-\dfrac{1}{2}+\dfrac{E}{2},x\right)}{2^{(-1/2+\frac{E}{2})}\,e^{\left(\frac{x^2}{2}\right)}}$$

There are four unknown coefficients, A, B, $\_C1$, and $\_C2$, so four equations are needed to determine them. Continuity of the amplitudes $\psi1$ and $\psi2$ at $x = 0$ is imposed in the first boundary (or matching) condition $bc1$. The latter waveform must be carefully handled. One cannot simply evaluate $\psi2$ at $x = 0$, because the error message "division by zero" will appear. One must take the limit as x approaches zero from the right, i.e., from $x > 0$. The quantity Γ appearing in the output is the *gamma* function.

> `bc1:=eval(psi1,x=0)=limit(psi2,x=0,right);`

$$bc1 := 1 + A = \frac{\_C2\,\sqrt{\pi}}{\Gamma\left(\dfrac{3}{4} - \dfrac{E}{4}\right)}$$

In the second boundary condition, $\psi2$ and $\psi3$ are equated at $x = L$.

> `bc2:=eval(psi2,x=L)=eval(psi3,x=L);`

$$bc2 := \_C1\,\text{WhittakerM}\left(\frac{E}{4}, \frac{1}{4}, 1\right) + \frac{\_C2\,\text{HermiteH}\left(-\dfrac{1}{2} + \dfrac{E}{2}, 1\right)}{2^{(-1/2 + \frac{E}{2})}\,e^{(1/2)}} = B\,e^{(\sqrt{E}\,I)}$$

Because the second derivative is finite, the first derivative (slope) of the wave forms must be continuous at $x = 0$. This condition is imposed in $bc3$, the limiting process again being used for the derivative $d\psi2/dx$.

> `bc3:=eval(diff(psi1,x),x=0)=limit(diff(psi2,x),x=0,right);`

$$bc3 := \sqrt{E}\,I - A\,\sqrt{E}\,I =$$
$$\frac{\_C1\Gamma\left(\dfrac{5}{4} - \dfrac{E}{4}\right)2^{(-1/2 + \frac{E}{2})} - \_C2\,2^{(-3/2 + \frac{E}{2})}\sqrt{\pi} + \_C2\,2^{(-3/2 + \frac{E}{2})}\sqrt{\pi}\,E}{\Gamma\left(\dfrac{5}{4} - \dfrac{E}{4}\right)2^{(-1/2 + \frac{E}{2})}}$$

The slope continuity condition is also imposed at $x = L$ (lengthy output not shown here).

> `bc4:=eval(diff(psi2,x),x=L)=eval(diff(psi3,x),x=L);`

The four boundary conditions are solved for the four unknown coefficients, and the solution is assigned.

> `sol3:=solve({bc1,bc2,bc3,bc4},{A,B,_C1,_C2}): assign(sol3):`

The reflection and transmission coefficients are calculated, the very lengthy expressions being suppressed with line-ending colons. Noting that the amplitudes are complex, these coefficients are given by $R = |A|^2 = A \times A^*$ and $T = |B|^2 = B \times B^*$, where the asterisk denotes the complex conjugate. The `conjugate` command is used to enter the complex conjugates.

> `R:=A*conjugate(A): T:=B*conjugate(B):`

The reflection and transmission coefficients are now plotted over the energy range $E = 0$ to 1, the two curves being colored blue and red on the computer screen. The black-and-white version is shown in Figure 3.10.

```
>  plot([R,T],E=0..1,color=[blue,red],tickmarks=[3,3],
   view=[0..1,0..1]);
```

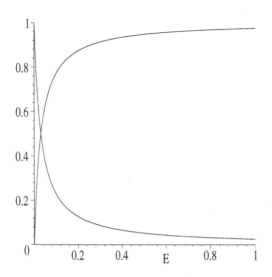

Figure 3.10: Reflection and transmission coefficients versus energy E.

The transmission coefficient rapidly increases from 0 and approaches 1 as E is increased. Conversely, the reflection coefficient rapidly decreases with increasing E. By energy conservation, the sum of the reflection and transmission coefficients should sum to 1 for all values of the energy E. From the figure, this appears to be the case, the confirmation being left for you as a problem. Note that the reflection coefficient is not equal to zero at $E = 1$, even though the energy of the incoming particle is equal to that at the top of the barrier.

PROBLEMS:

Problem 3-25: R+T=1

Confirm that the sum of the reflection and transmission coefficients is equal to 1 for all E. You may do this either graphically or analytically.

Problem 3-26: Equality

Determine the energy at which the reflection and transmission coefficients are equal by (i) using the mouse, (ii) using the fsolve command.

Problem 3-27: Exponential barrier

Determine the transmission and reflection coefficients as a function of energy E if the barrier has the form $V(x) = V e^{-x/L}$ between $x = 0$ and $L = 1$, with $V = 1$. Outside the barrier region, $V(x) = 0$. Plot the reflection and transmission coefficients in the same figure and discuss how the results compare with those obtained in the text recipe.

Chapter 4

Nonlinear ODE Models

The elegant body of mathematical theory pertaining to linear systems ... tends to dominate even moderately advanced university courses. The mathematical intuition so developed ill equips the student to confront the bizarre behavior exhibited by the simplest of ...nonlinear systems. Yet such nonlinear systems are surely the rule, not the exception .... Not only in research, but also in the everyday world of politics and economics, we would all be better off if more people realized that simple nonlinear systems do not necessarily possess simple dynamic properties.
Robert M. May, mathematical biologist, *Nature*, Vol. 261, 459 (1976)

In Chapter 1, phase-plane portraits were used to explore some simple nonlinear ODE models whose temporal evolution could not have been predicted, even qualitatively, before the portraits were numerically constructed. An example was the period-doubling route to chaos exhibited by the Duffing equation

$$\ddot{x} + 2\gamma\,\dot{x} + \alpha\,x + \beta\,x^3 = F\,\cos(\omega\,t) \tag{4.1}$$

when the amplitude F of the driving force was increased, the other parameters being held fixed. If the nonlinear term, $\beta\,x^3$, were not present, this "bizarre" period-doubling behavior would not even be possible. If we were to change the various coefficient values in (4.1), the response of the nonlinear system would in general be entirely different and not easily predicted on the basis of mathematical or physical intuition alone. To aid in the qualitative understanding of the behavior of nonlinear ODE systems such as this one, the concepts of fixed points and phase-plane analysis were discussed in Chapter 2.

One might well ask whether there are mathematical techniques for obtaining the exact analytic solutions to nonlinear ODEs, and if so, whether Maple can be used to find these solutions. The answer is that the vast majority of nonlinear ODEs of interest to physicists and engineers do not possess exact analytic solutions. Only a handful of ODEs of physical interest can be solved exactly, and for those equations Maple can be used to find the solutions. Some examples of first-order nonlinear ODEs for which this is the case are illustrated in the following section.

4.1 First-Order Models

4.1.1 An Irreversible Reaction

*I shall use the phrase "time's arrow" to express this one-way prop-
erty of time which has no analogue in space.*
Arthur Eddington, British astrophysicist (1882–1944)

As our first example, we consider the following problem, which the reader was
previously asked to solve numerically (see Problem 2-29).
   Consider the irreversible chemical reaction
$$2\,K_2Cr_2O_7 + 2\,H_2O + 3\,S \rightarrow 4\,KOH + 2\,Cr_2O_3 + 3\,SO_2$$
with initially N_1 molecules of potassium dichromate ($K_2Cr_2O_7$), N_2 molecules of
water (H_2O), N_3 atoms of sulphur (S), and 0 molecules of potassium hydroxide
(KOH). The number x of KOH molecules at time t seconds is given by the
nonlinear rate equation
$$\dot{x} = k\,(2\,N_1 - x)^2\,(2\,N_2 - x)^2\,(4\,N_3/3 - x)^3$$
with $k = 1.64 \times 10^{-20}$ s^{-1}. Taking $N_1 = 2000$, $N_2 = 2000$, and $N_3 = 3000$,
analytically determine the number of KOH molecules at arbitrary time $t > 0$
and plot the solution for the first second after the chemical reaction begins.
How many KOH molecules are present at 0.2 seconds? How many are present
in the limit $t \rightarrow \infty$?
   To observe Maple's method of solution, the `infolevel[dsolve]` command
is set equal to 2, and the rate equation entered.

```
>   restart: infolevel[dsolve]:=2:
>   de:=diff(x(t),t)=k*(2*N1-x(t))^2*(2*N2-x(t))^2
                 *(4*N3/3-x(t))^3;
```

$$de := \frac{d}{dt}x(t) = k\,(2\,N1 - x(t))^2\,(2\,N2 - x(t))^2\left(\frac{4\,N3}{3} - x(t)\right)^3$$

The rate coefficient k and the initial molecule numbers are specified.

```
>   k:=1.64*10^(-20): N1:=2000: N2:=2000: N3:=3000:
```

The differential equation de is analytically solved for $x(t)$, subject to $x(0) = 0$.

```
>   sol:=dsolve({de,x(0)=0},x(t));
```

Methods for first order ODEs:
— Trying classification methods —
trying a quadrature
trying 1st order linear
trying Bernoulli
trying separable
<— separable successful

$$sol := x(t) = -\frac{4000\,5^{(2/3)}}{(251904\,t + 625)^{(1/6)}} + 4000$$

In arriving at the answer, Maple has recognized that the first-order differential equation is *separable*. That is to say, it can be written in the form $dx/dt = f(t)/g(x)$, where $f(t)$ and $g(x)$ are given functions. So, on rearranging and integrating, one has $\int g(x)\,dx = \int f(t)\,dt$, the answer following if the integrals can be performed (which is the case here).

The right-hand side of the solution *sol* can be converted into a functional operator x using the `unapply` command. Then entering `x(t)` will evaluate x at the specified value of t.

> `x:=unapply(rhs(sol),t);`

$$x := t \rightarrow -\frac{4000\,5^{(2/3)}}{(251904\,t + 625)^{(1/6)}} + 4000$$

The number $x(t)$ of KOH molecules at time t is plotted over the time interval $t = 0$ to 1 second.

> `plot(x(t),t=0..1,labels=["t","x"]);`

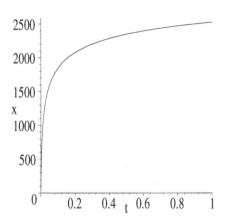

Figure 4.1: Number of KOH molecules as a function of time.

The number of KOH molecules present after 0.2 seconds is determined,

> `number:=evalf(x(0.2));`

$$number := 2079.400844$$

and rounded off to the nearest integer.

> `KOH:=round(number);`

$$KOH := 2079$$

So, 2079 KOH molecules are present 0.2 seconds after the reaction begins.

The number of KOH molecules appears to be leveling off in Figure 4.1 as the time increases. From the analytic form of the solution, clearly the limiting number N is 4000 KOH molecules as $t \rightarrow \infty$. This can be confirmed by applying the following `limit` command to $x(t)$:

> `N:=limit(x(t),t=infinity);`

$$N := 4000$$

PROBLEMS:

Problem 4-1: Falling basketball

A spherical object (diameter d meters and mass m kg) falling from rest experiences [FC99] a drag force $F_{drag} = -A\,v - B\,v^2$ newtons, where v is the velocity in m/s and $A = 1.55 \times 10^{-4}\,d$, $B = 0.22\,d^2$. Derive the nonlinear ODE governing the velocity of the falling sphere. If the sphere is a basketball of diameter 25 cm and mass 0.60 kg, analytically determine $v(t)$. Does Maple recognize the ODE as being separable? Plot $v(t)$ and show that the ball will reach a terminal velocity. Determine the terminal velocity. How long does it take the basketball to come within 1 percent of the terminal velocity?

Problem 4-2: It's separable

Consider the ODE
$$\frac{dy}{dx} = \frac{2\,x^3\,y - y^4}{x^4 - 2\,x\,y^3}, \quad y(1) = 5.$$

Does this ODE appear to be separable? Show that assuming $y(x) = x\,z(x)$ leads to a separable equation for $z(x)$. Making use of this transformation, determine $y(x)$ and plot the solution, starting at $x = 1$, over the range for which it remains real. What is the x value at the upper end of this real range?

4.1.2   The Struggle for Existence

The mathematics of uncontrolled growth are frightening. A single cell of the bacterium E. coli *would, under ideal circumstances, divide [in two] every twenty minutes .... it can be shown that in a single day, one cell of* E. Coli *could produce a super-colony equal in size and weight to the entire planet Earth.*
Michael Crichton, *The Andromeda Strain* (1969)

A classic experiment in microbiology is to grow yeast, or other microorganisms, in a nutrient broth inside a test tube or flask at a suitable fixed temperature.

As an assignment associated with her microbiology course in the premed program at MIT, Heather has been asked to create a Maple worksheet that illustrates the solution of simple model equations describing the growth of yeast in a test tube. She is to search the literature and find realistic numbers for the parameter values and use these to create suitable plots of the solutions. Consulting her older sister Jennifer who, recall, is a mathematics faculty member at MIT, Heather is guided to the text *Mathematical Models in Biology* by Leah Edelstein-Keshet [EK88]. This interesting and easy-to-read book describes some yeast-growing experiments and associated model equations.

After loading the library plots package,

```
> restart: with(plots):
```

Heather considers the simplest model of yeast growth, which would apply if there were an unlimited supply of nutrient. She lets $N(t)$ be the yeast population density at time $t > 0$ and $k > 0$ the rate of reproduction. If k is taken to be a constant, the following linear yeast equation, *YE*, results:

```
>  YE:=diff(N(t),t)=k*N(t);
```

$$YE := \frac{d}{dt} N(t) = k \, N(t)$$

This growth equation is historically known as *Malthus's law*, named in honor of Thomas Malthus (1766–1834), who published a pamphlet (entitled *Essay on Population*) on population growth in 1798. Although Heather can solve this simple linear ODE in her head, she lets Maple generate the solution, subject to the initial condition $N(0) = No$.

```
>  ic:=N(0)=No;
```

$$ic := N(0) = No$$

```
>  sol:=dsolve({YE,ic},N(t));
```

$$sol := N(t) = No \, e^{(k \, t)}$$

Malthus's law leads to exponential growth of the yeast population density.

Having read *The Andromeda Strain*, Heather is curious as to how many *Escherichia coli* there would be at the end of 24 hours if the exponential solution prevailed. According to the Crichton quotation, the doubling time is 20 minutes, or $1/3$ of an hour. In the following command line, Heather takes $N(t = 1/3)/No = 2$ in the exponential solution,

```
>  eq:=2=exp(k/3); #time in hours
```

$$eq := 2 = e^{\left(\frac{k}{3}\right)}$$

and numerically solves for the reproductive rate constant, labeled *k1*.

```
>  k1:=fsolve(eq,k);
```

$$k1 := 2.079441542$$

Using the Malthus solution, at the end of 24 hours the number of *E. coli* would have grown from a single bacterium $(No = 1)$,

```
>  Number:=1*exp(k1*24); #in 24 hours
```

$$Number := 0.4722366529 \, 10^{22}$$

to about 10^{22} bacteria. Talk about explosive growth! Heather wonders how accurate the Malthus solution is. After all, she need not have formulated the *E. coli* growth as a continuous-time ODE, but instead could calculate the growth directly. In a 24-hour period there would be $24 \times 3 = 72$ doublings. At the end of 24 hours, the number of *E. coli* is given by the output of the following line:

```
>  Number2:=2^(24.0*3);
```

$$Number2 := 0.4722366483 \, 10^{22}$$

The two numbers differ only in the eighth decimal place.

Although 10^{22} *E. coli* is a large number, Heather notes that since an *E. coli* bacterium weighs about 10^{-12} gm, their total weight is only 10^{10} gm, much less than the 6×10^{27} gm weight of the earth. Crichton's conclusion seems wrong.

This is interesting, but Heather realizes that she should get back to the yeast problem. Such unlimited growth as seen above clearly cannot occur in a test tube experiment with only a finite amount of nutrient available. It seems more reasonable that the growth would depend on the amount of nutrient left in the test tube. To this end, Heather modifies her original model by assuming that the reproductive rate k is not a constant but is proportional to the concentration $C(t)$ of nutrient, the proportionality constant being labeled K.

> k:=K*C(t);

$$k := K\ C(t)$$

The new yeast equation then is given by NYE.

> NYE:=YE;

$$NYE := \frac{d}{dt} N(t) = K\ C(t)\ N(t)$$

As the yeast density increases, the nutrient concentration must decrease. Heather assumes that the rate of decrease of $C(t)$ is proportional to the rate of increase of $N(t)$, the positive proportionality constant being labeled a. The concentration equation (CE) then is as follows.

> CE:=diff(C(t),t)=-a*diff(N(t),t);

$$CE := \frac{d}{dt} C(t) = -a \left(\frac{d}{dt} N(t) \right)$$

The problem now involves two coupled ODEs, NYE and CE, but the general solution of the latter is easily obtained, _C1 being the integration constant.

> dsolve(CE,C(t));

$$C(t) = -a\ N(t) + \_C1$$

The last output is then substituted into the new yeast equation.

> NYE:=subs(%,NYE);

$$NYE := \frac{d}{dt} N(t) = K\ (-a\ N(t) + \_C1)\ N(t)$$

It is traditional in mathematical biology to make the form of the equation notationally simpler, so Heather substitutes $\_C1 = a\,b$ and $K = r/(a\,b)$ into NYE, where b and r are new constants.

> subs({_C1=a*b,K=r/(a*b)},NYE);

$$\frac{d}{dt} N(t) = \frac{r\ (-a\ N(t) + a\ b)\ N(t)}{a\ b}$$

To cancel the constant a, the simplify command is applied.

> NYE:=simplify(%);

$$NYE := \frac{d}{dt} N(t) = \frac{r\ (-N(t) + b)\ N(t)}{b}$$

The resulting nonlinear differential equation is known as the *logistic equation* and was first studied by the mathematician Verhulst in 1838.

According to Edelstein-Keshet's text, the logistic equation can be solved in a straightforward way, and the solution is quoted in her book. If an analytic solution exists, Heather reasons that she should be able to use the `dsolve` command to find it. Setting the `infolevel[dsolve]` command to be 2, *NYE* is solved for $N(t)$, subject to the same initial condition as for Malthus's law.

```
>  infolevel[dsolve]:=2:
>  yeastdensity:=dsolve({NYE,ic},N(t));
```

Methods for first order ODEs:
  — Trying classification methods —
  trying a quadrature
  trying 1st order linear
  trying Bernoulli
  $<-$ Bernoulli successful

$$yeastdensity := N(t) = \frac{No\, b}{No - e^{(-r\,t)}\, No + e^{(-r\,t)}\, b}$$

In solving *NYE* for the yeast density $N(t)$, Maple has recognized the ODE as being a *Bernoulli* equation. A Bernoulli differential equation (named after Jakob Bernoulli (1654–1705)) is of the general structure[1]

$$\frac{dy}{dt} + f(t)\, y = g(t)\, y^n. \tag{4.2}$$

The logistic equation is a special case of the Bernoulli equation with $n = 2$ and f and g constant. For *NYE*, $f = -r$ and $g = r/b$.

The exponential terms in the solution are now collected.

```
>  yeastdensity:=collect(yeastdensity,exp);
```

$$yeastdensity := N(t) = \frac{No\, b}{(-No + b)\, e^{(-r\,t)} + No}$$

This mathematical form is referred to as the *logistic curve*. For positive r, $N(t)$ approaches the value b as $t \to \infty$. Unlike Malthus's law, the logistic equation predicts "saturation" of the yeast number to a constant value at large times, rather than uncontrolled growth.

Heather notes that by loading the DEtools package, the general Bernoulli solution to *NYE* can also be obtained using the `bernoullisol` command.

```
>  with(DEtools): bernoullisol(NYE,N(t));
```

  trying Bernoulli
  $<-$ Bernoulli successful

$$\left\{ N(t) = \frac{b}{1 + e^{(-r\,t)}\, \_C1\, b} \right\}$$

As part of her assignment, Heather was asked to find some realistic parameter values. In a book aptly entitled *The Struggle for Existence*, Gause [Gau69] has reported on test-tube experiments involving the growth of the two yeasts, *Schizosaccharomyces kephir* and *Saccharomyces cerevisiae*. The yeasts were grown separately in the first experiment and as a competing mixture in the

---

[1]If proceeding by hand, the standard approach to solving equation (4.2) is to introduce the new dependent variable $p = 1/y^{n-1}$, which reduces the equation to a linear ODE.

second. With the initial population density $N_0 = 0.5329$ known, the data from the first experiment were used to find the parameter values b and r to use in the logistic-model solution for each yeast variety. Using a least-squares fitting approach, Gause found that for the *kephir*, $r = 0.0607$ and $b = 5.80$, while for the *cerevisiae* $r = 0.2183$ and $b = 13.0$. The first experiment was carried out over a time period of 160 hours. Heather decides to plot the logistic curve for the *kephir*, so enters the parameter values,

>   `b:=5.80: No:=0.5329: r:=0.0607: yeastdensity;`

$$N(t) = \frac{3.090820}{5.2671\, e^{(-0.0607\, t)} + 0.5329}$$

and applies the `plot` command, adding a title to the picture.

>   `plot(rhs(yeastdensity),t=0..160,labels=["t","A"],`
>   `color=blue,thickness=2,tickmarks=[2,4],`
>   `title="Amount of yeast (A) vs. hours (t)");`

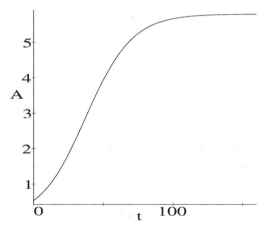

Figure 4.2: Growth curve for the yeast *kephir*.

The logistic growth curve of the *kephir* population is shown in Figure 4.2. The plot for the *cerevisiae* would be similar in appearance.

     Gause's second experiment studied the competition between the two yeasts for the available nutrient when they were grown together in the same test tube. To explore this case Heather labels the *kephir* population density and parameters with the subscript k, while the subscript c is used for the *cerevisiae*. To account for the interaction between the two species of yeast, Gause assumed that the probability of an interaction is proportional to the product of the population densities, i.e., to $N_k N_c$, and that the competition for nutrient was detrimental to both populations. Accordingly, Gause described the growth of the two interacting species by the following pair of coupled nonlinear equations:

$$\dot{N}_k = r_k \left(b_k - N_k - \beta_{kc} N_c\right) N_k/b_k, \quad \dot{N}_c = r_c \left(b_c - N_c - \beta_{ck} N_k\right) N_c/b_c, \quad (4.3)$$

with the interaction parameters experimentally determined to be $\beta_{kc} = 0.439$ and $\beta_{ck} = 3.15$. This set of nonlinear ODEs cannot be solved analytically, but the behavior of the two competing yeast populations can easily be determined by making a phase-plane portrait of N_k versus N_c.

Heather unassigns k so that it can be used as a subscript to label the *kephir* population and sets the infolevel[dsolve] command to zero.

```
>   unassign('k'): infolevel[dsolve]:=0:
```

The values determined by Gause for the coefficients in (4.3) are entered,

```
>   r[k]:=0.0607: b[k]:=5.8: beta[kc]:=0.439: r[c]:=0.2183:
    b[c]:=13.0: beta[ck]:=3.15:
```

as well as the system (*sys*) of equations.

```
> sys:=diff(N[k](t),t)=r[k]*(b[k]-N[k]-beta[kc]*N[c])*N[k]/b[k],
  diff(N[c](t),t)=r[c]*(b[c]-N[c]-beta[ck]*N[k])*N[c]/b[c];
```

$$sys := \frac{d}{dt} N_k(t) = 0.01046551724 \left(5.8 - N_k(t) - 0.439 N_c(t)\right) N_k(t),$$

$$\frac{d}{dt} N_c(t) = 0.01679230769 \left(13.0 - N_c(t) - 3.15 N_k(t)\right) N_c(t)$$

Heather takes $N_k(0) = N_c(0) = 0.5$ in the phaseportrait command,

```
>   phaseportrait([sys],[N[k](t),N[c](t)],t=0..250,
    [[N[k](0)=0.5,N[c](0)=0.5]],stepsize=0.1,linecolor=blue);
```

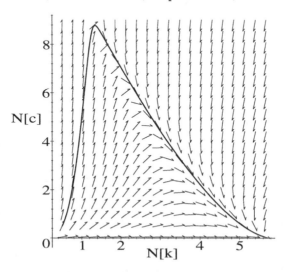

Figure 4.3: Phase portrait showing competition between *kephir* and *cerevisiae*.

producing the phase-plane picture shown in Figure 4.3. In the actual experiment, the time interval was only about 50 hours. On reducing the time range to be from $t = 0$ to 50 in the `phaseportrait` command line, Heather finds that the curve terminates in the vicinity of the maximum value for N_c. In this case, both species of yeast survived over the 50 hours of the experiment, with the *cerevisiae* dominating over the *kephir*. However, assuming that enough nutrient was available, she can see that ultimately only one species would survive over 250 hours and it isn't the *cerevisiae*, but rather the *kephir*. This is because the interaction coefficient β_{kc} is much smaller than β_{ck}. In this struggle for existence, there would be only the victorious *kephir* and the vanquished *cerevisiae*. To see the explicit time evolution of each yeast population, Heather uses the phase-plane portrait command again, but now creates two plots, one with the scene option `scene=[t,N[k]]`, the other with `scene=[t,N[c]]`.

```
>  plot1:=phaseportrait([sys],[N[k],N[c]],t=0..300,
           [[N[k](0)=0.5,N[c](0)=0.5]],stepsize=0.1,
           scene=[t,N[k]],color=red,linecolor=blue):
>  plot2:=phaseportrait([sys],[N[k],N[c]],t=0..300,
           [[N[k](0)=0.5,N[c](0)=0.5]],stepsize=0.1,
           scene=[t,N[c]],color=red,linecolor=red):
```

On superimposing the two plots with the `display` command, Figure 4.4 results.

```
>  display({plot1,plot2},tickmarks=[3,3]);
```

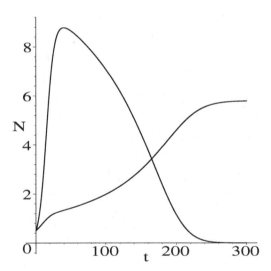

Figure 4.4: Time evolution of the *kephir* (saturable curve) and the *cerevisiae*.

By clicking with the mouse on the plot, the crossover point for the two population densities occurs at about 164 hours. With her assignment completed, Heather is grateful that her sister was able to recommend a useful text. She will have to treat her to a Starbucks latte the next time they get together.

PROBLEMS:

Problem 4-3: Nonlinear diode circuit
A linear capacitor with capacitance C is connected in series with a nonlinear diode that has a current (i)–voltage (v) relation of the form $i = a v + b v^2$, with the coefficients a and b both positive. The voltage across the capacitor at time $t = 0$ is $v = V$. Derive the nonlinear ODE governing this circuit. Introducing the dimensionless variables $y = v(t)/V$, $\tau = a t/C$, and $\beta = b V/a$, rewrite the ODE in dimensionless form. Analytically solve this ODE, demonstrating that Maple recognizes the ODE as a Bernoulli equation. For $\beta = 2$, plot the solution over the time interval $\tau = 0$ to 2.

Problem 4-4: Nonlinear diode revisited
Suppose that the current–voltage relation for the nonlinear diode in the previous problem is given by the more general relation $i = a v + b v^n$, where $n = 2, 3, 4, 5,$ ... Derive the corresponding general dimensionless ODE and analytically solve it for arbitrary n. For the same parameters and time range as in the previous problem, produce a single plot that shows the solutions for $n = 2$ to 5. Discuss the effect of increasing n.

Problem 4-5: A potpourri of Bernoulli equations
For each of the following nonlinear ODEs (with prime indicating an x-derivative):

- confirm that it is of the Bernoulli type;
- analytically solve the ODE for $y(0) = 1$;
- plot the solution $y(x)$ over the range $x = 0$ to 5.

(a) $y^3 y' + x^{-1} y^4 = x$; (b) $y' + y = x y^3$; (c) $y - y' = 3 y^3 e^{-2x}$;
(d) $y^3 + 3 y^2 y' = 4$.

Problem 4-6: General solution
Determine the general solution of the following Bernoulli equation:

$$x y' + y = x^3 y^6.$$

Problem 4-7: Laser beam competition
In the theory of *stimulated thermal scattering* [BEP71], the intensities I_L and I_S of two interacting collinear laser beams traveling in the z-direction are governed by the following pair of coupled first-order nonlinear ODEs:

$$\frac{dI_L}{dz} = -g I_L I_S - \alpha I_L, \quad \frac{dI_S}{dz} = g I_L I_S - \alpha I_S,$$

where the *gain coefficient* $g > 0$ and the *absorption coefficient* $\alpha \geq 0$.

(a) By adding the two equations and eliminating I_L, show that a single Bernoulli equation can be obtained for the intensity I_S.

(b) Analytically solve this Bernoulli equation for $I_S(z)$.

(c) Plot $\ln(I_S(z)/I_S(0))$ for $z = 0$ to 10 cm given that $I_S(0)/I_L(0) = 0.01$, $g I_L(0) = 1$ cm^{-1}, and (a) $\alpha = 0$, (b) $\alpha = 0.5$ cm^{-1}. Discuss the results.

Problem 4-8: Variable transformation
Explicitly show that the general Bernoulli equation (4.2) can be transformed into a linear ODE by making use of the transformation $p = 1/y^{n-1}$. Here n is an arbitrary positive integer greater than or equal to two.

Problem 4-9: Flu epidemic in Spuzzum
In the small town of Spuzzum (population $N_0 = 2000$ people), the number $N(t)$ of people infected with the flu after t weeks is governed by the logistic equation, $\dot{N} = 0.002\,N\,(N_0 - N)$. If two people are initially infected, derive the analytic formula for $N(t)$. How many weeks does it take before 75% of Spuzzum's population is infected with the flu? Plot $N(t)$ for the time that it takes 90% of the population to be infected.

Problem 4-10: Hare today, gone tomorrow
The growth of a snowshoe hare population is governed by the logistic equation, $\dot{N} = r\,N\,(N_0 - N)$, with the rate constant $r = 0.25/N_0$, time being measured in months. A viral epidemic of deadly *myxamatosis* strikes the population, suddenly reducing the hare number to 1% of the equilibrium number N_0. How many months does it take for the snowshoe hare population to climb back up to a level of 50% of its equilibrium number? Plot $N(t)/N_0$ over this time interval.

Problem 4-11: Harvesting of blue whales
In the absence of harvesting, the blue whale population number $N(t)$ is governed by the logistic equation. If a constant rate h of harvesting is included, the equation becomes
$$\dot{N}(t) = a\,(M - N(t))\,N(t) - h,$$
where M is the maximum number of whales in the absence of harvesting and a is a positive constant. With t in years, it has been estimated that $a = 6 \times 10^{-7}$ and $M=200,000$. The steady-state (equilibrium) population occurs when $\dot{N} = 0$.

(a) Plot h versus N for the steady state. Find the maximum value h_{max} of h by: (i) clicking the mouse on the maximum point of the curve, and (ii) working with the algebraic form and then substituting numbers. The quantity h_{max} is called the *maximum sustained yield*. What happens to the whale population for $h \gg h_{max}$?

(b) Using the `dsolve` command, solve the logistic equation for the given parameters and $N(0)=100,000$, $h = 7000$. How many years would it take for the blue whale population to go extinct? It is because of overharvesting that the blue whale population has been nearly wiped out.

Problem 4-12: Gompertz's law
Gompertz's law is an alternative nonlinear growth model to the logistic equation and has been used to model the growth of cancerous tumors. The Gompertz ODE is of the form
$$\dot{y} = k\,y\,\ln(y_m/y),$$
where k and y_m are positive constants. Derive the general analytic solution to the Gompertz ODE. If $k = 0.1$, $y_m = 0.5$, and $y(0) = 0.1$, determine the analytic form for $y(t)$ and plot the solution for $t = 0$ to 75. What is the interpretation of the constant y_m?

Problem 4-13: The von Bertalanffy model of growth

The limited-growth *von Bertalanffy model* is governed by the nonlinear ODE

$$\dot{y} = 3\,k\,y^{2/3}(y_m - y^{1/3}),$$

with k and y_m positive constants. Derive the general analytic solution to the von Bertalanffy ODE. If $k = 0.1$, $y_m = 0.5$, and $y(0) = 0.1$, determine the analytic form for $y(t)$ and plot the solution for $t = 0$ to 75.

Problem 4-14: The Michaelis–Menton equation

The nonlinear *Michaelis–Menton model* equation

$$\dot{y} = -a\,y/(b + y),$$

with a, b positive constants, describes the rate at which an enzyme reaction takes place. Here $y(t)$ is the amount of substrate that is being transformed by the enzyme at time t. Taking $a = b = y(0) = 1$, determine the analytic form of $y(t)$. Plot the analytic solution for $t = 0$ to 5.

4.1.3 The Bad Bird Equation

Everyone pushes a falling fence.
Chinese proverb about failure.

In an article by Peastrel and coworkers in the *The Physics of Sports* [PLA92], the experimental data for the distance y (in meters) that a badminton bird ("bad" bird, for short) falls in t seconds from rest is well described by

$$y = \frac{V^2}{g} \ln\left(\cosh\left(\frac{g\,t}{V}\right)\right), \tag{4.4}$$

where V is the terminal velocity and g is the acceleration due to gravity. Vectoria will now show you how equation (4.4) can be easily derived, using Newton's second law and assuming that the drag force is given by *Newton's law of air resistance*, viz., $F_{res} = -k\,m\,v^2$, with k the drag coefficient, m the mass of the bird, and v the speed. The quadratic dependence on v is appropriate for *turbulent* air flow. For the falling bird, the turbulence arises because of the bird's "feathers." Stokes's law of resistance ($F_{res} \propto v$) prevails for *laminar* (i.e., smooth) flow.[2]

Vectoria begins by loading the DEtools package,

```
>  restart: with(DEtools):
```

and entering Newton's law of air resistance.

```
>  F[res]:=-k*m*v(t)^2;
```

$$F_{res} := -k\,m\,v(t)^2$$

---

[2]The more general resistance law is a combination of linear and quadratic terms in the velocity. Which term dominates depends on the shape of the moving object and the magnitude of the velocity.

The equation of motion follows on applying Newton's second law to the motion
of the bad bird. Equating the net force on the bird, due to the pull of gravity
and the drag force, to the bird's mass times its acceleration produces eq.

> `eq:=m*g+F[res]=m*diff(v(t),t);`

$$eq := m\,g - k\,m\,v(t)^2 = m\,\left(\frac{d}{dt}v(t)\right)$$

The force equation is simplified by dividing eq through by the mass m.

> `eq2:=simplify(eq/m);`

$$eq2 := g - k\,v(t)^2 = \frac{d}{dt}v(t)$$

Since the terminal velocity V is reached when the upward and downward forces
balance, so that the acceleration is zero, then $g = k\,V^2$. Thus, the drag coefficient k may be eliminated by substituting $k = g/V^2$ into $eq2$.

> `ode:=subs(k=g/V^2,eq2); #Riccati equation`

$$ode := g - \frac{g\,v(t)^2}{V^2} = \frac{d}{dt}v(t)$$

Vectoria recognizes the resulting first-order nonlinear ode as a specific example
of $Riccati's\ equation$, which has the general structure

$$\dot{x} + a\,x^2 + f_1(t)\,x + f_2(t) = 0, \tag{4.5}$$

with a a constant. In the present case, Vectoria identifies $x \equiv v$, $a \equiv k = g/V^2$,
$f_1 = 0$, and $f_2 = -g$. She recalls that Riccati's equation can be reduced to a
linear ODE involving a new dependent variable $z(t)$ by introducing the transformation $z(t) = \exp(a \int_0^t x(t)\,dt)$. However, it is easier to derive the general
form of $v(t)$ by applying the `riccatisol` command to ode. The square brackets
are included at the end of the command line to remove the brackets that would
otherwise enclose the output.

> `sol:=riccatisol(ode,v(t))[];`

$$sol := v(t) = -\frac{1}{2}\,\frac{\tan\left(\dfrac{t\sqrt{-\dfrac{4\,g^2}{V^2}}}{2} - \dfrac{\_C1\sqrt{-\dfrac{4\,g^2}{V^2}}}{2}\right)\sqrt{-\dfrac{4\,g^2}{V^2}}\,V^2}{g}$$

The integration constant $\_C1$ is determined by evaluating the right-hand side
of the solution at $t = 0$, setting the result to zero (since the bird starts from
rest), and solving for the constant.

> `_C1:=solve(eval(rhs(sol),t=0)=0,_C1);`

$$\_C1 := 0$$

The integration constant is zero. The velocity expression follows on simplifying
the rhs of sol, assuming that $g > 0$, $V > 0$, and $t > 0$.

> `v:=simplify(rhs(sol)) assuming g>0,V>0,t>0;`

$$v := \tanh\left(\frac{t\,g}{V}\right) V$$

To determine $y(t)$, a second ODE is formed by equating dy/dt to v. To make the final expression appear in the desired form, the `convert` command is used to reexpress the hyperbolic tangent in terms of the hyperbolic sine and cosine.

```
>  ode2:=diff(y(t),t)=convert(v,sincos);
```

$$ode2 := \frac{d}{dt}y(t) = \frac{\sinh\left(\frac{t\,g}{V}\right) V}{\cosh\left(\frac{t\,g}{V}\right)}$$

Analytically solving *ode2* for $y(t)$, subject to the initial condition $y(0) = 0$,

```
>  dsolve({ode2,y(0)=0},y(t));
```

$$y(t) = \frac{V^2 \ln\left(\cosh\left(\frac{t\,g}{V}\right)\right)}{g}$$

produces an output that is identical in structure to equation (4.4), thus completing Vectoria's task.

PROBLEMS:

Problem 4-15: Return velocity
A ball of unit mass is thrown vertically upward near the earth's surface with an initial speed $v(0) = U$. Assuming that Newton's law of air resistance prevails, show that after the ball rises to its maximum height and begins to fall it passes its initial position with a velocity $v = (U\,V)/(\sqrt{U^2 + V^2})$, where V is the terminal velocity. *Hint:* Reexpress the acceleration as

$$\frac{dv}{dt} = \left(\frac{dv}{dx}\right)\left(\frac{dx}{dt}\right) = v(x)\left(\frac{dv(x)}{dx}\right)$$

and note that at the maximum height the speed is zero.

Problem 4-16: A Riccati equation
Consider the following Riccati equation:

$$y'(x) + y(x)/x + a\,y(x)^2 + b = 0,$$

with the initial condition $y(0) = A$, where a, b, and A are real constants.

(a) Analytically solve the ODE. Does the answer depend on the value of A?

(b) What does the general solution look like if no initial condition is specified? Explain what happens mathematically when $y(0) = A$ is imposed.

(c) Taking $a = 5$ and $b = 2$, plot the solution for $y(0) = A$ over the range $x = 0$ to 2, using the plot option `view=[0..2,-5..5]`.

(d) Explain the origin of the singular points in the graph in terms of the behavior of Bessel functions.

Problem 4-17: Nonlinear drag on Lake Ogopogo

A boat is launched on Lake Ogopogo with initial speed v_0. The water exerts a drag force $F(v) = -a\,e^{b\,v}$, with $a > 0$, $b > 0$, thus slowing the boat down.

(a) Find an analytic expression for the speed $v(t)$.

(b) Determine the time it takes for the boat to come to rest.

(c) How far does the boat travel along Lake Ogopogo before coming to rest?

4.2   Second-Order Models

The vast majority of second-order nonlinear models that physicists and engineers are interested in must be solved numerically. However, we begin this section with a few examples of models that lead to nonlinear ODEs having exact analytic solutions. Historically, these models have appeared in many equivalent guises, so keep this in mind when you read the stories.

4.2.1   Patches Gives Chase

Man ... cannot learn to forget, but hangs on the past:
however far or fast he runs, that chain runs with him.
Friedrich Nietzsche, German philosopher (1844–1900)

A loveable beagle, named Patches, is patrolling a flat farm field, sniffing contentedly at gopher holes, when she spots her mistress, Heather, walking at constant speed along a straight road at the edge of the field. Patches then runs at constant speed toward Heather in such a way as to aim always at her with her sensitive beagle nose. With distances in km, Patches is initially at $(x = 1, y = 0)$ and Heather at $(0, 0)$. The road is described by the equation $x = 0$, and the ratio of Heather's speed to Patches' speed is r.

(a) Derive the nonlinear ODE describing Patches' path $y(x)$.

(b) Analytically solve the ODE for $y(x)$. If Heather walks at 3 km/h and Patches runs at 8 km/h, plot $y(x)$ using constrained scaling.

(c) How many minutes does it take the dog to reach her mistress? How far has Heather walked in this time? How far has Patches run?

(d) Animate the motion of Patches and Heather, including a tangent line to Patches' instantaneous position, which points towards Heather. Again, use constrained scaling.

At some instant in time, Patches' coordinates are $(x, y(x))$, while Heather's are $(0, h)$. Patches runs towards Heather in such a way as to aim always at her. Therefore the slope of the tangent to Patches' path is $dy/dx = (h - y(x))/(0 - x)$, which is now entered in *eq*.

```
>   restart:
>   eq:=diff(y(x),x)=(h-y(x))/(0-x);
```

$$eq := \frac{d}{dx} y(x) = -\frac{h - y(x)}{x}$$

We solve *eq* for Heather's vertical coordinate *h*.

```
>   h:=solve(eq,h);
```

$$h := -\left(\frac{d}{dx} y(x)\right) x + y(x)$$

Heather's speed is r times that of Patches, so $dh/dt = r\,(ds/dt)$, where t is time and $ds = \sqrt{1 + (dy/dx)^2}\,dx$ is an element of arclength along Patches' path. But then $dh = r\,ds$, or $dh/dx = r\,(ds/dx) = r\,\sqrt{1 + (dy/dx)^2}$. Entering this last relation yields the relevant second-order nonlinear ODE for $y(x)$,

```
>   ode:=diff(h,x)=r*sqrt(1+diff(y(x),x)^2);
```

$$ode := -\left(\frac{d^2}{dx^2} y(x)\right) x = r \sqrt{1 + \left(\frac{d}{dx} y(x)\right)^2}$$

the expression for h having been automatically substituted. Although *ode* is nonlinear, it can be analytically solved. Patches is initially at $x = 1$, $y = 0$ and the tangent line there has zero slope. These starting conditions are used in the dsolve command.

```
>   sol:=dsolve({ode,y(1)=0,D(y)(1)=0},y(x));
```

$$sol := y(x) = -\frac{x}{2(-1+r)x^r} - \frac{x\,x^r}{2(1+r)} + \frac{r}{-1+r^2},$$

$$y(x) = \frac{x}{2(-1+r)x^r} + \frac{x\,x^r}{2(1+r)} - \frac{r}{-1+r^2}$$

Two forms of the solution are generated, the first corresponding to Heather moving downward (in the negative y direction) along the $x = 0$ line, the second corresponding to moving upward. The upward (second one here) solution is selected and simplified with respect to powers of x.

```
>   y:=simplify(rhs(sol[2]),power);
```

$$y := \frac{x^{(1-r)}}{2(-1+r)} + \frac{x^{(1+r)}}{2(1+r)} - \frac{r}{-1+r^2}$$

Evaluating y at $x = 0$ yield's Heather's vertical coordinate when Patches reaches her. Notice that as r approaches 1 from below, Y goes to infinity, i.e., Patches cannot catch Heather in a finite distance. Obviously, Patches must run faster than Heather can walk to catch her.

```
>   Y:=eval(y,x=0);
```

$$Y := -\frac{r}{-1+r^2}$$

The given ratio $r = \frac{3}{8}$ is entered, and y and Y are automatically evaluated.

```
>   r:=3/8: y:=y; Y:=evalf(Y);
```

$$y := -\frac{4\,x^{(5/8)}}{5} + \frac{4\,x^{(11/8)}}{11} + \frac{24}{55}$$

$$Y := 0.4363636364$$

Heather has walked about 0.44 km before Patches catches her. The times T in hours and $T2$ in minutes for Patches to reach Heather are calculated, along with the distance $Sdog$ that Patches runs. The absolute value is taken for T, because Y will be negative if the downward solution branch is chosen.

```
>  T:=abs(Y)/3; T2:=T*60*minutes; Sdog:=8*T;
```

$$T := 0.1454545455 \qquad T2 := 8.727272730\,minutes \qquad Sdog := 1.163636364$$

The time for Patches to catch Heather is about 0.15 hours, or 8.73 minutes. (Patches is rather slow!) Patches has run about 1.16 km.

A labeled plot of y is now created with constrained scaling,

```
>  pl:=plot(y,x=0..1,thickness=2,tickmarks=[3,3],
       labels=["x","y"],scaling=constrained):
```

and, loading the plots package, displayed in Figure 4.5.

```
>  with(plots): display(pl);
```

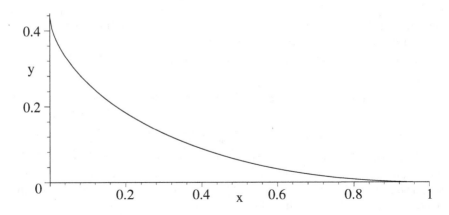

Figure 4.5: Patches' path is the curved line.

Finally, an animation of Heather's and Patches' motion is produced. The animation will contain $N = 40$ frames, the frames labeled from $i = 0$ to $N - 1$. The Student[Calculus1] library package is loaded so that the tangent line to Patches' position on the path can be drawn for each frame.

```
>  N:=40: with(Student[Calculus1]):
```

The following functional operator X will calculate Patches' horizontal coordinate for the ith frame. For $i = 0, 1, 2$, etc., $X(0) = 1$, $X(1) = 1 - 1/40 = 0.975$, $X(2) = 0.95$, etc.

```
>  X:=i->1-i/N:
```

The operator hh calculates Heather's y-coordinate for the ith frame. The sign command is included to pick up the correct sign for the coordinate, whether Heather's motion is upward or downward.

```
>   hh:=i->sign(-Y)*evalf(int(r*sqrt(1+diff(y,x)^2),x=1..X(i))):
```

The pointplot command is used to graph Heather's and Patches' positions for the ith frame as size-16 blue circles.

```
>   gr:=i->pointplot([[X(i),eval(y,x=X(i))],[0,hh(i)]],
            symbol=circle,symbolsize=16,color=blue):
```

The command Tangent(y,x=X(i),0..1) generates the equation of the line tangent to y at the point $x = X(i)$, the third argument giving the range of x. A plot of the tangent line is produced by including the arguments showtangent and output=plot.

```
>   gr2:=i->Tangent(y,x=X(i),0..1,showtangent,output=plot,
            thickness=2):
```

All the graphs are superimposed for the ith frame, using the display command.

```
>   Gr:=i->display([pl,gr(i),gr2(i)],view=[0..1,0..Y]):
```

The motion is animated using the display command again, with the option insequence=true.

```
>   display(seq(Gr(i),i=0..N-1),insequence=true,
        scaling=constrained);
```

On execution of the above command line, the opening frame of the animation appears on the computer screen. Click on the plot and on the start arrow to see the animation. The equation of the tangent curve is displayed at the top of each frame, along with the coordinates at which the tangent line is calculated.

PROBLEMS:

Problem 4-18: Heather slows down

On seeing Patches begin to run toward her, Heather slows down, her speed being $3\,x$ km/h, where x is Patches' horizontal coordinate. Thus, when Patches is at $x = 1$, Heather's speed is 3 km/h, but when Patches reaches her $(x = 0)$ Heather is standing still. If Patches still runs at a constant speed of 8 km/h, analytically determine the equation $y(x)$ describing Patches' path and plot it. How far has Heather walked when Patches catches her? How far has Patches run? How long does it take Patches to reach Heather?

Problem 4-19: Curves of pursuit

Do an Internet search on *curves of pursuit* and discuss the history and types of pursuit curves that have been studied in the mathematical literature. Also see Davis [Dav62]. Create a recipe that generates the pursuit curve of your choice and motion along it, including the instantaneous tangent line. Depending on your choice, you may have to solve the relevant ODE numerically.

4.2.2   Oh What Sounds We Hear!

A man falls in love through his eyes, a woman through her ears.
Woodrow Wyatt, British journalist, *To the Point,* "*The Ears Have It,*" 1981
Wyatt argues that what is said to a woman by a man, and what she believes
about his status, is usually more important than the superficiality of good looks.

At the end of the nineteenth century, the famous German scientist Heinrich
Helmholtz knew that the ear perceives frequencies that are not present in the
incident acoustic waves. He attributed this to an asymmetric nonlinear response
of the eardrum's tympanic membrane and developed a simple mechanical model
for the eardrum's vibrations.

In the present century, a nervous young mathematics major by the name of
Mike has an hour or so to fret over his upcoming date with Vectoria. Why is
Mike nervous? This is the night that he plans to reveal his serious intentions for
her hand, and he is worried about possible rejection. To calm him down, let's
try to take his mind off Vectoria by asking him to reproduce Helmholtz's derivation. Helmholtz's model involves applying Newton's second law to generate a
phenomenological nonlinear ODE describing the one-dimensional displacement
$x(t)$ of the freely vibrating eardrum about the equilibrium position $x = 0$.

Reluctantly acquiescing to our request, Mike begins by loading the plots
library package, and then formally Taylor expanding the restoring force per
unit mass, $F(x)$, about $x = 0$ out to third order in x.

```
>  restart: with(plots):
>  Force:=taylor(F(x),x=0,3);
```

$$Force := F(0) + D(F)(0)\,x + \frac{1}{2}\,(D^{(2)})(F)(0)\,x^2 + O(x^3)$$

The coefficients $D(F)(0)$ and $D^{(2)}(F)(0)$ in the above output stand for the first
and second derivatives of F with respect to x, evaluated at $x = 0$, respectively.
Mike then assumes that the displacement x is sufficiently small that only terms
to second order in x have to be retained. So he removes the $O(x^3)$ term.

```
>  Force:=convert(Force,polynom);
```

$$Force := F(0) + D(F)(0)\,x + \frac{1}{2}\,(D^{(2)})(F)(0)\,x^2$$

Why keep the second-order term? The term that is linear in x corresponds to
Hooke's law, while the quadratic term is the first nonlinear correction to Hooke's
law and is responsible for the asymmetric response noted by Helmholtz. As the
eardrum vibrates about the equilibrium position $x = 0$, the linear (x) term in
the force changes sign with x, but the quadratic (x^2) term does not change sign.

In equilibrium the restoring force must vanish, so Mike sets $F(0) = 0$. He
also makes the symbolic substitutions $D(F)(0) = -\omega_0^2$ and $D^{(2)}(F)(0) = -2\beta$,

```
>  Force:=subs({F(0)=0,D(F)(0)=-omega[0]^2,(D@@2)(F)(0)=
             -2*beta},%);
```

$$Force := -\omega_0^2\,x - \beta\,x^2$$

with ω_0 and β both positive. Using Newton's second law and substituting the explicit time dependence into the force yields the relevant second-order nonlinear ODE.

> ode:=diff(x(t),t,t)-subs(x=x(t),Force)=0;

$$ode := \left(\frac{d^2}{dt^2} x(t) \right) + \omega_0{}^2 x(t) + \beta x(t)^2 = 0$$

If β is set equal to zero, then the familiar linear simple harmonic oscillator equation results, the eardrum vibrating at the characteristic frequency ω_0. In this case, the period (time for one complete oscillation) is $T = 2\pi/\omega_0$. With $\beta \neq 0$, the above ODE is nonlinear, but can be solved analytically, as Mike will demonstrate. Since he is already furtively looking at his watch, we will not insist that Mike look for realistic parameter values for the eardrum, but instead allow him to set $\omega_0 = 1$ and $\beta = 1$, so the ode has the following form.

> omega[0]:=1: beta:=1: ode:=ode;

$$ode := \left(\frac{d^2}{dt^2} x(t) \right) + x(t) + x(t)^2 = 0$$

Applying the dsolve command to ode,

> solution:=dsolve(ode,x(t));

$$solution := \int^{x(t)} \frac{3}{\sqrt{-9\_a^2 - 6\_a^3 + 9\_C1}} \, d\_a - t - \_C2 = 0,$$
$$\int^{x(t)} - \frac{3}{\sqrt{-9\_a^2 - 6\_a^3 + 9\_C1}} \, d\_a - t - \_C2 = 0$$

produces two general *implicit solutions* for the time, with positive and negative square roots. To obtain a positive period, Mike chooses to work with the positive square root, which is the first solution for this particular run.

> answer:=solution[1]; #choose positive square root

$$answer := \int^{x(t)} \frac{3}{\sqrt{-9\_a^2 - 6\_a^3 + 9\_C1}} \, d\_a - t - \_C2 = 0$$

The constant $\_C2$ in the answer merely shifts the origin of time and so can be disregarded or set equal to zero. The other arbitrary constant $\_C1$ can be evaluated by specifying an initial amplitude $x(0) = A$ with zero initial velocity there. The amplitude sets the total energy E of the system. For a one-dimensional mechanical system having mass m and speed v, the total energy is the sum of the kinetic energy and potential energy $U(x)$, i.e.,

$$E = \frac{1}{2} m v^2 + U(x). \tag{4.6}$$

But this can be rewritten as

$$v = \frac{dx}{dt} = \pm \sqrt{\frac{2}{m}(E - U(x))}, \tag{4.7}$$

or, on integrating,

$$t = \pm \int_0^{x(t)} \frac{dx}{\sqrt{(2/m)(E - U(x))}}.\tag{4.8}$$

Comparing the result in *answer* with the positive square root in (4.8), the potential energy $U(x)$ for the present problem can be taken (to within an overall proportionality constant) as

$$U(x) = 9\,x^2 + 6\,x^3.\tag{4.9}$$

Now Mike differentiates the answer with respect to time,

> `eq:=diff(answer,t);`

$$eq := \frac{3\left(\frac{d}{dt}x(t)\right)}{\sqrt{-9\,x(t)^2 - 6\,x(t)^3 + 9\_C1}} - 1 = 0$$

thus removing the unwanted coefficient _C2 from the analysis. Next, he solves *eq* for dx/dt, i.e., for the velocity *vel*.

> `vel:=solve(eq,diff(x(t),t));`

$$vel := \frac{1}{3}\sqrt{-9\,x(t)^2 - 6\,x(t)^3 + 9\_C1}$$

and removes the time dependence for later integration purposes, replacing $x(t)$ with the dummy variable y.

> `vel:=subs(x(t)=y,vel);`

$$vel := \frac{\sqrt{-9\,y^2 - 6\,y^3 + 9\_C1}}{3}$$

The unknown constant _C1 can be determined by imposing the condition that the velocity must be zero at the turning point $y = A$,

> `condition:=eval(vel,y=A)=0;`

$$condition := \frac{\sqrt{-9\,A^2 - 6\,A^3 + 9\_C1}}{3} = 0$$

and solving for the constant. The velocity expression is then determined.

> `_C1:=solve(condition,_C1); vel:=vel;`

$$\_C1 := A^2 + \frac{2}{3}\,A^3$$

$$vel := \frac{\sqrt{-9\,y^2 - 6\,y^3 + 9\,A^2 + 6\,A^3}}{3}$$

To proceed any further, the amplitude A must be given a specific value. Mike enters the nominal value $A = \frac{1}{3}$, which is automatically substituted into *vel*.

> `A:=1/3: vel:=vel;`

$$vel := \frac{\sqrt{-9\,y^2 - 6\,y^3 + \frac{11}{9}}}{3}$$

In *vel*, the total energy is identified to be $E = 11/9$. To get a qualitative feeling for the eardrum motion, Mike plots the potential energy U as a thick red line over the range $x = -1.6$ to 0.6,

```
>   U:=plot(9*x^2+6*x^3,x=-1.6..0.6,color=red,thickness=2):
```

and the total energy E as a thick green horizontal line between the points $(x = -2, E = 11/9)$ and $(x = 1, E = 11/9)$.

```
>   E:=plot([[-2,11/9],[1,11/9]],style=line,color=green,
        thickness=2):
```

The two energy curves are superimposed in Figure 4.6.

```
>   display({U,E},tickmarks=[3,3],labels=["x","Energy"]);
```

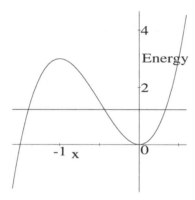

Figure 4.6: Asymmetric potential well for the eardrum model equation.

The horizontal total energy line in Figure 4.6 intersects the potential energy curve at three x values. The two intersection points inside the potential well centered at $x = 0$ are the *turning points*, the eardrum system oscillating between these two points if it is placed at one of these two points with zero velocity. The turning point on the right is the input value $x(0) = A = \frac{1}{3}$. Because the potential well is slightly asymmetric, the other turning point on the left is not at $x = -\frac{1}{3}$. The other turning point is easily determined by first finding all the intersection points using the `solve` command.

```
>   points:=solve(vel=0,y);
```

$$points := \frac{1}{3}, \ -\frac{11}{12} + \frac{\sqrt{33}}{12}, \ -\frac{11}{12} - \frac{\sqrt{33}}{12}$$

Three solutions are produced, corresponding to the three intersection points of the total energy line with the potential energy curve. The first answer corresponds to the turning point on the far right, and the third answer to the intersection on the far left. So the second answer is selected as corresponding to the second turning point for motion inside the potential well. This turning point is labeled B.

```
>  B:=points[2];
```

$$B := -\frac{11}{12} + \frac{\sqrt{33}}{12}$$

If $\beta = 0$, the potential well would have been symmetric about $x = 0$, and the second turning point would have been at $x = -\frac{1}{3} \approx -0.333$. For $\beta = 1$, the potential well is asymmetric and the second turning point is at $B = (-11 + \sqrt{33})/12 \approx -0.438$. Clearly, the period of oscillation will be somewhat longer than the period $2\pi/\omega_0 \approx 6.28$ for the linear case.

Mike calculates the period T for the nonlinear situation by multiplying the time it takes the system to go from B to A by 2, i.e, $T = 2\int_B^A (1/vel)\, dy$.

```
>  T:=2*int(1/vel,y=B..A);
```

$$T := \frac{24\,\mathrm{EllipticK}\left(\sqrt{\dfrac{15 - \sqrt{33}}{15 + \sqrt{33}}}\right)}{\sqrt{30 + 2\sqrt{33}}}$$

The period is expressed in terms of the *complete elliptic integral* $K(k)$ of the first kind, which is defined as

$$K(k) \equiv \int_0^{\pi/2} \frac{d\alpha}{\sqrt{1 - k^2 \sin^2 \alpha}}. \tag{4.10}$$

Maple expresses the complete elliptic integral in the form $\mathrm{EllipticK}(k)$, so in this case $k = \sqrt{(15 - \sqrt{33})/(15 + \sqrt{33})}$. The elliptic integral can be evaluated numerically,

```
>  T:=evalf(T);
```

$$T := 6.747679332$$

so the period is $T \approx 6.75$, which is about $7\frac{1}{2}\%$ longer than for the linear case.

To perform the necessary integration to obtain the analytic solution, it is necessary to assume that the displacement satisfies $x > B$ and $x < A$.

```
>  assume(x>B,x<A):
```

An implicit solution expressing t in terms of x is obtained by integrating the inverse velocity from the turning point B to an arbitrary point x (less than A).

```
>  sol:=t=int(1/vel,y=B..x); #implicit solution
```

$$sol := t = \frac{12\,\mathrm{EllipticF}\left(2\sqrt{\dfrac{45x + 33 - \sqrt{33} + 3\sqrt{33}\,x}{(15 - \sqrt{33})(11 + \sqrt{33} + 12x)}},\ \sqrt{\dfrac{15 - \sqrt{33}}{15 + \sqrt{33}}}\right)}{\sqrt{30 + 2\sqrt{33}}}$$

The implicit solution is expressed in terms of the *incomplete elliptic integral of the first kind*, $u \equiv F(\phi, k)$, which is defined as follows:

$$u \equiv F(\phi, k) = \int_0^\phi \frac{d\alpha}{\sqrt{1 - k^2 \sin^2 \alpha}} = \int_0^{\sin\phi} \frac{dy}{\sqrt{1 - y^2}\sqrt{1 - k^2 y^2}}. \tag{4.11}$$

Maple expresses the incomplete elliptic integral in the form $\mathrm{EllipticF}(z, k)$ with $z \equiv \sin\phi$. The complete elliptic integral $K(k)$ corresponds to taking $\phi = \pi/2$.

An explicit solution is obtained by solving *sol* for *x*.

> `x:=solve(sol,x);`

$$x := -\frac{1}{3} \frac{88\,\%1 + 8\,\%1\,\sqrt{11}\,\sqrt{3} - 77 - 3\sqrt{33}}{32\,\%1 - 43 - 5\sqrt{33}}$$

$$\%1 := \text{JacobiSN}\left(\frac{t\sqrt{30 + 2\sqrt{33}}}{12}, \sqrt{\frac{15 - \sqrt{33}}{15 + \sqrt{33}}}\right)^2$$

The formidable-looking answer, which is challenging to derive with pen and paper, is given in terms of the *Jacobian elliptic sine function*, which Maple expresses in the form JacobiSN(u, k). When written by hand, the elliptic sine function is often written as sn(u), the argument k not being explicitly expressed.

How is sn(u) defined? The upper limit ϕ in the first integral of equation (4.11) is referred to as the "amplitude" of u and is written as $\phi = $ am(u). Then, for a given k value, the elliptic sine function, sn(u), is defined by

$$\text{sn}(u) = \sin\phi = \sin(\text{am}(u)). \tag{4.12}$$

For $k = 0$, it is easy to verify that sn(u) reduces to sin(u). As k is increased, the elliptic sine function deviates away from the "ordinary" sine function shape. To learn more about the Jacobian elliptic sine function, Mike suggests that you try some of the relevant problems that follow this recipe.

The solution is now plotted over the time interval $t = 0$ to 30, displaying asymmetric oscillation about the equilibrium point.

> `plot(x,t=0..30,thickness=2);`

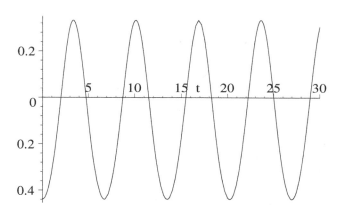

Figure 4.7: Asymmetric oscillations of the eardrum.

Unfortunately, Mike has to leave to pick Vectoria up, so let's wish him luck in his quest for this fair damsel's hand and hope that he will return soon to explore some more interesting recipes with us.

PROBLEMS:
Problem 4-20: Different amplitude
Taking the initial conditions $x(0) = -\frac{4}{5}$, $\dot{x}(0) = 0$ in the text recipe, determine the analytic form of the period and its numerical value. Determine the analytic solution and plot it over a suitable time range.

Problem 4-21: Maximum energy for bounded motion
What is the maximum total energy E for which oscillatory motion of the eardrum can occur in the text recipe? What are the numerical values of the turning points?

Problem 4-22: Higher-order correction to the period
For the freely vibrating eardrum, also keep the cubic term in the Taylor expansion of the force law and express the restoring force per unit mass as $F = -\omega_0^2 x - \beta x^2 - \gamma x^3$. Taking $\omega = 1$, $\beta = \frac{3}{4}$, $\gamma = \frac{1}{2}$, and $A = \frac{1}{3}$ determine the period of vibrations of the eardrum. By how much is the period changed with respect to that predicted by the linear Hooke's law?

Problem 4-23: Plotting the elliptic sine function
Plot the Jacobian elliptic sine function $\mathrm{sn}(u)$ over the range $u = 0$ to 20 in the same graph for $k = 0.1$, $k = 0.5$, $k = 0.9$, and $k = 0.995$. Discuss the behavior of $\mathrm{sn}(u)$ as k is increased.

Problem 4-24: Plotting the elliptic cosine function
In analogy to the elliptic sine function, the *Jacobian elliptic cosine function* of u is defined as $\mathrm{cn}(u) = \cos \phi = \cos(\mathrm{am}(u))$. Noting that the Maple command for $\mathrm{cn}(u)$ is `JacobiCN(u,k)`, plot the elliptic cosine function over the range $u = 0$ to 20 in the same graph for $k = 0.1$, $k = 0.5$, $k = 0.9$, and $k = 0.995$. Discuss the behavior of $\mathrm{cn}(u)$ as k is increased.

Problem 4-25: Properties of elliptic functions
Using the definition of $\mathrm{cn}(u)$ in the previous problem and defining another elliptic function $\mathrm{dn}(u) = \sqrt{1 - k^2 \sin^2 \phi} = \sqrt{1 - k^2 \,\mathrm{sn}^2(u)}$, prove the following:

(a) $\dfrac{d}{du}(\mathrm{cn}(u)) = -\,\mathrm{sn}(u)\,\mathrm{dn}(u)$;

(b) $\dfrac{d^2}{du^2}(\mathrm{cn}(u)) = (2\,k^2 - 1)\,\mathrm{cn}(u) - 2\,k^2\,\mathrm{cn}^3(u)$;

(c) $\displaystyle\int \mathrm{cn}(u)\,du = \frac{1}{k}\arccos(\mathrm{dn}(u))$.

Problem 4-26: Vibrating hard spring
The equation of motion for a "hard" spring is
$$\ddot{x}(t) + (1 + a^2\,x(t)^2)\,x(t) = 0.$$
Analytically determine the period and the solution $x(t)$ for the hard spring, given the initial conditions $x(0) = A$, $\dot{x}(0) = 0$. Plot the solution over several cycles for $a = A = 1$.

4.2.3 Vectoria Feels the Force and Hits the Bottle

Man your ships, and may the force be with you.
Film line from *Star Wars*, written by George Lucas, 1977

Our fair damsel, Vectoria, has felt the force of Mike's love and can't wait to tell her girlfriends tonight about becoming engaged. Perhaps a bottle of sparkling Asti Spumante wine would be in order then, but in the meantime a homework assignment in her electromagnetics course beckons and an entirely different sort of bottle awaits her immediate consideration.

She has been asked to create a recipe that illustrates the movement of a charged particle in a *magnetic bottle*. The phrase "magnetic bottle" refers to the carefully designed arrangements of magnetic fields that are used for the confinement of extremely hot plasmas studied in experiments on thermonuclear fusion. Basically, the magnetic field is approximately uniform in the center of the "bottle" but gets stronger and pinches inward at the ends of the bottle. A charged particle starting in the central region, and given an initial velocity toward one end of the bottle, spirals around a field line traveling toward that end. As it approaches the end, the orbits become smaller and smaller, and finally at some point the charge reflects off the "end" of the bottle and reverses direction. It then travels to the other end of the bottle and repeats the scenario.

In her recipe, Vectoria will make use of the *Lorentz force law* and a magnetic field configuration of her own mathematical design. A particle of charge q moving with velocity $\vec{v}$ in an external electric field $\vec{E}$ and magnetic field $\vec{B}$ will experience the Lorentz force $\vec{F} = q\vec{E} + q(\vec{v} \times \vec{B})$. If present, the electric field produces a contribution to the Lorentz force in the direction of $\vec{E}$, while the magnetic force contribution is perpendicular to the velocity vector. So, the electric force will accelerate the charge, while the magnetic force deflects it.

Vectoria begins her recipe by loading the VectorCalculus and plots packages, the former being needed for creating and manipulating vectors.

```
>  restart: with(VectorCalculus): with(plots):
```

Using Cartesian coordinates, with $\hat{e}_x$, $\hat{e}_y$, and $\hat{e}_z$ pointing along the x-, y-, and z-axes respectively, the position vector $\vec{r} = x(t)\,\hat{e}_x + y(t)\,\hat{e}_y + z(t)\,\hat{e}_z$ of the charge at time t is entered, and its velocity $\vec{v} = d\vec{r}/dt$ and acceleration $\vec{a} = d\vec{v}/dt$ calculated. The position vector is entered here with the "shorthand" syntax, the longer form being Vector([x(t),y(t),z(t)]). The default output is in terms of Cartesian unit vectors, although Maple does not place "hats" on them.

```
>  r:=<x(t),y(t),z(t)>; v:= diff(r,t); a:=diff(v,t);
```

$$r := x(t)\,\mathrm{e_x} + y(t)\,\mathrm{e_y} + z(t)\,\mathrm{e_z}$$

$$v := \left(\frac{d}{dt}\,x(t)\right)\mathrm{e_x} + \left(\frac{d}{dt}\,y(t)\right)\mathrm{e_y} + \left(\frac{d}{dt}\,z(t)\right)\mathrm{e_z}$$

$$a := \left(\frac{d^2}{dt^2}\,x(t)\right)\mathrm{e_x} + \left(\frac{d^2}{dt^2}\,y(t)\right)\mathrm{e_y} + \left(\frac{d^2}{dt^2}\,z(t)\right)\mathrm{e_z}$$

The radial distance $R = \sqrt{x(t)^2 + y(t)^2}$ at time t is entered. The coordinates and initial conditions will be chosen so that spirals are traced out in the radial direction with translation of the spirals along the z-axis.

```
>  R:=sqrt(x(t)^2+y(t)^2);
```

$$R := \sqrt{x(t)^2 + y(t)^2}$$

For the magnetic bottle fields, Vectoria takes the electric field to be zero, and starts to build up the magnetic field structure, inputting the x-, y-, and z-components but with the mathematical forms of the radial function Br and longitudinal function Bz not yet specified.

```
>  E:=<0,0,0>; B:=<-Br*x(t)/R,-Br*y(t)/R,Bz>;
```

$$E := 0 \, e_x$$

$$B := -\frac{Br \, x(t)}{\sqrt{x(t)^2 + y(t)^2}} \, e_x - \frac{Br \, y(t)}{\sqrt{x(t)^2 + y(t)^2}} \, e_y + Bz \, e_z$$

The Lorentz force on the charge q is calculated, the short-hand form v &x B being used to enter the cross product $\vec{v} \times \vec{B}$. The long form is CrossProduct(v,B).

```
>  F:=q*E+q*(v &x B);
```

$$F := q \left(\left(\frac{d}{dt} y(t) \right) Bz + \frac{\left(\frac{d}{dt} z(t) \right) Br \, y(t)}{\sqrt{x(t)^2 + y(t)^2}} \right) e_x$$

$$+ q \left(-\frac{\left(\frac{d}{dt} z(t) \right) Br \, x(t)}{\sqrt{x(t)^2 + y(t)^2}} - \left(\frac{d}{dt} x(t) \right) Bz \right) e_y$$

$$+ q \left(-\frac{\left(\frac{d}{dt} x(t) \right) Br \, y(t)}{\sqrt{x(t)^2 + y(t)^2}} + \frac{\left(\frac{d}{dt} y(t) \right) Br \, x(t)}{\sqrt{x(t)^2 + y(t)^2}} \right) e_z$$

Taking the mass of the charge to be m, Newton's second law gives the system of three component equations. Only the first ODE is shown here in the text.

```
>  sys:=seq(m*a[i]=F[i],i=1..3);
```

$$sys := m \left(\frac{d^2}{dt^2} x(t) \right) = q \left(\left(\frac{d}{dt} y(t) \right) Bz + \frac{\left(\frac{d}{dt} z(t) \right) Br \, y(t)}{\sqrt{x(t)^2 + y(t)^2}} \right), \dots$$

Vectoria uses the piecewise command to build up the functional forms of Br and Bz. The structure of Br is such that there is no radial component of the magnetic field between $z = -1$ and $z = +1$. In this region, the magnetic field is completely in the z-direction. Outside this central region, the radial field component is allowed to vary with z according to a hyperbolic cosine function. The function Bz is taken to be constant inside the region $z = -0.5$ to 0.5 and allowed to grow linearly stronger outside this region.

```
>  Br:=piecewise(z(t)>1,cosh(z(t)),z(t)<-1,-cosh(z(t)),0);
```

$$Br := \begin{cases} \cosh(z(t)) & 1 < z(t) \\ -\cosh(z(t)) & z(t) < -1 \\ 0 & otherwise \end{cases}$$

```
>  Bz:=piecewise(abs(z(t))<.5,4,4+4*abs(abs(z(t))-.5));
```

$$Bz := \begin{cases} 4 & |z(t)| < 0.5 \\ 4 + 4\,||z(t)| - 0.5| & otherwise \end{cases}$$

To see whether her piecewise functional forms produce a suitable field configuration for the occurrence of magnetic bottle behavior, Vectoria will plot the magnetic field in the $y = 0$ plane. To accomplish this she first substitutes $y(t) = 0$ and removes the time dependence from x and z in the first and third components of the magnetic field.

```
>  B1:=subs({x(t)=x,z(t)=z,y(t)=0},B[1]):
   B3:=subs({x(t)=x,z(t)=z},B[3]):
```

She uses the `fieldplot` command to represent the direction and magnitude of the field with thick red arrows. A cross-sectional plot of the symmetrical magnetic field is shown in Figure 4.8. Note that the size of the arrows is an indication of field strength, bigger arrows for stronger fields. The field configuration shown in the figure should allow magnetic bottle behavior.

```
>  fieldplot([B3,B1],z=-1.5..1.5,x=-1..1,arrows=THICK,grid=
   [15,15],color=red,scaling=constrained,tickmarks=[3,3]);
```

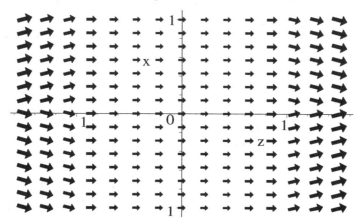

Figure 4.8: Magnetic field configuration to produce a magnetic bottle.

To animate the motion of a charge in her magnetic bottle, Vectoria chooses the nominal mass and charge values $m = 1$ and $q = 1$ and the initial conditions

$x(0) = 0.5$, $y(0) = 0$, $z(0) = 0$, $\dot{x}(0) = 0$, $\dot{y}(0) = -2$, and $\dot{z}(0) = -0.2$.

```
>  m:=1: q:=1:
```

```
> ic:=(x(0)=0.5,y(0)=0,z(0)=0,D(x)(0)=0,D(y)(0)=-2,D(z)(0)=0.2):
```

Maple is not able to produce an explicit closed-form analytic solution for the nonlinear ODE system, *sys*, so a numerical solution is generated, the output being given as a "listprocedure."

```
>  sol:=dsolve({sys,ic},{x(t),y(t),z(t)},numeric,
        output=listprocedure):
```

Evaluating $x(t)$, $y(t)$, and $z(t)$ with *sol* will allow Vectoria to calculate the charge's position at an arbitrary time t.

```
>  X:=eval(x(t),sol): Y:=eval(y(t),sol): Z:=eval(z(t),sol):
```

For example, the charge's x-coordinate at $t = 2$ time units is now evaluated.

```
>  X(2);
```

$$-0.0727500262205896948$$

Vectoria will now animate the motion of the charge in the magnetic bottle. She takes the total time to be $T = 24.0$ time units and will have $N = 200$ time steps in her animation. The time step size T/N is then calculated.

```
>  T:=24.0: N:=200: step:=T/N;
```

$$step := 0.1200000000$$

The `spacecurve` command is used to plot (but not display) the entire trajectory over the time interval $t = 0$ to T, the trajectory being colored with the `zhue` option. To obtain a smooth curve, the minimum number of plotting points is taken to be 1000.

```
>  sc:=spacecurve([X(t),Y(t),Z(t)],t=0..T,shading=zhue,
        numpoints=1000):
```

Using a do loop running from $n = 0$ to N, the charge is plotted on each time step in 3 dimensions as a size-18 red circle superimposed on the entire trajectory.

```
>  for n from 0 to N do
```

```
>  t:=step*n;
```

```
>  pp[n]:=pointplot3d([X(t),Y(t),Z(t)],symbol=circle,
        symbolsize=18,color=red);
```

```
>  gr[n]:=display({sc,pp[n]});
```

```
>  end do:
```

The motion of the charge in the magnetic bottle is animated with the `display` command, using the `insequence=true` option. To see the animation execute the worksheet, click on the plot, and on the start arrow in the tool bar.

```
>  display(seq(gr[n],n=0..N),insequence=true,axes=frame,
        labels=["x","y","z"]);
```

Vectoria is quite pleased that her piecewise model mimics magnetic bottle behavior. But now she has to forsake this bottle for another one, as she rushes off to join her girlfriends and share a bottle of Asti Spumante.

PROBLEMS:

Problem 4-27: Crossed electric and magnetic fields

Consider a charge q and mass m moving in the crossed electric and magnetic fields described by $\vec{E} = E_0 \sin(\omega t)\,\hat{e}_z$, $\vec{B} = B_0\,\hat{e}_x$, where E_0, B_0, and ω are real constants. Using the Lorentz force law, derive the system of ODEs governing the motion of the charge. Is the system linear or nonlinear? Analytically derive the solution of the system, given the initial conditions $x(0) = 0$, $y(0) = 0$, $z(0) = 0$, $\dot{x}(0) = 0$, $\dot{y}(0) = 0$, and $\dot{z}(0) = 0$. Taking $m = 1$, $q = 1$, $\omega = 5.1$, $E_0 = 2$, and $B_0 = 2.3$, animate the motion of the charge (represented as a circle), superimposing the motion on the entire trajectory.

Problem 4-28: Altering the magnetic bottle field

In the text recipe, experiment with different functional forms of Br and Bz and discuss the effect of your alterations on magnetic bottle behavior. Optional: Create a continuous field configuration that produces magnetic bottle behavior and check that $\mathrm{div}\vec{B} = 0$.

4.2.4 Golf Is Such an "Uplifting" Experience

If you watch a game, it's fun. If you play it, it's recreation.
If you work at it, it's golf.
Bob Hope, American comedian, *Readers Digest*, October 1958.

Heather's girlfriends have invited her to go golfing, but they still need another woman to make up a foursome for the Thursday afternoon ladies' day at the Metropolis Country Club. Heather assures them that she can probably twist her older sister Jennifer's arm to go with them. Jennifer, who has been busy preparing lectures for her applied mathematics course, has had little time to play golf recently, but is persuaded to go on the basis that a gourmet buffet supper will be available in the clubhouse after the 18 holes are completed.

Although it won't help her to play golf any better, Jennifer's inborn curiosity about how she would go about realistically modeling the flight of a golf ball is piqued, so she trots over to the university science library to look up some articles on the physics of golf [Erl83], [MA88], [MH91]. She recalls from her undergraduate physics courses that the trajectory of a golf ball was calculated assuming that there was no atmospheric drag on the ball, and also not taking any possible uplift due to backspin of the ball into account. Jennifer wonders what effect the inclusion of these two contributions might have on the ball's flight path? For example, can the uplift compensate sufficiently for the aerodynamic drag to enable the ball to fly farther? After photocopying the cited references, she makes a beeline back to her office, where she commences to simulate the flight of the golf ball with both drag and uplift included. Fortunately, the references also provide actual parameter values that can be used in her computer modeling.

As she was taught way back in first-year physics, Jennifer begins by making a free-body diagram showing all the forces acting on the golf ball. This diagram is reproduced in Figure 4.9.

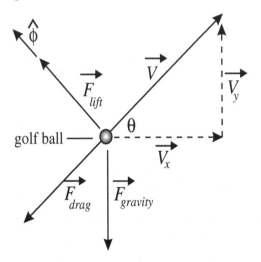

Figure 4.9: Free-body diagram for a golf ball, including drag and lift.

Three forces are included, the pull of gravity $\vec{F}_{gravity}$ downward, the drag force $\vec{F}_{drag}$ in the direction opposite to the instantaneous velocity vector $\vec{V}$, and the lift force $\vec{F}_{lift}$ in a direction perpendicular to $\vec{V}$. For simplicity, Jennifer assumes that the ball's spin axis is in the z-direction and that the translational motion is in the x-y plane with x horizontal and y vertical. She makes a call to the plots and VectorCalculus packages,

> `restart: with(plots): with(VectorCalculus):`

and enters the velocity vector $\vec{V} = V_x\,\hat{e}_x + V_y\,\hat{e}_y$.

> `Velocity:=<V[x],V[y]>;`

$$Velocity := V_x\,e_x + V_y\,e_y$$

Since the speed V will be needed, it is calculated by taking the dot product of the velocity vector with itself and then taking the square root. The dot product is entered with the shorthand[3] dot syntax. The unit vector (assigned the name u_v) in the direction of $\vec{V}$ is then equal to the velocity divided by the speed.

> `Speed:=sqrt(Velocity . Velocity); u[v]:=Velocity/Speed;`

$$Speed := \sqrt{V_x{}^2 + V_y{}^2}$$

$$u_v := \frac{V_x}{\sqrt{V_x{}^2 + V_y{}^2}}\,e_x + \frac{V_y}{\sqrt{V_x{}^2 + V_y{}^2}}\,e_y$$

---

[3]The long form is `DotProduct(Velocity,Velocity)`.

Referring to the freebody diagram, $\sin(\theta) = V_y/V$ and $\cos(\theta) = V_x/V$.

```
>  sin(theta):=V[y]/Speed; cos(theta):=V[x]/Speed;
```

$$\sin(\theta) := \frac{V_y}{\sqrt{V_x{}^2 + V_y{}^2}} \qquad \cos(\theta) := \frac{V_x}{\sqrt{V_x{}^2 + V_y{}^2}}$$

The unit vector $\hat{\phi}$ pointing in the lift direction is related to the Cartesian unit vectors $\hat{e}_x$ and $\hat{e}_y$ pointing along the x- and y-directions, respectively, by the relation $\hat{\phi} = -\sin\theta\,\hat{e}_x + \cos\theta\,\hat{e}_y$. This relation is entered and labeled u_ϕ.

```
>  u[phi]:=<-sin(theta),cos(theta)>;
```

$$u_\phi := -\frac{V_y}{\sqrt{V_x{}^2 + V_y{}^2}}\, e_x + \frac{V_x}{\sqrt{V_x{}^2 + V_y{}^2}}\, e_y$$

The gravitational force $\vec{F}_{gravity} = -m\,g\,\hat{e}_y$, where m is the mass of the ball and g is the acceleration due to gravity.

```
>  F[gravity]:=<0,-m*g>;
```

$$F_{gravity} := -m\,g\,e_y$$

According to the photocopied references, both the drag force $\vec{F}_{drag}$ and the lift force $\vec{F}_{lift}$ are proportional to the square of the speed. Jennifer labels the proportionality constants (per unit mass) as K_{drag} and K_{lift}, respectively, and enters the two forces.

```
>  F[drag]:=-K[drag]*m*Speed^2*u[v];
```

$$F_{drag} := -K_{drag}\, m\, \sqrt{V_x{}^2 + V_y{}^2}\, V_x\, e_x - K_{drag}\, m\, \sqrt{V_x{}^2 + V_y{}^2}\, V_y\, e_y$$

```
>  F[lift]:=K[lift]*m*Speed^2*u[phi];
```

$$F_{lift} := -K_{lift}\, m\, \sqrt{V_x{}^2 + V_y{}^2}\, V_y\, e_x + K_{lift}\, m\, \sqrt{V_x{}^2 + V_y{}^2}\, V_x\, e_y$$

The net force $\vec{F}$ acting on the golf ball is the vector sum of the three forces.

```
>  F:=F[gravity]+F[drag]+F[lift]:
```

Now, $V_x = dx(t)/dt$ and $V_y = dy(t)/dt$, where $x(t)$ and $y(t)$ are the horizontal and vertical coordinates of the golf ball at time t.

```
>  V[x]:=diff(x(t),t): V[y]:=diff(y(t),t):
```

Newton's second law of motion is applied in the x- and y-directions.

```
>  xeq:=diff(V[x],t)=simplify(F[1]/m);
```

$$xeq := \frac{d^2}{dt^2}x(t) = -\sqrt{\left(\frac{d}{dt}x(t)\right)^2 + \left(\frac{d}{dt}y(t)\right)^2}\left(K_{drag}\left(\frac{d}{dt}x(t)\right) + K_{lift}\left(\frac{d}{dt}y(t)\right)\right)$$

```
>  yeq:=diff(V[y],t)=simplify(F[2]/m);
```

$$yeq := \frac{d^2}{dt^2} y(t) = -g - K_{drag} \sqrt{\left(\frac{d}{dt}x(t)\right)^2 + \left(\frac{d}{dt}y(t)\right)^2} \left(\frac{d}{dt}y(t)\right)$$

$$+ K_{lift} \sqrt{\left(\frac{d}{dt}x(t)\right)^2 + \left(\frac{d}{dt}y(t)\right)^2} \left(\frac{d}{dt}x(t)\right)$$

Jennifer's task is now to solve the coupled nonlinear ODEs, xeq and yeq, numerically, since they do not have analytic solutions. To accomplish this, the parameter values must be entered. From the references, she finds that the drag and lift parameters for a British golf ball of mass $m = 0.046$ kg and radius $r = 0.0207$ m launched at an initial speed $V0 = 61$ m/s and with a backspin of about 60 rev/s are approximately equal to each other with a value 0.28. The density ρ of dry air at 1 atm pressure and 20°C is $\rho = 1.21$ kg/m^3. Jennifer takes $g = 9.81$ m/s and the initial launch angle of the ball to be 16°.

>   `rho:=1.21: m:=0.046: r:=0.0207: g:=9.81: V0:=61: Angle:=16:`
     `Drag:=0.28: Lift:=0.28:`

According to the references, the drag and lift coefficient are given by $K_{drag} = \rho \pi r^2 \, \text{Drag}/(2\,m)$ and $K_{lift} = \rho \pi r^2 \, \text{Lift}/(2\,m)$. These two coefficients are numerically evaluated and temporarily labeled $KDrag$ and $KLift$.

>   `KDrag:=evalf(rho*Pi*r^2*Drag/(2*m));`
     `KLift:=evalf(rho*Pi*r^2*Lift/(2*m));`
        $KDrag := 0.004957310686$ $KLift := 0.004957310686$

The initial angle is converted from degrees into radians,

>   `Theta:=convert(Angle*degrees,radians);`

$$\Theta := \frac{4\pi}{45}$$

and the initial velocity of the golf ball calculated.

>   `Vo:=evalf(V0*<cos(Theta),sin(Theta)>);`

$$Vo := 58.63696345\, e_x + 16.81387871\, e_y$$

The ball starts with a horizontal velocity component of 58.6 m/s and a vertical component of 16.8 m/s. Taking the golf ball to be initially positioned at the origin of the coordinate system, the initial conditions are then entered.

>   `ic:=x(0)=0,D(x)(0)=Vo[1],y(0)=0,D(y)(0)=Vo[2]:`

Jennifer wishes to compare the path followed by the ball when both drag and lift are included with the trajectory that results when both contributions are absent. So in the following do loop, $n = 2$ corresponds to the former, while $n = 1$ produces the latter situation.

>   `for n from 1 to 2 do`

For $n = 1$ in the following command line, $K_{drag} = 0$ and $K_{lift} = 0$, while $n = 2$ produces $K_{drag} = KDrag$ and $K_{lift} = KLift$.

>   `K[drag]:=(n-1)*KDrag: K[lift]:=(n-1)*KLift:`

The nonlinear ODEs xeq and yeq are solved numerically for $x(t)$ and $y(t)$,

```
> sol:=dsolve({xeq,yeq,ic},{x(t),y(t)},type=numeric);
```
and the `odeplot` command used to plot the trajectory of the ball. The time interval for the golf ball to strike the ground again is different for the two trajectories. Without drag and lift, the time interval is about 3.42 s, while the time interval lengthens to about 7 s when these contributions are both included. Jennifer allows for this in the following command line. The linestyle option, `linestyle=3-n`, is used to produce a solid curve for the situation that both lift and drag are included ($n = 2$), and a dashed line when they are not.

```
> pl[n]:=odeplot(sol,[x(t),y(t)],0..3.42+3.6*(n-1),
             labels=["x","y"],linestyle=3-n,thickness=2):
> end do:
```
On ending the do loop, Jennifer muses about what usually happens when she goes golfing. Invariably, she ends up in the rough, in a sand trap, or has to shoot over a tree. So in her simulation, she places a thick brown "tree" 30 meters high at a distance of 120 meters from where the ball is struck.

```
> tree:=plot([[120,0],[120,30]],color=brown,thickness=3):
```
The two possible trajectories are superimposed with the tree plot to produce

```
> display([seq(pl[i],i=1..2),tree],scaling=constrained);
```

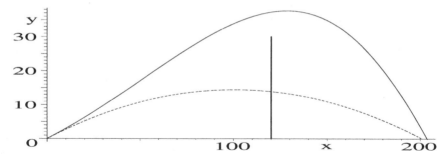

Figure 4.10: Solid curve: drag and lift included. Dashed curve: no drag or lift.

Figure 4.10. Jennifer notes a number of important differences between the two trajectories. Despite the effect of drag, her simulation indicates that a golf ball in the real world rises higher (thus, clearing the "tree" here) and consequently travels farther because of the lift on the golf ball due to backspin. This is not too surprising, since the ball stays in the air about twice as long as when both drag and lift are neglected. She further notices that the ball's path actually bends upward before reaching the zenith of its trajectory, a phenomenon that she has seen on the golf course, especially when the club pro is playing. In her model calculation the ball has traveled about 203 m (222 yards), hardly a long shot by the standards of someone like Tiger Woods, who averages about 300 yards a drive. However, she is confident that she could modify the initial conditions to mimic one of Tiger's shots. Undoubtedly Heather, who is a big Tiger Woods fan, will ask her to do it when she learns about the simulation.

As much fun as she has had, Jennifer suddenly realizes that she had better spend a little time hitting balls down at the local driving range before she goes out with Heather and her friends. Otherwise, she might end up proving to be an embarrassment to her sister. Now where did she put those golf clubs?

PROBLEMS:

Problem 4-29: The Boffo brothers go golfing

Syd and Benny Boffo have taken an afternoon off from their realty business and gone golfing at the Metropolis Country Club. After a good approach shot on a level fairway, Benny Boffo has placed his golf ball 100 m from the hole. He selects a 9-iron and hits the ball perfectly, landing practically at the hole. If both lift and drag are included and all other parameters of the text recipe prevail, at what angle to the horizontal (in degrees) was the ball hit? How high did the ball rise?

Problem 4-30: Syd Boffo makes a hole in one

Syd and Benny Boffo's golf match continues and they move onto the next hole, a par 3. The tee-off point is elevated 20 m above the hole, the ball having to travel over an intervening lily pond. The horizontal distance from the tee-off position to the hole is 160 m. Syd miraculously scores a hole in one, which means that drinks are on him at the clubhouse after the match. Assuming that both lift and drag are included and all other parameters of the text recipe prevail, at what angle did Syd's ball ascend in order for him to make the hole in one? How high did the ball rise relative to the hole? If the green is circular (radius 5 m) with the hole at its center, and the pond stretches up to the edge of the green, would the ball have ended up in the pond if lift had been neglected? Or would the ball have overshot the green?

Problem 4-31: Rocky Mountain high

At a Rocky Mountain golf resort, the elevation is such that the density of air is $\rho = 1.0$ kg/m^3 and $g = 9.8$ m/s^2. How much farther would the ball travel than for the sea-level data given in the text recipe?

Problem 4-32: Horizontal drive

A golf ball that has been teed up 3 cm above the ground is hit badly and leaves the tee horizontally. Including lift and drag and taking all other parameters as in the text recipe, plot the trajectory of the ball and determine where it strikes the ground, assuming that the fairway is level. How does the distance compare to the situation in which there is no lift?

Problem 4-33: Vertical shot

As part of its testing routine, a golf ball manufacturing company tests the performance of one of its golf balls by shooting it vertically into the air with the same spin and other parameters as in the text recipe. If lift and drag are included, where would the golf ball land?

4.3 Variational Calculus Models

The models in this section are based on the following mathematical framework developed by the mathematicians Euler and Lagrange. Consider a function $y(x)$ with fixed values $y(x_0) = y_0$ and $y(x_1) = y_1$ at two distinct points A and B, respectively. Form a specific function F of x, $y(x)$ and $y' \equiv dy(x)/dx$, i.e., $F = F(x, y, y')$. In our examples, F will be determined by the specification of the problem. Among all functions $y(x)$ connecting A and B, find those that give an extremum (minimum or maximum, usually) to the integral

$$I[y] = \int_{x_0}^{x_1} F(x, y(x), y'(x))\, dx. \tag{4.13}$$

The value I of the integral depends on the functional form chosen for $y(x)$. Rather than trying different $y(x)$ ad infinitum, it turns out that the $y(x)$ that yields an extremum value for I must satisfy the *Euler–Lagrange* equation [FC99]

$$\frac{\partial F}{\partial y} - \frac{d}{dx}\left(\frac{\partial F}{\partial y'}\right) = 0. \tag{4.14}$$

The resulting ODE is often nonlinear in nature, but may be susceptible to analytic solution. If not, it can be solved numerically.

4.3.1 Dress Design, the Erehwonese Way

Haute Couture should be fun, foolish and almost unwearable.
Christian Lacroix, French couturier (1951–)

On the planet Erehwon the leading dress designer, Gino Spoofia, has created a new design, called the lampshade look,[4] as shown in Figure 4.11, to accommo-

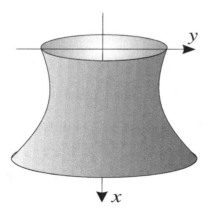

Figure 4.11: The lampshade look in dress design.

---

[4]The supporting straps are not displayed.

date the shapes of its most beautiful (by Erehwon standards) female citizens. Being an amateur scientist on the side and also to keep down material cost, he wants the surface area of cloth in the one-piece dress to be a minimum. The smaller upper ring at $x = 0$ is of radius $y = 1$ sretem (the local unit of length), and the lower, larger, ring at $x = 1$ sretem is of radius $y = 2$ sretem. To answer the question of what the shape of the surface should be to minimize the area, Gino consults one of Erehwon's most prominent mathematicians, Greg Arious Nerd, who says that this problem was solved many millennia ago. Greg then proceeds to outline the solution to the problem for his dress-designing friend. For the benefit of readers who may not understand the peculiar Northern Ere-hwonese dialect spoken by Greg, we shall translate what he has to say. The authors are not totally fluent in this language, so bear with us if we occasionally are in doubt about the exact words and thus resort to paraphrasing what Greg says instead of reporting his remarks verbatim.

"It's all very simple," Greg starts out. "The surface area of the dress can be generated by revolving the curve $y(x)$ about the x-axis. So what[5] is the shape of $y(x)$?

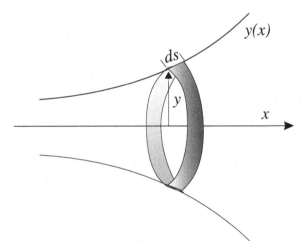

Figure 4.12: Determining the surface area of the dress.

Referring to Figure 4.12, an element dA of surface area is given by

$$dA = (2\pi y)\, ds = 2\pi y \sqrt{(dx)^2 + (dy)^2} = 2\pi y \sqrt{1 + (y')^2}\, dx, \qquad (4.15)$$

where the prime indicates a derivative with respect to x. So the surface area between $x = 0$ and $x = 1$ is

$$A = 2\pi \int_0^1 y \sqrt{1 + (y')^2}\, dx. \qquad (4.16)$$

[5]Author's note: This appears to be a rhetorical question on Greg's part, since Gino hasn't opened his mouth yet.

The goal is to find the shape $y(x)$ that minimizes A."

"I know that, Greg," Gino interjects, "but get to the point. As my business associates from Earth say, time is money. What is the bottom line? How many square sretem of cloth will I need?"

A little perturbed by this sign of impatience, Greg's dialect has become stronger and his response is difficult for us to precisely translate, but the gist of it goes something like this. Long ago, some ancestors on his mother's side had adopted a trial-and-error approach to finding $y(x)$. This consisted in choosing some form for $y(x)$ and performing the integration to calculate A. Then the venerable ancestors would choose another form and see whether it generated a smaller value for A. With a touch of hyperbole, we think, Greg said that some kept repeating this process until either they became residents in the Erehwon asylum or inductees into the Erehwon Academy of Mathematics. According to his unbiased (?) version of history, a better approach was discovered long ago by one of Greg's illustrious ancestors on his father's side, Hil Arious Nerd. Evidently, Hil independently discovered the same method as developed by Euler and Lagrange on far-distant Earth. The method is as follows.

Label the integrand of the area integral as F, i.e., $F = 2\pi y \sqrt{1 + (y')^2}$. The curve $y(x)$ that minimizes the area is the solution of the Euler–Lagrange equation (known as the Nerd equation on Erehwon)

$$\frac{\partial F}{\partial y} - \frac{d}{dx}\frac{\partial F}{\partial y'} = 0. \tag{4.17}$$

Solving this type of problem by hand has been a standard assignment given in the past by perverse professors on Erehwon[6] to their students.

It appears that Greg has calmed down so that we can understand him better, so let's pick up Greg's detailed narrative once again.

"It's now frowned upon to inflict mental pain on students (and dress designers) by performing hand calculations when computers can do all the algebra for you with no errors. At the Erehwon Institute of Technology we use the Elpam computer algebra system, so let's use it to crack your problem.

I will begin by loading the VariationalCalculus library package which contains the `EulerLagrange` command for generating the left-hand side of the Euler–Lagrange (Nerd) equation (4.17).

```
>   restart: with(VariationalCalculus):
```
Next let's enter the integrand, $F = y(x)\sqrt{1 + (dy(x)/dx)^2}$, omitting the factor of 2π, which would ultimately cancel out of the equation,

```
>   F:=y(x)*sqrt(1+diff(y(x),x)^2);
```

$$F := y(x)\sqrt{1 + \left(\frac{d}{dx}y(x)\right)^2}$$

The `EulerLagrange` command is applied to F, the second argument (x) being the independent variable, the third argument the dependent variable. The result is then simplified.

---

[6] Author's note: This quaint, but archaic, custom is still widely practiced on Earth.

```
>  eq:=simplify(EulerLagrange(F,x,y(x)));
```

$$eq := \left\{ \frac{y(x)}{\sqrt{1+\left(\frac{d}{dx}y(x)\right)^2}} = K_1, \ \frac{1+\left(\frac{d}{dx}y(x)\right)^2 - y(x)\left(\frac{d^2}{dx^2}y(x)\right)}{\left(1+\left(\frac{d}{dx}y(x)\right)^2\right)^{(3/2)}} \right\}$$

Two results are generated in *eq*. The expression not involving the constant K_1 is just the lhs of the Euler–Lagrange equation. It can be extracted using the **remove** command to remove the expression containing K_1 from *eq*. Inclusion of the brackets at the end of the command line removes the brackets that would otherwise enclose the following answer.

```
>  eq2:=remove(has,eq,K[1])[];
```

$$eq2 := \frac{1+\left(\frac{d}{dx}y(x)\right)^2 - y(x)\left(\frac{d^2}{dx^2}y(x)\right)}{\left(1+\left(\frac{d}{dx}y(x)\right)^2\right)^{(3/2)}}$$

The relevant governing nonlinear ODE that must be solved for $y(x)$ follows on setting the numerator of *eq2* equal to zero. This ODE would have been tedious to derive by hand.

```
>  ode:=numer(eq2)=0; #relevant ODE
```

$$ode := 1+\left(\frac{d}{dx}y(x)\right)^2 - y(x)\left(\frac{d^2}{dx^2}y(x)\right) = 0$$

If proceeding by hand, one could reduce the second-order *ode* to a first-order ODE by setting $p = dy/dx$, so that $d^2y/dx^2 = dp/dx = (dy/dx)(dp/dy) = p(dp/dy)$. The variables can then be separated and the integrations performed, yielding $p \equiv dy/dx$ with one arbitrary constant. But this *first integral* is already contained in *eq*, being the term involving the constant K_1. This first integral can be extracted from *eq* using the following **select** command:

```
>  eq3:=select(has,eq,K[1])[]; #first integral
```

$$eq3 := \frac{y(x)}{\sqrt{1+\left(\frac{d}{dx}y(x)\right)^2}} = K_1$$

The first-order ODE *eq3* is solved for $y(x)$, a second constant *_C1* appearing in the solution, which contains two equivalent answers (not displayed).

```
>  sol:=dsolve(eq3,y(x));
```

I will select the first answer in *sol* and simplify the rhs of the result.

```
>  y:=simplify(rhs(sol[1]));
```

$$y := \frac{1}{2}\left(K_1{}^2 + e^{\left(\frac{2(x-\_C1)}{K_1}\right)}\right) e^{\left(-\frac{x-\_C1}{K_1}\right)}$$

We now have the general curve y that will minimize the area, but we have to determine the two constants K_1 and $\_C1$. Evaluating y at $x = 0$ and setting the result to 1 yields the first boundary condition, while evaluating y at $x = 1$ and setting the result to 2 yields the second one. The two boundary conditions are then solved numerically for K_1 and $\_C1$ and the solution is assigned.

```
>   bc1:=eval(y,x=0)=1: bc2:=eval(y,x=1)=2:
>   sol2:=fsolve({bc1,bc2},{K[1],_C1}); assign(sol2):
```

$$sol2 := \{K_1 = 0.9499888273, \_C1 = -0.2581775581\}$$

On expanding and simplifying, y is now completely determined,

```
>   y:=simplify(expand(y));
```

$$y := 0.3438580549\, e^{(-1.052643959\, x)} + 0.6561419450\, e^{(1.052643959\, x)}$$

and can be plotted, the vertical view being controlled."

```
>   plot(y,x=0..1,view=[0..1,0..2.5],tickmarks=[4,4]);
```

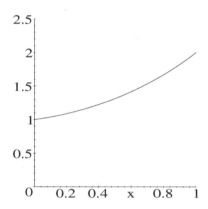

Figure 4.13: The curve $y(x)$ that minimizes the surface area of the dress.

"That's O.K., Greg," Gino interrupts, "but a 3-dimensional picture of the dress would be nicer, and more importantly, what is the minimum surface area?"

"Since you're getting impatient, let's load the Student[Calculus1] package,

```
>   with(Student[Calculus1]):
```

and use the `SurfaceOfRevolution` command to obtain the minimum area A.

```
>   A:=evalf(SurfaceOfRevolution(y,x=0..1));
```

$$A := 13.06167468$$

The minimum surface area of your lampshade dress is about 13 square sretem.

Using the same command, but with the option `output=plot` included,

```
>   SurfaceOfRevolution(y,x=0..1,output=plot);
```

will yield the desired surface of revolution, i.e., your "dress," with the form of $y(x)$ included. The dress will be oriented horizontally on the computer screen but can be 'dragged' into a vertical position."

PROBLEMS:
Problem 4-34: Other curves
Calculate the surface areas for dresses generated with some other curves $y(x)$ joining the endpoints $(x = 0, y = 1)$ and $(x = 1, y = 2)$. How do these areas compare with the area obtained in the text recipe? Plot the surfaces of revolution generated by the curves.

Problem 4-35: Brachistochrone
Consider the smooth curve $y(x)$ joining the two points A and B in Figure 4.14.

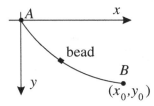

Figure 4.14: Brachistochrone problem.

(a) If a small bead slides without friction between A and B, what $y(x)$ corresponds to the shortest time? This famous *brachistochrone* ("shortest time" in Greek) problem was proposed and solved by Jakob and Johann Bernoulli. *Hint:* First show that the time is given by

$$t = \int_0^{x_0} \sqrt{(1 + (y')^2)/(2\,g\,y)}\, dx,$$

where g is the acceleration due to gravity.

(b) Let the coordinates of B be $x_0 = 0.5$ m, $y_0 = 0.5$ m for the remainder of this problem. What is the shortest time for the bead to slide from point A at the origin to point B?

(c) Plot the path of the bead from A to B.

(d) What is the speed of the bead when half the shortest time has elapsed?

(e) What is the position of the bead at this latter time?

Problem 4-36: Fermat's principle and mirages
Fermat's principle states that a ray of light in a medium with a variable refractive index will follow the path that requires the shortest traveling time. For a two-dimensional situation, write down the integral that must be minimized to obtain such a path. Note that the speed of light in a medium of refractive index n is c/n, where c is the vacuum speed of light. If the refractive index varies vertically from ground level according to the formula $n = n_0(1 + \alpha y)$, with $n_0 > 0$ and $\alpha > 0$, determine the equation for the path taken by a light

ray. By plotting a light ray path for physically reasonable parameter values, discuss how your answer may be related to the phenomenon of mirages.

Problem 4-37: A different refractive index

Making use of Fermat's principle, prove that a light ray will follow a semicircular path in a medium whose refractive index $n(x, y)$ equals $1/y$. Plot a typical path.

Problem 4-38: Geodesic

The *geodesic* between two points is the curve that gives the shortest distance. Show that the geodesics on the surface of a sphere are the arcs of great circles. (A great circle is a curve resulting from the intersection of a sphere with a plane passing through the center of that sphere.) Create a three-dimensional plot of the geodesic between New York City and Sydney, Australia, assuming that the Earth is spherical. You will have to look up the necessary parameter values.

4.3.2 Queen Dido Wasn't a Dodo

Mathematics is the queen of the sciences.
Carl Friedrich Gauss, German mathematician (1777–1855)

According to the *Aeneid*, written by the Roman poet Virgil in the first century BC, the Phoenician Queen Dido was able to convince the North African ruler King Jambas to give her as much land as she could enclose with an ox-hide. Being rather clever, she had the hide cut into very thin strips, the ends stitched together, and was able to stake out a sizable area along the Mediterranean coast on which she built the city of Carthage (now Tunis). Queen Dido's problem was to lay out the joined oxhide strips that had a total fixed length in such a way as to maximize the area enclosed. Can you suggest what shape the perimeter traced out by the joined strips might take? Problems of this type that involve maximizing an area enclosed by a perimeter of fixed length are called *isoperimetric* (constant perimeter) problems.

A recipe is now given to solve Queen Dido's problem. A strip of oxhide of fixed length $L > a$ is connected at its ends to the points $(x = 0, y = 0)$ and $(x = a, y = 0)$. The area enclosed by a strip of shape $y(x)$ and the x-axis between $x = 0$ and $x = a$ is $A = \int_0^a y(x)\, dx$ and the length of the strip is given by $L = \int_0^a \sqrt{1 + (y')^2}\, dx$. If, say, $a = 1$ milion[7] and $L = 1.5$ milion, what is the shape $y(x)$ that maximizes the area subject to the length constraint? Plot $y(x)$ and determine the value of the maximum area.

If the integrands of A and L are labeled as F and G, respectively, the shape $y(x)$ may be found by solving the Euler–Lagrange equation of the previous subsection with F replaced by $FF = F + \lambda G$, where λ is an undetermined parameter. See, e.g., references [MW71] and [Boa83].

The VariationalCalculus package is loaded,

```
>  restart: with(VariationalCalculus):
```

---

[7] A milion is a Roman mile, so this recipe involve one very big oxhide!

and the integrands F and G are entered.

> F:=y(x); G:=sqrt(1+diff(y(x),x)^2);

$$F := y(x) \qquad G := \sqrt{1 + \left(\frac{d}{dx}y(x)\right)^2}$$

Then $FF = F + \lambda G$ is formed.

> FF:=F+lambda*G;

$$FF := y(x) + \lambda\sqrt{1 + \left(\frac{d}{dx}y(x)\right)^2}$$

The `EulerLagrange` command is applied to FF.

> eq:=EulerLagrange(FF,x,y(x));

$$eq := \left\{ 1 + \frac{\lambda\left(\frac{d}{dx}y(x)\right)^2\left(\frac{d^2}{dx^2}y(x)\right)}{\left(1 + \left(\frac{d}{dx}y(x)\right)^2\right)^{(3/2)}} - \frac{\lambda\left(\frac{d^2}{dx^2}y(x)\right)}{\sqrt{1 + \left(\frac{d}{dx}y(x)\right)^2}}, \right.$$

$$\left. y(x) + \lambda\sqrt{1 + \left(\frac{d}{dx}y(x)\right)^2} - \frac{\left(\frac{d}{dx}y(x)\right)^2\lambda}{\sqrt{1 + \left(\frac{d}{dx}y(x)\right)^2}} = K_1 \right\}$$

The first integral is extracted from eq by selecting the term that has K_1 in it, and removing the brackets that would otherwise appear in ode.

> ode:=select(has,eq,K[1])[];

$$ode := y(x) + \lambda\sqrt{1 + \left(\frac{d}{dx}y(x)\right)^2} - \frac{\left(\frac{d}{dx}y(x)\right)^2\lambda}{\sqrt{1 + \left(\frac{d}{dx}y(x)\right)^2}} = K_1$$

A general analytic solution to the first-order nonlinear ode is sought, yielding two answers that differ by the sign in front of the square root.

> sol:=dsolve(ode,y(x));

$$sol := y(x) = K_1 + \sqrt{\lambda^2 - \_C1^2 - x^2 + 2x\_C1},$$
$$y(x) = K_1 - \sqrt{\lambda^2 - \_C1^2 - x^2 + 2x\_C1}$$

The rhs of, say, the negative square root result (second one) is chosen.

> Y:=rhs(sol[2]);

$$Y := K_1 - \sqrt{\lambda^2 - \_C1^2 - x^2 + 2x\_C1}$$

The boundary conditions that $Y = 0$ at $x = 0$ and at $x = a$ are imposed in *bc1* and *bc2*, which are then solved for K_1 and *_C1* in *sol2*.

```
>   bc1:=eval(Y,x=0)=0; bc2:=eval(Y,x=a)=0;
```

$$bc1 := K_1 - \sqrt{\lambda^2 - \_C1^2} = 0$$

$$bc2 := K_1 - \sqrt{\lambda^2 - \_C1^2 - a^2 + 2\,a\,\_C1} = 0$$

```
>   sol2:=solve({bc1,bc2},{K[1],_C1});
```

$$sol2 := \{\_C1 = \frac{a}{2}, K_1 = \frac{\sqrt{4\lambda^2 - a^2}}{2}\}$$

Assigning *sol2*, we obtain Y in the following form.

```
>   assign(sol2): Y:=Y;
```

$$Y := \frac{\sqrt{4\lambda^2 - a^2}}{2} - \sqrt{\lambda^2 - \frac{1}{4}a^2 - x^2 + x\,a}$$

To determine λ, the constraint condition $\int_0^a G(y(x) = Y)\,dx = L$ is imposed.

```
>   eq2:=simplify(int(eval(G,y(x)=Y),x=0..a))=L;
```

$$eq2 := \sqrt{-\frac{\lambda^2}{-4\lambda^2 + a^2}}\,\sqrt{-4\lambda^2 + a^2}\,(-\ln(-a + \sqrt{-4\lambda^2 + a^2})$$

$$+\ln(a + \sqrt{-4\lambda^2 + a^2})) = L$$

The values $a = 1$ and $L = 1.5$ are entered, and *eq2* numerically solved for λ, the search range $\lambda = 0$ to 0.6 being specified.

```
>   a:=1: L:=1.5: lambda:=fsolve(eq2,lambda=0..0.6);
```

$$\lambda := 0.5014101096$$

The final form of Y is then displayed, and $|Y|$ plotted in Figure 4.15.

```
>   Y:=Y; plot(abs(Y),x=0..a,scaling=constrained);
```

$$Y := 0.03757789244 - \sqrt{0.0014120980 - x^2 + x}$$

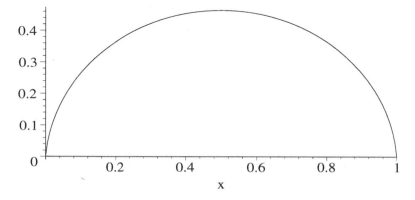

Figure 4.15: Shape of the oxhide strip that maximizes the enclosed area.

The absolute value of Y was plotted, because in some executions of the work sheet Y will be negative because the ordering of the answers in *sol* is reversed. The shape of the strip that maximizes the area is a circular arc.

The area $\int_0^a Y\, dx$ is now calculated, the absolute value being taken to avoid a possible negative area if the "wrong" solution is selected in *sol*.

> `A:=abs(int(Y,x=0..a));`

$$A := 0.3572686358$$

The maximum area is about 0.36 square milions. Without doing any mathematical calculation, can you suggest why this is a maximum and not a minimum?

PROBLEMS:

Problem 4-39: Maximum volume of a solid

A curve $y(x)$ of length 2 is drawn between the points (0, 0) and (1, 0) in such a way that the solid obtained by rotating the curve about the x-axis has the largest possible volume.

(a) Determine $y(x)$.

(b) Plot $y(x)$ over the range $x = 0$ to 1.

(c) What is the value of y at $x = 0.5$?

(d) Make a three-dimensional plot of the solid.

(e) What is the volume of the solid?

Problem 4-40: The catenary curve

Consider a uniform cable of length $L = 1.5$ km and mass per unit length $\epsilon = 1$ kg/m suspended between the two endpoints $(-a/2, b)$ and $(a/2, b)$, where $a = b = 1$ km.

(a) Determine the equilibrium shape (referred to as a *catenary* curve) of the cable. *Hint*: The potential energy of the cable will be at a minimum when the cable has its equilibrium shape. Take $g = 10$ m/s^2.

(b) Plot the equilibrium shape of the cable.

(c) If the cable crosses a very deep Himalayan gorge with the river located a distance b below the endpoints of the cable, what is the distance between the minimum in the cable and the river?

(d) What is the distance down to the river from a point one-third of the way along the cable?

(e) What force is exerted on the supports at the endpoints of the cable?

(f) What length of cable should be used if the sag in the middle of the cable is not to exceed 50 m?

4.3.3 The Human Fly Plans His Escape Route

A product of the untalented, sold by the unprincipled to the utterly bewildered.
Al Capp, American cartoonist, comment on abstract art (1907–1979)

Mr. X, whom the popular press have dubbed the "human fly," plans to scale the outside vertical wall of one of Metropolis's tallest skyscrapers, the Metropolis Stock Exchange, using a combination of his rock-climbing skills and suction cups. Knowing that once the climb has begun, the police will quickly learn of his exploit and try to capture him, Mr. X plans to execute a daring escape. He intends to evade the law enforcement officers by sliding down a thin, but strong, wire to a yet-undetermined window at a lower elevation in the slanted roof of the Museum of Modern Art directly across the street (see Figure 4.16). The window will be left open so that Mr. X can zoom through to the room in-

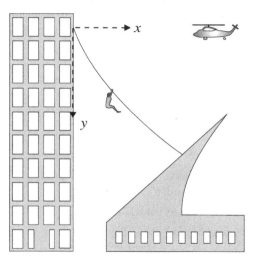

Figure 4.16: Mr. X sliding down wire with a police helicopter hovering nearby.

side, brake rapidly, and make a (hopefully) relatively soft landing on the floor. Catching the police by surprise and taking advantage of the museum's many exits, Mr. X hopes once again to avoid capture and expand his ever-growing legend. However, being somewhat cautious, despite an otherwise harebrained scheme, Mr. X decides to check the details of the escape route by conferring with his mathematician friend Mike.

"Mike, the roof of the stock exchange is slightly over 200 meters above street level and the police will surely be waiting for me on the rooftop and in the street below. They will probably even have a police helicopter hovering nearby. If I were to connect a wire to the stock exchange outer wall at the 200 meter point, what shape should the wire have, and what point on the slanting museum surface should it be connected to, in order to minimize the time of descent? What is the time of descent and what speed would I have acquired on reaching the slanted museum roof? Fortunately, the museum roof has many windows to let in light and I could arrange for the appropriate window to be left open. A few relevant facts are as follows. The museum roof slants at 45° to the horizontal. The street is 50 meters wide, and the vertical section of the museum wall adjoining the slanted roof is 50 meters tall."

"You're crazy," Mike replies, "but you do pose an interesting mathematical problem. I can tackle the solution using the Euler–Lagrange equation. Referring to Figure 4.16, let's choose to measure x to the right and y downward from the point at which the wire is attached to the stock exchange building. We need to find a general expression for the time of descent along the wire. Let the equation of the wire be $y(x)$. Neglecting friction and equating the increase in kinetic energy of a falling mass to its decrease in potential energy yields a speed $v = \sqrt{2gy}$, where g is the acceleration due to gravity. But $v = ds/dt$, where $ds = \sqrt{1 + (dy/dx)^2}\, dx$ is an element of arc length along the wire. If the (unknown) coordinates of the contact point on the museum roof are $(x1, y1)$, the time of descent will be given by

$$T = \int_0^{x1} \frac{\sqrt{1 + (dy/dx)^2}}{\sqrt{2gy}}\, dx.$$

Since the factor $\sqrt{2g}$ will cancel out, the integrand to use in the Euler–Lagrange equation can be taken to be

$$F = (\sqrt{1 + (dy/dx)^2})/\sqrt{y}.$$

I will now use Maple to solve the problem. Loading the necessary library packages, I enter the integrand F.

```
>   restart: with(VariationalCalculus): with(plots):
>   F:=sqrt((1+diff(y(x),x)^2)/y(x));
```

$$F := \sqrt{\dfrac{1 + \left(\dfrac{d}{dx}y(x)\right)^2}{y(x)}}$$

The EulerLagrange command is applied to F, and the first integral result, involving the constant K_1, is selected from the output (not shown) and simplified in ode.

```
>   EL:=EulerLagrange(F,x,y(x));
>   ode:=simplify(select(has,EL,K[1])[]);
```

$$ode := \dfrac{1}{\sqrt{\dfrac{1 + \left(\dfrac{d}{dx}y(x)\right)^2}{y(x)}}\, y(x)} = K_1$$

A general solution to the first-order nonlinear ode is sought.

```
>  sol:=dsolve(ode,y(x)); #implicit solution
```

$$sol := x + \frac{\sqrt{y(x) - K_1{}^2 y(x)^2}}{K_1} - \frac{1}{2} \frac{\arctan\left(\frac{\sqrt{K_1{}^2}\left(y(x) - \frac{1}{2}\frac{1}{K_1{}^2}\right)}{\sqrt{y(x) - K_1{}^2 y(x)^2}}\right)}{K_1\sqrt{K_1{}^2}} - \_C1 = 0,$$

$$x - \frac{\sqrt{y(x) - K_1{}^2 y(x)^2} + 1}{K_1} - \frac{1}{2}\frac{\arctan\left(\frac{\sqrt{K_1{}^2}\left(y(x) - \frac{1}{2}\frac{1}{K_1{}^2}\right)}{\sqrt{y(x) - K_1{}^2 y(x)^2}}\right)}{K_1\sqrt{K_1{}^2}} - \_C1 = 0$$

A formidable-looking implicit solution has been produced, with two forms present. It turns out that we must select the form having the positive square root (the first solution for this particular run). For later convenience, I will also replace the awkward constant symbols K_1 and $\_C1$ with $1/\sqrt{A}$ and B.

```
>  ans:=subs({K[1]=1/sqrt(A),_C1=B},sol[1]); #select + root
```

$$ans := x + \sqrt{A}\sqrt{y(x) - \frac{y(x)^2}{A}} - \frac{1}{2}\frac{\sqrt{A}\arctan\left(\frac{\sqrt{\frac{1}{A}}\left(y(x) - \frac{A}{2}\right)}{\sqrt{y(x) - \frac{y(x)^2}{A}}}\right)}{\sqrt{\frac{1}{A}}} - B = 0$$

To proceed any further with the implicit answer, its square-root structure suggests a trigonometric substitution of the form $y = A\sin^2\theta$, where θ is an angular parameter not connected to any geometrical feature in our picture. Here A and B are assumed to be positive, as are $\sin\theta$ and $\cos\theta$ as well.

```
>  assume(A>0,B>0,sin(theta)>0,cos(theta)>0):
>  Y:=y(x)=A*sin(theta)^2;
```

$$Y := y(x) = A\sin(\theta)^2$$

Then Y is substituted into the answer, and simplified with the trig option.

```
>  sol1:=simplify(subs(Y,ans),trig);
```

$$sol1 := x + A\sin(\theta)\cos(\theta) + \frac{1}{2}A\arctan\left(\frac{1}{2}\frac{-1 + 2\cos(\theta)^2}{\sin(\theta)\cos(\theta)}\right) - B = 0$$

Then we solve *sol1* for x and combine with respect to trig terms.

```
>  x:=combine(solve(sol1,x),trig);
```

$$x := -\frac{1}{2}A\sin(2\theta) - \frac{1}{2}A\arctan\left(\frac{\cos(2\theta)}{\sin(2\theta)}\right) + B$$

So, now we have expressions for the coordinates x and y of our sought-after curve in terms of A, B, and θ. If A and B can be found and the range of θ

determined, the curve that minimizes the time of descent will be known.

At the starting point $x = 0$, we shall choose to set the parameter θ equal to zero. Some care must be taken in evaluating the arctan term at $\theta = 0$. The limit must be taken from the positive side (i.e., from the "right") of $\theta = 0$.

>   `eq:=limit(x,theta=0,right)=0;`

$$eq := B - \frac{A\,\pi}{4} = 0$$

The resulting *eq* is easily solved for the constant B.

>   `sol2:=B=solve(eq,B);`

$$sol2 := B = \frac{A\,\pi}{4}$$

The solution *sol2* is assigned and x displayed.

>   `assign(sol2); x:=x;`

$$x := -\frac{1}{2}\,A\sin(2\,\theta) - \frac{1}{2}\,A\arctan\left(\frac{\cos(2\,\theta)}{\sin(2\,\theta)}\right) + \frac{A\,\pi}{4}$$

We still have to determine the constant A and the range of θ. Now it gets a bit tricky. We have no idea yet what the coordinates $(x1,\ y1)$ of the endpoint on the museum roof should be.

Hand me that copy of *Mathematical Methods in Physics* [MW71]. Ah, here we go. The case in which one endpoint of the sought-after curve is fixed and the other endpoint is allowed to vary along a line $g(x,y) = 0$ is considered. It is shown that if the Euler–Lagrange function F is of the structure[8] $F = f(x,y)\sqrt{1 + (y')^2}$, which it is in our case, then the condition for determining the unknown endpoint is that the slope of $y(x)$ must satisfy the condition $y' = (\partial g/\partial y)/(\partial g/\partial x)$ at that point. But this is just a mathematical statement that the curve $y(x)$ of quickest descent must intersect the destination curve $g(x,y) = 0$ at right angles. Here the equation for the slanting museum roof is the straight line $g(x,y) = y + x - 200 = 0$. But both $\partial g/\partial y$ and $\partial g/\partial x$ are equal to 1, so we have $y' = 1$ at the museum roof. Since the parameter θ has been introduced, the slope of $y(x)$ must be calculated in terms of θ.

>   `slope:=simplify(diff(rhs(Y),theta)/diff(x,theta));`

$$slope := \frac{\cos(\theta)}{\sin(\theta)}$$

Setting the slope to 1, the value Θ that the parameter θ must have at the museum roof is determined, assuming that θ is positive.

>   `Theta:=solve(slope=1,theta) assuming theta>0;`

$$\Theta := \frac{\pi}{4}$$

---

[8] If F is not of this structure, the endpoint condition is more complicated, taking the form

$$\left(F - y'\frac{\partial F}{\partial y'}\right)\frac{\partial g}{\partial y} - \frac{\partial F}{\partial y'}\frac{\partial g}{\partial x} = 0.$$

Evaluating x at $\theta = \Theta$ must yield $x1$,

> `eq1:=x1=eval(x,theta=Theta);`

$$eq1 := x1 = -\frac{1}{2}A + \frac{1}{4}A\pi$$

while evaluating the rhs of Y at $\theta = \Theta$ must yield $y1$.

> `eq2:=y1=eval(rhs(Y),theta=Theta);`

$$eq2 := y1 = \frac{A}{2}$$

Finally, the straight-line equation for the museum roof at $(x1, y1)$ is entered,

> `eq3:=x1+y1=200;`

$$eq3 := x1 + y1 = 200$$

and the three equations solved for A, $x1$, and $y1$.

> `sol3:=fsolve({eq1,eq2,eq3},{A,x1,y1}); assign(sol3):`

$$sol3 := \{A = 254.6479090,\ y1 = 127.3239545,\ x1 = 72.67604557\}$$

Thus, the endpoint of the wire on the museum roof has coordinates $x1 \approx 72.7$, $y1 \approx 127.3$. The endpoint is about 127 meters below the starting point on the stock exchange wall. I hope you realize, X, that this means that you will be dropping more than 40 stories as you slide along the wire!

I will plot the curve (colored blue in graph **gr1**) of minimum descent to the museum roof (colored red in **gr2**). The wire and roof are labeled in **gr3**.

> `gr1:=plot([x,-rhs(Y)+200,theta=0..Theta],color=blue):`
> `gr2:=plot([[50,0],[50,50],[125,125]],color=red,thickness=2):`
> `gr3:=textplot({[40,140,"wire"],[100,85,"roof"]}):`
> `display({gr1,gr2,gr3},scaling=constrained,tickmarks=[3,3]);`

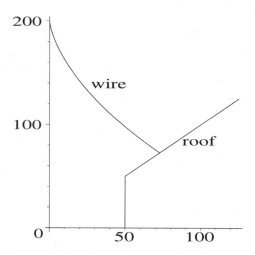

Figure 4.17: Shape of wire that minimizes the descent time to the roof.

OK, X, there's your curve shown in Figure 4.17. You can arrange for a window to be left open at the point where the curve intersects the museum roof.

Now that we know the distance through which you will be dropping, we can easily calculate your velocity from the expression $v = \sqrt{2\,g\,y1}$. Taking the gravitational acceleration to be $g = 9.81$ m/s^2,

> g:=9.81: vel:=sqrt(2*g*y1);

$$vel := 49.98095625$$

you will be traveling at almost 50 m/s as you pass through the window on the museum roof, if you haven't been braking. This is an upper bound on your speed, of course, because we have completely neglected friction and air resistance. I can see by the look on your face that you don't quite appreciate how fast this is. So, let me convert the velocity to kilometers per hour.

> vel2:=convert(vel,units,m/s,km/h)*km/h;

$$vel2 := \frac{179.9314425\,km}{h}$$

Your theoretical speed would be almost 180 km/hr but, as I said, in reality it would be somewhat less. The time of descent can now be calculated from

$$T = \int_{\theta=0}^{\Theta} \frac{\sqrt{(dx/d\theta)^2 + (dy/d\theta)^2}}{2\,g\,y(\theta)}\, d\theta, \quad \text{viz.,}$$

> T:=int(sqrt(diff(x,theta)^2+diff(rhs(Y),theta)^2)
    /sqrt(2*g*rhs(Y)),theta=0..evalf(Theta));

$$T := 5.659009626$$

The minimum time of descent is about 5.7 seconds. Again, friction, air resistance, and any braking on your part would lengthen this time. Given the estimates that we have come up with, do you still want to go through with your bizarre escape route?"

"Sure, Mike," Mr. X replies, "and I am counting on you to help attach the wire to the appropriate window in the museum roof and leave it open for me."

"Oh no," Mike groans. "If I get caught, Vectoria's parents are sure to call off our recently announced engagement."

PROBLEMS:

Problem 4-41: A different escape route

Determine Mr. X's escape route if the museum roof slanted at 30° to the horizontal, the street is 50 m wide, and the vertical section of the museum wall adjoining the slanted roof is 25 m tall. Make a plot similar to that in Figure 4.17. How close to the edge of the roof would Mr. X land? What would be his speed at this point? Neglect friction and air resistance.

Problem 4-42: In search of reality

Suppose that Mr. X decides that sliding down a steel wire at nearly 180 km/hr is a little too insane even for his bizarre tastes. His friend Mike is unavailable for technical advice because he is busy with preparations for his upcoming wedding to Vectoria. So Mr. X comes to you and asks what speed he would reach and how long would his slide take if air resistance and friction are included, the same wire shape being used. He tells you that his mass is about 60 kg. You may assume that the force due to air resistance is $0.5\,v^2$ newtons, where v is Mr. X's speed, and that the coefficient of kinetic friction is about 0.6 (appropriate for steel on steel). Mr. X would also like to know the maximum speed and trip time if a braking action is applied. You may assume that the brakes exert a force of 400 newtons on the wire. *Hint*: You may find that using a do loop to compute the tangential velocity and position of Mr. X at discrete time intervals is the easiest way to attack this problem.

4.3.4   This Would Be a Great Amusement Park Ride

To gyre is to go around and round like a gyroscope.
To gimble is to make holes like a gimlet.
Lewis Carroll, English writer and mathematician (1832–1898)

Consider the following possible amusement park ride consisting of a small cage of mass m (including the screaming victims inside) being swung around at the end of a light, but strong, connecting arm of fixed radius r as in Figure 4.18. The cage can trace out various spherical surface trajectories depending on the

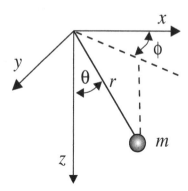

Figure 4.18: Configuration of the amusement park ride.

conditions imposed by the slightly sadistic ride operator. What kind of trajectories can the victims be subjected to?

Our task is to investigate this question using *Lagrange's equations* of motion, an alternative approach to the Newtonian formulation. The Lagrangian L is defined as $L = T - V$, where T is the kinetic energy and V is the potential energy of the system being studied. Formulating the Lagrangian is often much easier than determining all the forces and their components that are required to apply Newton's second law. If L depends on the coordinates $q_1, q_2, \ldots, q_i, \ldots, q_N$ and the "velocities" $\dot{q}_i \equiv \partial q_i / \partial t$, it is shown in standard mechanics texts [FC99] that the equations of motion are given by *Lagrange's equations*,

$$\frac{\partial L}{\partial q_i} - \frac{d}{dt}\left(\frac{\partial L}{\partial \dot{q}_i}\right) = 0, \tag{4.18}$$

with $i = 1, 2, \ldots, N$. Since each component equation is of the Euler–Lagrange structure, it should come as no surprise that the `EulerLagrange` command can be used to derive the relevant ODEs, given a specified form of L. Let's now use this approach for the amusement park ride.

After we load the plots and VariationalCalculus library packages,

> `restart: with(plots): with(VariationalCalculus):`

the Cartesian coordinates of the mass m are expressed in terms of the spherical polar coordinates r (which is fixed), $\theta(t)$, and $\phi(t)$, as in Figure 4.18.

> `x:=r*sin(theta(t))*cos(phi(t));`

$$x := r\sin(\theta(t))\cos(\phi(t))$$

> `y:=r*sin(theta(t))*sin(phi(t));`

$$y := r\sin(\theta(t))\sin(\phi(t))$$

> `z:=r*cos(theta(t));`

$$z := r\cos(\theta(t))$$

The kinetic energy of the mass at time t is calculated and simplified.

> `T:=simplify((m/2)*(diff(x,t)^2+diff(y,t)^2+diff(z,t)^2));`

$$T := -\frac{1}{2}m\,r^2\left(-\left(\frac{d}{dt}\phi(t)\right)^2 + \left(\frac{d}{dt}\phi(t)\right)^2\cos(\theta(t))^2 - \left(\frac{d}{dt}\theta(t)\right)^2\right)$$

The potential energy $V = -m\,g\,r\,\cos(\theta(t))$, with g the gravitational acceleration, is entered and the Lagrangian $L = T - V$ calculated.

> `V:=-m*g*r*cos(theta(t)): L:=T-V;`

$$L := -\frac{1}{2}m\,r^2\left(-\left(\frac{d}{dt}\phi(t)\right)^2 + \left(\frac{d}{dt}\phi(t)\right)^2\cos(\theta(t))^2 - \left(\frac{d}{dt}\theta(t)\right)^2\right)$$
$$+m\,g\,r\cos(\theta(t))$$

The `EulerLagrange` command is applied to L, the independent variable t being specified, and the two angular coordinates $\phi(t)$ and $\theta(t)$ being entered as a list. The lengthy output is suppressed here in the text.

> `eq:=EulerLagrange(L,t,[phi(t),theta(t)]);`

The first integral in *eq* containing the constant K_1 is selected.

> `ode1:=select(has,eq,K[1])[];`

$$ode1 := -\frac{1}{2}m\,r^2\left(-2\left(\frac{d}{dt}\phi(t)\right) + 2\left(\frac{d}{dt}\phi(t)\right)\cos(\theta(t))^2\right) = K_1$$

A second ODE relating $\theta(t)$ to $\phi(t)$ is required. To this end, we select the expression in *eq* containing $d^2\theta(t)/dt^2$ and set the result to zero.

> ode2:=select(has,eq,diff(theta(t),t,t))[]=0;

$$ode2 := m\,r^2\left(\frac{d}{dt}\phi(t)\right)^2 \cos(\theta(t))\sin(\theta(t)) - m\,g\,r\sin(\theta(t)) - m\,r^2\left(\frac{d^2}{dt^2}\theta(t)\right) = 0$$

With *ode1* and *ode2*, we have two coupled nonlinear ODEs. However, we can solve the former for $d\phi/dt$ by isolating this term on the lhs of the equation,

> ode1b:=isolate(ode1,diff(phi(t),t));

$$ode1b := \frac{d}{dt}\phi(t) = -\frac{2\,K_1}{m\,r^2\,(-2 + 2\cos(\theta(t))^2)}$$

and then simplifying with the algebraic substitution $\cos(\theta(t))^2 = 1 - \sin(\theta(t))^2$.

> ode1c:=algsubs(cos(theta(t))^2=1-sin(theta(t))^2,ode1b);

$$ode1c := \frac{d}{dt}\phi(t) = \frac{K_1}{m\,r^2\,\sin(\theta(t))^2}$$

Then substituting *ode1c* and $g = r\,\omega^2$, where ω is a frequency, into $-ode2/(m\,r^2)$ and expanding,

> ode2b:=expand(subs({ode1c,g=r*omega^2},-ode2/(m*r^2)));

$$ode2b := -\frac{K_1^{\,2}\cos(\theta(t))}{m^2\,r^4\,\sin(\theta(t))^3} + \omega^2\sin(\theta(t)) + \left(\frac{d^2}{dt^2}\theta(t)\right) = 0$$

yields the second-order nonlinear ODE *ode2b* entirely in terms of $\theta(t)$. This ODE cannot be solved analytically. Before seeking a numerical solution, it is instructive to consider some simpler special cases first.

Consider the situation in which the ride operator is feeling in a rare mellow mood and allows the cage to swing to and fro at a constant angle ϕ, so $\dot{\phi} = 0$ and, from *ode1c*, $K_1 = 0$. Then *ode2b* reduces to the well-known linear ODE for the simple undamped plane pendulum with characteristic frequency ω. Those readers who have gone to an amusement park lately may have seen a "boat" ride that behaves as a driven simple pendulum. This ride is much too tame for our ride operator and the large teenage market that he is after.

Another special case corresponds to the cage orbiting in a horizontal circle at a fixed angle $\theta(t) = \Theta$. (A popular young children's ride does exactly this.) We evaluate *ode2b* at this fixed angle.

> eq2:=eval(ode2b,theta(t)=Theta);

$$eq2 := -\frac{K_1^{\,2}\cos(\Theta)}{m^2\,r^4\,\sin(\Theta)^3} + \omega^2\sin(\Theta) = 0$$

At the fixed angle Θ, K_1 is given from *ode1c* by the following command line, where we have set the angular velocity $d\phi(t)/dt = \Omega$.

> K[1]:=Omega*m*r^2*sin(Theta)^2;

$$K_1 := \Omega\,m\,r^2\,\sin(\Theta)^2$$

The angular velocity needed to maintain the horizontal circular motion at the fixed angle $\theta(t) = \Theta$ is then obtained by solving $eq2$ for Ω.

> `Omega:=solve(eq2,Omega);`

$$\Omega := \frac{\omega}{\sqrt{\cos(\Theta)}}, \quad -\frac{\omega}{\sqrt{\cos(\Theta)}}$$

The two angular velocity solutions given above correspond to rotations in the opposite sense.

The two rides described so far are not sufficiently exciting to the teenage generation. Taking the nominal values $r = 1$, $m = 1$, and $\omega = 1$ for the parameters (the reader can experiment with more realistic numbers), can one create a more interesting trajectory? For a horizontal circular orbit at an inclination to the vertical of $\Theta = 60°$, the constant K_1 is given by

$$K_1 = \frac{\omega\,m\,r^2\,\sin^2\Theta}{\sqrt{\cos\Theta}} = \sqrt{\frac{9}{8}} \approx 1.06.$$

To create a more interesting ride let's take $K_1 = 0.5$. The trajectory of the ride will be plotted over $T = 20$ time units.

> `r:=1: m:=1: omega:=1: K[1]:=0.5: T:=20:`

Let's suppose that initially, $\phi(0) = 0$, $\theta(0) = \pi/3$ radians, and $\dot{\theta}(0) = 0$,

> `ic:=phi(0)=0,theta(0)=Pi/3,D(theta)(0)=0:`

and numerically solve $ode1c$ and $ode2b$ subject to this initial condition.

> `sol:=dsolve({ode1c,ode2b,ic},{theta(t),phi(t)},numeric):`

To get a feeling for the three-dimensional motion, the spherical surface on which the cage can move is created with the **sphereplot** command. The radius $r = 1$ of the sphere is specified as well as the angular ranges $\theta = 0$ to 2π and $\phi = 0$ to π. A **wireframe** style is used to represent the spherical surface.

> `gr1:=sphereplot(r,theta=0..2*Pi,phi=0..Pi,style=wireframe):`

The **odeplot** command is now used to plot the numerically determined trajectory of the mass m over the time interval $t = 0$ to T. Normal axes are chosen, and the minimum number of plotting points is taken to be 500 to obtain a smooth curve.

> `gr2:=odeplot(sol,[x,y,z],0..T,axes=normal,numpoints=500,`
>   `thickness=3):`

The **display** command is used to overlay the two graphs. Axis labels are added, and the orientation and size of the 3-dimensional viewing box controlled, as well as the number of tickmarks.

> `display(gr1,gr2,labels=["x","y","z"],orientation=[50,-100],`
>   `view=[-1..1,-1..1,-1..1], tickmarks=[2,2,4]);`

Figure 4.19 shows the 3-dimensional trajectory. The viewing box can be rotated to observe the trajectory from different angles by dragging the box on the computer screen with your mouse.

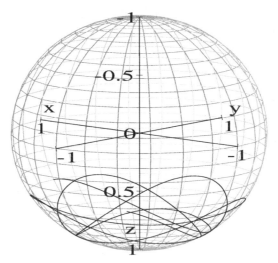

Figure 4.19: A wild amusement park ride trajectory on a spherical surface.

If this trajectory isn't wild enough, feel free to create your own crazy ride. You could also try using more realistic values for the parameters instead of the nominal values used in the above recipe.

Have a stomach-churning ride!

PROBLEMS:

Problem 4-43: A twirling-loop ride

A vertically oriented circular loop of radius ℓ rotates with angular velocity ω about the z-axis as shown in Figure 4.20. A cage of unit mass ($m = 1$) is allowed to slide along the frictionless loop. If the plane of the loop is oriented along the

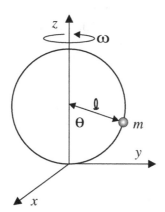

Figure 4.20: Rotating circular loop with cage (mass m) free to slide.

y-axis at $t = 0$, what are the x-, y-, and z-coordinates of the cage at time t? Using the Lagrangian approach, show that the cage's motion is described by

$$\ddot{\theta} + \omega_0^2 \sin\theta - \frac{1}{2}\omega^2 \sin(2\theta) = 0,$$

with $\omega_0 = \sqrt{g/\ell}$. Numerically solve the equation of motion for $\omega_0 = 1$ and varying values of ω and discuss the behavior to which the cage is subjected.

Problem 4-44: Ride into the jaws of chaos
The pivot point O for the simple pendulum is undergoing vertical oscillations given by $A\sin(\omega t)$ as indicated in Figure 4.21.

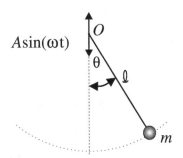

Figure 4.21: An example of parametric excitation.

Show that the relevant equation of motion is

$$\ddot{\theta} + \omega_0^2 \left(1 - \frac{A\omega^2}{g}\sin(\omega t)\right)\sin\theta = 0,$$

with $\omega_0 = \sqrt{g/\ell}$. This nonlinear ODE with a time-dependent coefficient is referred to in the mathematics literature as an example of *parametric excitation*. Taking $\omega_0 = 1$ and $\omega = 1$, numerically study the effect of changing the ratio A/g. Then take this ratio equal to 1, and study the effect of changing ω.

Problem 4-45: Horizontally oscillating pivot point
The pivot point O in the previous problem is undergoing horizontal oscillations given by $A\sin(\omega t)$. Derive the relevant equation of motion. Taking $\omega_0 = 1$, $\omega = 1$, and $g = 10$, numerically study the effect of changing the amplitude A.

Chapter 5

Linear PDE Models. Part 1

Because linear partial differential equations play such an important role in the mathematical description of electromagnetic waves, heat flow, elastic vibrations, and many other scientific phenomena, there is an abundance of wonderful examples that can be solved using computer algebra. For this reason, this topic is split over two chapters. We begin with examples of checking PDE solutions, either obtained by intelligent guessing or quoted, without derivation, in some scientific reference. Diffusion and Laplace's equation models are then presented.

5.1 Checking Solutions

5.1.1 The Palace of the Governors

The knowledge of the world is only acquired in the world,
and not in a closet.
Phillip Dormer Stanhope, Earl of Chesterfield (1694–1773)

While driving back to Phoenix from Los Alamos, where he attended an engineering conference at the Los Alamos National Laboratory, Russell stops in Sante Fe to have something to eat and to tour the historic Plaza area of town. After treating himself to a gourmet lunch, consisting of a tasty rattlesnake burger washed down with a Corona beer, he strolls around the Plaza. As an engineer, Russell is particularly impressed by the massively thick walls of the Palace of the Governors. This long, low adobe structure, which was built in 1610 by the Spanish, is the oldest continuous seat of government in the United States. The thick walls, which is a design feature of many historic buildings in the American Southwest, not only offered protection from attackers but helped to keep the heat of the summer sun at bay as well as excluding the cold breath of winter.

On completing his tour of the Plaza, Russell is intrigued by the question of how effective a very thick adobe wall is in cutting down the incident solar radiation from the summer sun. Pulling out his laptop computer, and finding a shady spot, he formulates the following relevant model. For simplicity, he considers a semi-infinite solid medium whose bounding planar surface is periodically heated by the sun. The direction transverse to the surface is taken to be the x-direction, with the surface located at $x = 0$, and positive x taken to be inside the surface. In order to animate the temperature profile inside the surface, Russell loads the plots library package.

>   restart: with(plots):

From undergraduate thermodynamics, he knows that the time-dependent temperature distribution $T(x, t)$ obeys the one-dimensional heat diffusion equation

$$\frac{\partial T}{\partial t} = d\,\frac{\partial^2 T}{\partial x^2}, \tag{5.1}$$

with the heat diffusion coefficient $d = K/(\rho C)$, where K is the thermal conductivity, ρ the density, and C the specific heat. Russell enters the heat equation.

>   heateq:=diff(T(x,t),t)-d*diff(T(x,t),x,x)=0;

$$heateq := \left(\tfrac{\partial}{\partial t}\, T(x,\, t)\right) - d\left(\tfrac{\partial^2}{\partial x^2}\, T(x,\, t)\right) = 0$$

To account for the periodic heating of the planar surface due to the sun, he takes the temperature variation at $x = 0$ to be $T(0, t) = T_0 \cos(\omega t)$. Rather than derive the temperature variation $T(x, t)$ for $x > 0$, Russell decides to make an intelligent guess as to its mathematical form and check it by substituting the form back into the diffusion equation. Since he is looking for a steady-state response and the temperature should decrease as x increases inside the surface, Russell conjectures that the solution should be of the structure $T(x, t) = T_0\, e^{-\alpha x} \cos(\omega t - \beta x)$, with the parameters α and β as yet undetermined. At $x = 0$, the boundary condition is satisfied, and an exponentially decaying cosine solution seems reasonable inside the surface. Both α and β must be positive for a waveform propagating in the positive x-direction.

What forms should α and β have? From the structure of the heat diffusion equation, the units of d are m^2/s, so $1/\sqrt{d}$ has units of $\mathrm{s}^{1/2}{\cdot}\mathrm{m}^{-1}$. Since the argument in the exponential function must be dimensionless, α must have the dimensions of m^{-1}. The only other parameter in the problem involving a time unit is the frequency ω with units s^{-1}. So, noting that the combination $\sqrt{\omega/d}$ has the units m^{-1}, Russell takes $\alpha = a\sqrt{\omega/d}$, where a is a numerical factor, yet to be determined. By similar reasoning, the constant β is set equal to $b\sqrt{\omega/d}$, with b another numerical factor. Both a and b must be positive.

To check the postulated solution and determine a and b, Russell enters $T(x, t)$, which will be automatically substituted into the heat equation.

>   T(x,t):=T[0]*exp(-a*sqrt(omega/d)*x)
             *cos(omega*t-b*sqrt(omega/d)*x);

$$T(x,\, t) := T_0\, e^{\left(-a\,\sqrt{\frac{\omega}{d}}\, x\right)} \cos\left(-\omega t + b\,\sqrt{\frac{\omega}{d}}\, x\right)$$

Russell then simplifies the heat equation,

>   eq:=simplify(heateq);

and collects the coefficients of the cosine and sine terms.

```
>  collect(eq,{cos,sin});
```

$$-T_0\, w\, e^{\left(-a\, \sqrt{\frac{w}{d}}\, x\right)} (a^2 - b^2) \cos\left(-w\, t + b\, \sqrt{\frac{w}{d}}\, x\right)$$

$$- T_0\, w\, e^{\left(-a\, \sqrt{\frac{w}{d}}\, x\right)} (-1 + 2\, a\, b) \sin\left(-w\, t + b\, \sqrt{\frac{w}{d}}\, x\right) = 0$$

For the left-hand side of the above output to be equal to zero for arbitrary x and t, one must have $b = a$ and $-1 + 2\, a\, b = 0$, so that $a = 1/\sqrt{2} = b$. These values of a and b are entered, numerically evaluated, and labeled aa and bb.

```
>  aa:=evalf(1/sqrt(2)): bb:=aa:
```

Adobe brick is made up of a mixture of dried clay and straw, so Russell consults a handbook of physical constants that he just happens to have brought along with him. He ascertains that for dried clay, the major component of the bricks, $K = 0.4$ W/(m·K), $\rho = 2000$ kg/m^3, and $C = 920$ J/(kg·K). Russell then is able to calculate the diffusion coefficient d_{clay} for clay, which he will use as an estimate for the adobe brick.

```
>  K:=0.4: rho:=2000: C:=920: d[clay]:=K/(rho*C);
```

$$d_{clay} := 0.2173913043\, 10^{-6}$$

The rotational *period* of the earth is about 24 hours, or $24 \times 60 \times 60 = 86{,}400$ seconds, so its rotational frequency, *freq*, can be calculated.

```
>  period:=24*60*60; freq:=evalf(2*Pi/period);
```

$$period := 86400 \qquad freq := 0.00007272205218$$

Summer temperatures in the American Southwest can vary from daily highs of $100°$F or more, to overnight minima in the low to mid-60s. For his calculation, Russell takes a mean daily temperature of $80°$F with a variation of $20°$ on either side. Converting to degrees Celsius, this translates into a mean temperature of $26.6°$C and $T_0 = 11°$ for the amplitude of the temperature variation. Now $T(x, t)$ is evaluated using the parameter values that have been obtained.

```
>  T:=eval(T(x,t),{a=aa,b=bb,T[0]=11,d=d[clay],omega=freq});
```

$$T := 11.\, e^{(-12.93293161\, x)} \cos(-0.00007272205218\, t + 12.93293161\, x)$$

Russell creates a plot of the mean daily temperature, representing the temperature by a dashed (linestyle=3) blue line. The spatial range of the plot is from $x = 0$ to 0.5 meters.

```
>  meantemp:=plot(26.6,x=0..0.5,color=blue,linestyle=3):
```

The total temperature profile over a 0.5-meter spatial range is animated over a time interval of one complete period (day). Fifty frames are included in the animation. The mean temperature line is included with the background option.

```
>  animate(plot,[T+26.6,x=0..0.5],t=0..period,frames=50,
     thickness=2,background=meantemp);
```

The initial temperature profile in the animation is similar to the solid curve shown in Figure 5.1, the horizontal dashed line being the mean temperature.

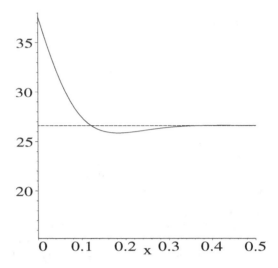

Figure 5.1: Initial temperature profile inside the adobe wall.

On running the animation, the temperature curve will oscillate up and down as the externally imposed temperature on the surface varies during the day. However, the temperature variation drops to zero within a penetration depth of less than 0.4 m or 40 cm (16 in). Well within the surface, the temperature is in thermal equilibrium with the mean exterior temperature. Since the adobe walls of many of the historic buildings of Arizona and New Mexico are substantially thicker than 16 inches, the interior temperature of these buildings is quite insensitive to external temperature variations.

PROBLEMS:
Problem 5-1: A house of wood
In the text recipe, make the following modifications:

(a) The region $x > 0$ is composed of solid wood for which $K = 0.15$ W/(m·K), $\rho = 700$ kg/m^3, and $C = 1800$ J/(kg·K).

(b) The mean daily temperature is 30°C and the amplitude of the temperature variation is 20°.

Run the animation and determine the approximate depth in centimeters at which the temperature variation is essentially zero.

Problem 5-2: Pulsating sphere
The surface of a sphere of radius $r = a$, surrounded by an ideal compressible fluid, pulsates radially with frequency ω. The radial velocity of the surface is given by $V = U \cos(\omega t)$. It is stated in an advanced calculus text that the steady-state fluid velocity at an arbitrary point $r > a$ is of the form

$$V = \frac{U a^2}{(c^2 + a^2\omega^2) r^2} \left[(c^2 + a r \omega^2) \cos(\theta) + c \omega (r - a) \sin(\theta) \right],$$

where $\theta \equiv \omega(r-a)/c - \omega t$ and c is the speed of sound in the fluid.

(a) Check that the solution satisfies the boundary condition.

(b) The velocity V is related to the velocity potential Φ by $V = -\partial\Phi/\partial r$. Determine the radial dependence of the velocity potential.

(c) Since Φ depends only on the distance r from the center of the sphere, it satisfies the wave equation in the form

$$\frac{\partial^2 (r\,\Phi)}{\partial r^2} = \frac{1}{c^2} \frac{\partial^2 (r\,\Phi)}{\partial t^2}.$$

Verify that Φ satisfies the wave equation, thus ensuring that it is the correct solution to the pulsating sphere problem.

(d) Taking the nominal values $U = 1$, $a = 1$, $c = 1$, and $\omega = 1$, animate the analytic formula for V in the region outside the spherical surface.

(e) How far from the surface does the velocity oscillation amplitude drop to 5% of the value at the surface?

5.1.2 Play It, Sam

You just pick a chord, go twang, and you've got music.
Syd Vicious, British rock musician (1957–1979)

In a famous scene from the classic movie *Casablanca*, Humphrey Bogart is annoyed by the musical piece that the nightclub pianist is playing and doesn't want to hear it. But Ingrid Bergman turns to the piano player, and says "Play it, Sam." Humphrey Bogart then echoes her, by saying "If she can stand it, I can. Play it."

To a certain undergraduate physics student watching this old movie, an individual who tends to look for deeper understanding rather than simply enjoying the movie and the music, the sounds that emanate are of course due to the transverse vibrations of the piano strings as each is successively struck by a piano hammer. This student is our old friend Vectoria, who is spending a lonely evening by herself, since her fiancé Mike is out of town at a mathematics conference. Hoping that Mike will phone before it gets too late, Vectoria decides in the meantime to look at the mathematical vibrations of an elastic string fixed at its ends.

Not yet having taken the necessary mathematics prerequisites to study the topic of vibrating strings in depth, Vectoria consults a physics text that gives no derivation but simply the formula for the transverse displacement of a light, horizontal, elastic string of length a fixed at both ends when struck and given a certain initial velocity profile. The information that Vectoria gleans from this particular text is as follows.

Neglecting stiffness, we can model the transverse displacement $\psi(x,t)$ of a light, initially horizontal $(\psi(x,0) = 0)$ piano string by the wave equation

$$\frac{\partial^2 \psi}{\partial x^2} - \frac{1}{c^2}\frac{\partial^2 \psi}{\partial t^2} = 0, \tag{5.2}$$

where the wave velocity c equals $\sqrt{T/\epsilon}$, T being the tension in the string and ϵ the mass per unit length. If a string of length a is held fixed at both ends $(\psi(0,t) = \psi(a,t) = 0)$ for all times, and is given an initial transverse velocity

$$\dot{\psi}(x,0) = \begin{cases} 4\,v\,x/a, & 0 < x < a/4, \\ (4\,v/a)((a/2) - x), & a/4 < x < a/2, \\ 0, & a/2 < x < a, \end{cases} \tag{5.3}$$

the solution to the initial value problem is given by the *Fourier series* expansion $\psi(x,t) = \sum_{n=1}^{\infty} \psi_n(x,t)$, where a representative Fourier term has the form

$$\psi_n(x,t) \equiv \frac{8\,v\,a}{\pi^3\,n^3 c}\left(2\sin\left(\frac{n\,\pi}{4}\right) - \sin\left(\frac{n\,\pi}{2}\right)\right)\sin\left(\frac{n\,\pi\,x}{a}\right)\sin\left(\frac{n\,\pi\,ct}{a}\right). \tag{5.4}$$

Not knowing whether the formula is correct and not yet having the mathematical tools to derive it, Vectoria at least knows how to check the alleged solution to see whether it is indeed valid. The formula for the total displacement must, of course, satisfy the wave equation, as must each individual Fourier term ψ_n.

After loading the plots package, Vectoria checks that this is the case by entering the left-hand side of the PDE (5.2) applied to $\psi_n(x,t)$,

```
>   restart: with(plots):
>   PDE:=diff(psi[n](x,t),x,x)-(1/c^2)*diff(psi[n](x,t),t,t);
```

$$PDE := \left(\frac{\partial^2}{\partial x^2}\,\psi_n(x,\,t)\right) - \frac{\frac{\partial^2}{\partial t^2}\,\psi_n(x,\,t)}{c^2}$$

and the given mathematical form of this term.

```
>   term:=psi[n](x,t)=(8*v*a/(Pi^3*n^3*c))*(2*sin(n*Pi/4)
        -sin(n*Pi/2))*sin(n*Pi*x/a)*sin(n*Pi*c*t/a);
```

$$term := \psi_n(x,\,t) = \frac{8\,v\,a\left(2\sin\left(\frac{n\,\pi}{4}\right) - \sin\left(\frac{n\,\pi}{2}\right)\right)\sin\left(\frac{n\,\pi\,x}{a}\right)\sin\left(\frac{n\,\pi\,ct}{a}\right)}{\pi^3\,n^3\,c}$$

The pdetest command is used to test whether $\psi_n(x,t)$ satisfies *PDE*.

```
>   pdetest(term,PDE);
```

$$0$$

The zero result confirms that the quoted *term* does satisfy the wave equation.

Next, Vectoria looks at the boundary conditions at $x = 0$ and $x = a$ to see whether each Fourier term vanishes at these points for arbitrary t. For integer n, the spatial part $\sin(n\,\pi\,x/a)$ correctly goes to zero at the endpoints.

What about satisfying the initial conditions? At $t = 0$, the time part $\sin(n\,\pi\,ct/a)$ equals 0 as required. The initial velocity distribution, which is a piecewise function, is slightly harder to verify. One approach is to plot the velocity profile predicted by the series solution at $t = 0$ and compare it with the analytic piecewise form. Adding, say, the first five terms in the series solution,

```
> psi:=add(psi[n],n=1..5);
```

$$\psi := \frac{8 \, v\, a \, (\sqrt{2}-1) \sin\left(\frac{\pi x}{a}\right) \sin\left(\frac{\pi ct}{a}\right)}{\pi^3 c} + \frac{2\, v\, a \sin\left(\frac{2\pi x}{a}\right) \sin\left(\frac{2\pi ct}{a}\right)}{\pi^3 c}$$

$$+ \frac{8}{27} \frac{v\, a\, (\sqrt{2}+1) \sin\left(\frac{3\pi x}{a}\right) \sin\left(\frac{3\pi ct}{a}\right)}{\pi^3 c}$$

$$+ \frac{8}{125} \frac{v\, a\, (-\sqrt{2}-1) \sin\left(\frac{5\pi x}{a}\right) \sin\left(\frac{5\pi ct}{a}\right)}{\pi^3 c}$$

and differentiating with respect to time, yields the velocity in the transverse direction (output suppressed here).

```
> vel:=diff(psi,t);
```

To plot the transverse velocity at $t = 0$, some representative values must be substituted for the parameters. Vectoria chooses $v = 5$, $a = 2$, and $c = 1$.

```
> t:=0: v:=5: a:=2: c:=1:
```

To compare the series representation at $t = 0$ with the given piecewise initial velocity distribution, the latter is entered,

```
> V:=piecewise(x<a/4,4*v*x/a,x<a/2,(4*v/a)*(a/2 -x),x<a,0):
```

and plotted together with *vel*, the result being shown in Figure 5.2.

```
> plot([vel,V],x=0..a,color=[red,green],thickness=2,
    labels=["x","vel"],tickmarks=[2,4]);
```

As seen in Figure 5.2 the agreement on keeping five terms in the Fourier series

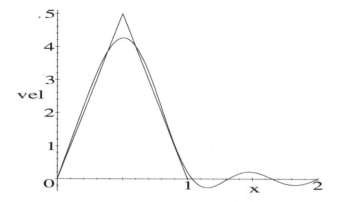

Figure 5.2: Comparing the sum of five Fourier terms to the exact input velocity.

solution isn't too bad, but not great either. Increasing the n value, Vectoria observes an increasingly better fit. The reader should be the judge of how many terms suffice to give a good fit for the parameters chosen.

Although reasonably satisfied that the quoted Fourier series expansion is correct, Vectoria still isn't entirely happy until the actual motion predicted by the formula is observed. After all, it is not obvious from the series solution exactly what the behavior of the string is after being struck. So the displacement $\psi(x, t)$ is animated for the time interval $t = 0$ to 50. First, the time variable t must be unassigned. Otherwise, Maple will remember the value $t = 0$ used to check the initial velocity profile.

```
>  unassign('t'):
```
Finally, the animation command is given with 100 frames being used.

```
>  animate(plot,[psi,x=0..a],t=0..50,frames=100,thickness=2);
```
On running the animation, Vectoria observes that the wave form begins to grow in the region where the string was struck. This makes intuitive sense. Because the wave form is created on the left side of the string, it then moves with wave velocity c to the right and reflects off the boundary at $x = a$. On reflection the wave form is inverted, a characteristic feature for a fixed-ends boundary condition. The wave then propagates to the left boundary at $x = 0$ before inverting again and repeating the oscillatory behavior.

Finally, with a feeling of accomplishment, Vectoria is able to appreciate the deeper content underlying the simple remarks made by Ingrid Bergman and Humphrey Bogart. She will be even more content if Mike phones soon.

PROBLEMS:

Problem 5-3: Plucked string

An elastic string fixed between $x = 0$ and L, and initially at rest, is "plucked," its initial shape being given by the following symmetric triangular profile,

$$\psi(x, 0) = \begin{cases} 2\,h\,x/L, & 0 \le x \le L/2, \\ 2\,h\,(L - x)/L, & L/2 \le x \le L. \end{cases}$$

Verify that the motion for $t > 0$ may be described by the Fourier series solution

$$\psi(x, t) = \frac{8\,h}{\pi^2} \sum_{n=1}^{\infty} \frac{\sin(n\,\pi/2)}{n^2} \sin(n\,\pi\,x/L) \cos(n\,\pi\,c\,t/L)$$

and animate the solution for parameter values of your own choosing.

Problem 5-4: A striking piano hammer

A piano string fixed between $x = 0$ and a is struck by a piano hammer in a region of width d centered at $x = x_0$. Its initial velocity distribution is

$$\dot{\psi}(x, 0) = \begin{cases} v\,\cos(\pi\,(x - x_0)/d), & |x - x_0| < d/2, \\ 0, & |x - x_0| > d/2. \end{cases}$$

Neglecting stiffness and assuming $\psi(x,0) = 0$,

$$\psi(x,t) = \frac{4\,v\,d}{\pi^2 c} \sum_{n=1}^{\infty} \frac{1}{n} \frac{\sin(n\,\pi\,x_0/a)\cos(n\,\pi\,d/(2\,a))}{(1 - (n\,d/a)^2)} \sin(n\,\pi\,x/a)\sin(n\,\pi\,c\,t/a)$$

is the shape of the string at time t. Verify that this series solution is correct and animate it for parameter values of your own choosing.

5.1.3 Three Easy Pieces

I would advise you Sir, to study algebra, if you are not already an adept in it: your head would be less muddy...
Samuel Johnson, English writer and lexicographer (1709–1784)

Spurred by her earlier success, and feeling happier after her long phone conversation last night with Mike, Vectoria is pursuing another vibrating string example, which she has entitled *Three Easy Pieces*. This does not refer to the old Jack Nicholson movie with a similar title (cf. *Five Easy Pieces*), but to the fact that the algebraic manipulations involved in dealing with plane-wave propagation along a three-piece string are easy if one uses computer algebra.

When an intermediate section of a very long horizontal string has a greater mass density ϵ than the remaining two identical portions of the string, a transverse plane wave incident on that section will in general experience partial reflection and transmission. Recall that the velocity of the transverse wave is given by $c = \sqrt{T/\epsilon}$, where T is the tension in the string. The wave number is $k = \omega/c = 2\pi/\lambda$, where ω is the angular frequency and λ is the wavelength. Since the frequency of the wave and the tension must remain the same in each region, the ratio r of wave numbers in two different regions of mass density ϵ_2 and ϵ_1 is given by $r = k_2/k_1 = \sqrt{\epsilon_2/\epsilon_1}$. Vectoria reads in a certain physics text that for the case $k_1 = K$, the ratio $r = 3$, and the more massive segment (labeled 2) stretches from $x = 0$ to $x = L$, the energy transmission (T) and reflection (R) coefficients are given by,

$$T = \frac{9}{17 - 8\cos(6\,K\,L)}, \qquad R = \frac{8 - 8\cos(6\,K\,L)}{17 - 8\cos(6\,K\,L)}.$$

Here T and R measure the fraction of the incident power that is transmitted and reflected. The power here is proportional to the square of the string amplitude.

Vectoria's objective is to verify the cited reflection and transmission coefficients and plot them for $K = 1$ as a function of L. Her method of attack is to write down plane-wave expressions in each region, determine the coefficients by matching the solutions at $x = 0$ and $x = L$, and then calculate T and R.

She begins by entering $r = 3$, $k_1 = K$, and $k_2 = r\,k_1$.

```
>   restart: r:=3: k[1]:=K; k[2]:=r*k[1];
```

$$k_1 := K \qquad k_2 := 3\,K$$

The plane wave is assumed to be traveling in the positive x-direction, its time part being $e^{-I\omega t}$, where $I = \sqrt{-1}$. In the first region, $x < 0$, labeled by the subscript 1, the spatial part of the wave will be $\psi_1 = e^{I k_1 x} + a\, e^{-I k_1 x}$.

```
>   psi[1]:=exp(I*k[1]*x)+a*exp(-I*k[1]*x);
```

$$\psi_1 := e^{(K\,x\,I)} + a\,e^{(-I\,K\,x)}$$

The first term is the incident plane wave, while the second term is the reflected wave with amplitude a. Since the transmission and reflection coefficients involve ratios of squared amplitudes, the amplitude of the incident wave has been set equal to one without loss of generality. The energy reflection coefficient then is given by $R = |a|^2/1 = |a|^2$.

In region 2 ($0 \le x \le L$), the wave form must be made up of waves traveling in the positive and negative x-directions, viz., $\psi_2 = b\,e^{I k_2 x} + c\,e^{-I k_2 x}$, with undetermined amplitudes b and c.

```
>   psi[2]:=b*exp(I*k[2]*x)+c*exp(-I*k[2]*x);
```

$$\psi_2 := b\,e^{(3\,I\,K\,x)} + c\,e^{(-3\,I\,K\,x)}$$

In the third region, $x > L$, there will be only a transmitted plane wave, with spatial part $\psi_3 = d\,e^{I k_1 x}$, the wave number being $k_1 = K$ since the string density is the same as in the first region. The fraction of the energy incident in region 1 that is transmitted into region 3 is given by the transmission coefficient $T = |d|^2/1 = |d|^2$.

```
>   psi[3]:=d*exp(I*k[1]*x);
```

$$\psi_3 := d\,e^{(K\,x\,I)}$$

To evaluate the four unknown coefficients a, b, c, and d, four independent equations are needed. The first two equations, $eq1$ and $eq2$, follow from the physical continuity of the string. The string segment in region 2 is joined to the segment in region 1 at $x = 0$ and to the segment in region 3 at $x = L$.

```
>   eq1:=eval(psi[1]=psi[2],x=0);
```

$$eq1 := 1 + a = b + c$$

```
>   eq2:=eval(psi[2]=psi[3],x=L);
```

$$eq2 := b\,e^{(3\,I\,K\,L)} + c\,e^{(-3\,I\,K\,L)} = d\,e^{(K\,L\,I)}$$

Since the wave equation, and therefore the second spatial derivative, remains finite everywhere along the string, the first derivative of ψ with respect to x must be continuous. So, continuity of the slope at $x = 0$ and $x = L$ yields the third and fourth equations.

```
>   eq3:=eval(diff(psi[1],x)=diff(psi[2],x),x=0);
```

$$eq3 := K\,I - a\,K\,I = 3\,I\,b\,K - 3\,I\,c\,K$$

```
>   eq4:=eval(diff(psi[2],x)=diff(psi[3],x),x=L);
```

$$eq4 := 3\,I\,b\,K\,e^{(3\,I\,K\,L)} - 3\,I\,c\,K\,e^{(-3\,I\,K\,L)} = d\,K\,e^{(K\,L\,I)}\,I$$

The system of four equations is now solved for the four unknown amplitudes, and the solution is assigned.

> `sol:=solve({eq1,eq2,eq3,eq4},{a,b,c,d}); assign(sol):`

$$sol := \{b = -\frac{2\,e^{(-3\,I\,K\,L)}}{e^{(3\,I\,K\,L)} - 4\,e^{(-3\,I\,K\,L)}}, \ c = -\frac{e^{(3\,I\,K\,L)}}{e^{(3\,I\,K\,L)} - 4\,e^{(-3\,I\,K\,L)}}$$

$$a = -\frac{2\,(e^{(3\,I\,K\,L)} - e^{(-3\,I\,K\,L)})}{e^{(3\,I\,K\,L)} - 4\,e^{(-3\,I\,K\,L)}}, \ d = -\frac{3\,e^{(-3\,I\,K\,L)}\,e^{(3\,I\,K\,L)}}{e^{(K\,L\,I)}\,e^{(3\,I\,K\,L)} - 4\,e^{(-3\,I\,K\,L)}}\}$$

Since the amplitudes are complex, the transmission coefficient is given by $T = |d|^2 = d \times d^*$, where the asterisk denotes the complex conjugate. The command `conjugate(d)` is used to enter d^*. The complex evaluation command, `evalc`, which breaks a complex quantity into real and imaginary parts, is used to generate a completely real answer. Finally, the result is simplified.

> `T:=simplify(evalc(d*conjugate(d)));`

$$T := -\frac{9}{-25 + 16\cos(3\,K\,L)^2}$$

Applying the `combine` command, with the `trig` option, produces the desired form of the transmission coefficient.

> `T:=combine(T,trig);`

$$T := -\frac{9}{-17 + 8\cos(6\,K\,L)}$$

Similarly, the reflection coefficient is given by $R = a \times a^*$, and the desired form follows on using the same command structure as for deriving T.

> `R:=simplify(evalc(a*conjugate(a)));`

$$R := \frac{16\,(-1 + \cos(3\,K\,L)^2)}{-25 + 16\cos(3\,K\,L)^2}$$

> `R:=combine(R,trig);`

$$R := \frac{-8 + 8\cos(6\,K\,L)}{-17 + 8\cos(6\,K\,L)}$$

By energy conservation the sum of the reflection and transmission coefficients should add up to one. This is checked in the next command line. The value $K = 1$ is also entered for plotting purposes.

> `check:=simplify(R+T); K:=1:`

$$check := 1$$

Finally, Vectoria plots the reflection and transmission coefficients in the same figure over the range $L = 0$ to 5, using a solid (`linestyle=1`) red curve for R, and a dashed (`linestyle=3`) blue curve for T. To maintain order, lists are used here for the energy coefficients, the colors, and the line styles. For better visualization, thick curves are employed. Labels are also added to the plot, entered as strings.

> `plot([R,T],L=0..5,color=[red,blue],linestyle=[1,3],`
>   `thickness=2,labels=["L","R,T"]);`

The resulting picture is shown in Figure 5.3.

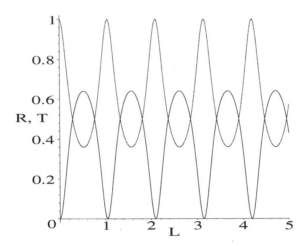

Figure 5.3: Reflection (upper curve) and transmission (lower) coefficients vs. L.

When $L = n\pi/3$, with $n = 0$ or a positive integer, the reflection coefficient vanishes and there is 100% transmission. Looking back at the expression for R, and recalling that $k_2 = 3K$ here, Vectoria notes that this situation will occur whenever $k_2 L = n\pi$ or $L = n(\lambda_2/2)$, i.e., when the length L of the thick portion exactly accommodates an integer number of half-wavelengths in that region. Does this condition make sense to you?

PROBLEMS:

Problem 5-5: Variation on the text example
If $\epsilon_1 = 1$, $\epsilon_2 = 16$, $k_1 = 1$, and $L = 2$, what fraction of the incident plane wave energy is transmitted through the middle segment into the third region?

Problem 5-6: As simple as 1-2-3
Consider a plane wave of frequency w traveling from $x = -\infty$ along an infinitely long string with three regions of different density, the middle one being located between $x = 0$ and L. Suppose that the wave number in region 1 $(x < 0)$ is $k_1 = 1$, in region 2 $(0 < x < L)$ is $k_2 = 2$, and in region 3 $(x > L)$ is $k_3 = 3$.

(a) Confirm that the reflection coefficient is

$$R = \frac{17 + 15 \cos(4L)}{113 + 15 \cos(4L)}.$$

(b) Because $k_3 \neq k_1$, the transmission coefficient T is equal to $k_3|d|^2/(k_1|a|^2)$, where a and d are the incident and transmitted amplitudes. Determine the analytic form of T and confirm that $R + T = 1$.

(c) Plot R and T on the same graph.

(d) Is there an L value for which 100% transmission is possible?

(e) What is the maximum value of T, and for what L value does this transmission occur?

Problem 5-7: Five easy pieces (not the movie)
Consider an infinitely long string that has a linear density $\epsilon_1 = 1$ in region 1 ($x < 0$), density $\epsilon_2 = 4$ in region 2 ($0 < x < L$), density $\epsilon_3 = 1$ in region 3 ($L < x < 2L$), density $\epsilon_4 = 4$ in region 4 ($2L < x < 3L$), and density $\epsilon_5 = 1$ in region 5 ($x > 3L$). For a plane wave of frequency ω coming from $x = -\infty$:

(a) Show that the transmission coefficient T into region 5 is

$$T = 2048 / \left(\sum_{n=0}^{5} b_n \cos(2\,n\,L) \right)$$

with $b_0 = 3686$, $b_1 = -882$, $b_2 = -1800$, $b_3 = 1611$, $b_4 = 162$, $b_5 = -729$.

(b) Show that the sum $T + R$ equals 1, where R is the reflection coefficient.

(c) Plot the transmission and reflection coefficients in the same graph.

(d) At what values of L is there 100% transmission? Discuss your answer.

Problem 5-8: Quantum-mechanical tunneling revisited
Schrödinger's equation describing one-dimensional motion of a particle of mass m and energy E in a potential $V(x)$ is

$$-\frac{\hbar^2}{2\,m} \frac{d^2\psi(x)}{dx^2} + V(x)\,\psi(x) = E\,\psi(x).$$

Here $\psi(x)$ is the probability amplitude for finding the particle at x and $\hbar = h/(2\pi)$, where h is Planck's constant. Consider a particle with energy $E < V_0$ incident on a rectangular barrier $V(x) = V_0 > 0$ located in the region $x = 0$ to L. The potential $V(x)$ equals 0 outside the barrier.

(a) Show that the transmission coefficient T, which may be calculated in a similar manner to that for the 3-piece string, is

$$T = \left(1 + \frac{\sinh^2(K\,L)}{(4\,E/V_0)\,(1 - E/V_0)} \right)^{-1},$$

with $K = \sqrt{2\,m\,(V_0 - E)/\hbar^2}$.

(b) Plot T as a function of KL for some representative values of $E/V_0 < 1$.

(c) Discuss the behavior of the transmission coefficient and contrast it with what would be expected classically.

5.1.4   Complex, Yet Simple

I adore simple pleasures. They are the last refuge of the complex.
Oscar Wilde, Anglo-Irish playwright (1854–1900)

Consider the complex function $w = u(x,y) + I\,v(x,y) = z + 1/z$, where $z = x + I\,y$ with $I = \sqrt{-1}$. It is stated in a certain mathematical physics text that because the real functions u and v satisfy the *Cauchy–Riemann conditions*,

$$\frac{\partial u}{\partial x} = \frac{\partial v}{\partial y}, \quad \frac{\partial v}{\partial x} = -\frac{\partial u}{\partial y}, \tag{5.5}$$

then both $u(x,y)$ and $v(x,y)$ satisfy *Laplace's* equation,

$$\nabla^2 u = \frac{\partial^2 u}{\partial x^2} + \frac{\partial^2 u}{\partial y^2} = 0, \quad \nabla^2 v = \frac{\partial^2 v}{\partial x^2} + \frac{\partial^2 v}{\partial y^2} = 0, \tag{5.6}$$

and therefore can represent possible real potentials. It is further stated that the constant-v curves can be used to represent the equipotentials outside an infinitely long grounded conducting cylinder of unit radius placed in a previously uniform electric field oriented perpendicular to the cylinder. The electric field lines outside the cylinder are given by the constant-u curves, and the electric field vectors by $\vec{E} = -\nabla v$.

Our goal in this recipe is to illustrate how simple it is to check these statements using Maple. Loading the plots and VectorCalculus packages,

```
>   restart: with(plots): with(VectorCalculus):
```
both $z = x + I\,y$ and $w = z + 1/z$ are entered.

```
>   z:=x+I*y; w:=z+1/z;
```

$$z := x + y\,I \qquad w := x + y\,I + \frac{1}{x + y\,I}$$

The complex function w is separated into real and imaginary parts.

```
>   w:=evalc(w);
```

$$w := x + \frac{x}{x^2 + y^2} + \left(y - \frac{y}{x^2 + y^2} \right) I$$

The term involving I is removed from w, thus yielding u. The term containing I is selected and divided by I to produce v.

```
>   u:=remove(has,w,I); v:=select(has,w,I)/I;
```

$$u := x + \frac{x}{x^2 + y^2} \qquad v := y - \frac{y}{x^2 + y^2}$$

We now check that both Cauchy–Riemann conditions (5.5) are satisfied.

```
>   CR1:=simplify(diff(u,x)-diff(v,y));
```

$$CR1 := 0$$

```
>   CR2:=simplify(diff(v,x)+diff(u,y));
```

$$CR2 := 0$$

A functional operator L is formed to calculate and simplify the Laplacian (∇^2) of a function f expressed in terms of the Cartesian coordinates x and y.

> `L:=f->simplify(Laplacian(f,'cartesian'[x,y])):`

Applying L to u and to v yields zero, confirming that these functions satisfy Laplace's equation and can be regarded as real potentials.

> `L(u); L(v);`

$$0 \qquad 0$$

To confine our attention to the region outside the cylinder, two piecewise functions, *pw1* and *pw2*, are formed which are equal to zero for $x^2 + y^2 < 1$ and u and v, respectively, outside this circular region.

> `pw1:=piecewise(x^2+y^2<1,0,x^2+y^2>=1,u);`

$$pw1 := \begin{cases} 0 & x^2 + y^2 < 1 \\ x + \dfrac{x}{x^2 + y^2} & 1 \le x^2 + y^2 \end{cases}$$

> `pw2:=piecewise(x^2+y^2<1,0,x^2+y^2>=1,v):`

A contour plot operator CP is formed to plot the equipotentials for a given potential function f. The color C must also be specified. The contours are drawn for potentials equal to $0.2\,n$, with n ranging from -11 to $+11$. The grid spacing is taken to be 90×90, and constrained scaling is imposed.

> `CP:=(f,C)->contourplot(f,x=-2..2,y=-2..2,contours=`
> `   [seq(0.2*n,n=-11..11)],grid=[90,90],color=C,`
> `   scaling=constrained,thickness=2):`

The curves corresponding to constant u are colored red, those corresponding to constant v colored blue. The two sets of curves are superimposed. A black-and-white version of the plot is shown in Figure 5.4.

> `display({CP(pw1,red),CP(pw2,blue)});`

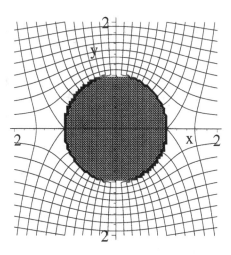

Figure 5.4: Equipotentials and electric field lines outside a conducting cylinder.

The horizontal curves are the equipotentials (constant v) and the vertical curves are the electric field lines (constant u). The two sets of curves intersect at right angles, as would be expected. The electric field lines also intersect the cylindrical surface perpendicularly, because the conducting surface is also an equipotential.

Instead of showing the electric field lines, one can plot the electric field vectors. The electric field $\vec{E} = -\nabla v$ can be calculated in Cartesian coordinates using the `Gradient` command.

```
>  E:=-Gradient(v,'cartesian'[x,y]);
```

$$E := -\frac{2\,y\,x}{(x^2 + y^2)^2}\,e_x + \left(-1 + \frac{1}{x^2 + y^2} - \frac{2\,y^2}{(x^2 + y^2)^2}\right)\,e_y$$

Since we are interested only in the electric field outside the cylinder, two piecewise functions are formed to calculate the x and y components of the electric field in the region $x^2 + y^2 \geq 1$.

```
>  Ex:=piecewise(x^2+y^2<1,0,x^2+y^2>=1,E[1]):
>  Ey:=piecewise(x^2+y^2<1,0,x^2+y^2>=1,E[2]):
```

The `fieldplot` command is used to plot the electric field vectors as thick red arrows, the grid being 12×12. The arrows will point in the direction of $\vec{E}$, their size being a measure of the magnitude, $|\vec{E}|$.

```
>  FP:=fieldplot([Ex,Ey],x=-2..2,y=-2..2,arrows=THICK,
         grid=[12,12],color=red):
```

The equipotentials and electric field vectors are superimposed in the same picture, a black-and-white rendition being shown in Figure 5.5.

```
>  display({CP(pw2,blue),FP},labels=["x","y"],tickmarks=[2,2]);
```

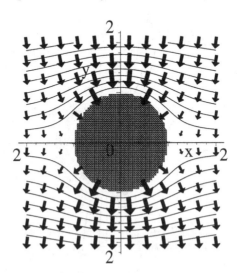

Figure 5.5: Equipotentials and electric field vectors outside the cylinder.

PROBLEMS:

Problem 5-9: Intersecting conducting plates

In the text recipe replace the complex function with $w = u + I\,v = z^2$, with $z = x + I\,y$. Confirm that $u(x, y)$ and $v(x, y)$ satisfy the Cauchy–Riemann conditions and Laplace's equations. Confirm by making a suitable plot that the equipotentials and electric field lines are appropriate to the quarter-space bounded by two semi-infinite conducting plates intersecting at right angles. Make another plot that shows the electric field vectors and equipotentials.

Problem 5-10: Fluid flow around a plate

Consider the function $w = u + i\,v = \sqrt{z^2 - 1}$, with $z = x + i\,y$. By creating a suitable figure in the x-y plane, show that the constant-u curves can represent the *streamlines* (tracks of the fluid particles) for fluid flow around an infinitely long plate of finite width, lying between $x = -1$ and $+1$, inserted perpendicular to a previously uniform fluid flow in the y direction. Include the plate and the equipotentials (constant v curves) in your figure.

5.2 Diffusion and Laplace's Equation Models

In this section, diffusion and Laplace's equation models are solved in Cartesian and other common coordinate systems, using some of the standard methods of mathematical physics. Of course, we will let Maple do the heavy slogging, concentrating on the underlying physics.

5.2.1 Freeing Excalibur

A sleeping presence is always a mystery ... seemingly peaceful,
yet in reality off on wild adventures in strange landscapes.
A. Alvarez, British novelist (1929–)

Russell works as a control systems engineer for an aerospace company in the Phoenix area. Earlier in the day, several of his old ASU engineering classmates dropped by his workplace and persuaded him to go to the Monastery, a local outdoor pub, after work. There they ate chicken wings and pizza, drank pitchers of Mexican beer, and swapped stories about their undergraduate days and what had happened to each of them since graduating from engineering school.

Later that night, probably triggered by the day's events, Russell is having a wild dream in which he has been transported back to the mythical times of Merlin the magician. Merlin has wrapped the fabled sword Excalibur in thin insulating material and has embedded it in a large rock with only the tip and hilt protruding. The legend is that whoever manages to pull out the sword will become ruler of the Kingdom with all its associated wealth. Many have tried, but all[1] have failed to pull the sword out of the rock. Having taken a course in thermodynamics as an engineering undergraduate, Russell speculates as to whether the sword could be pulled out by cooling the ends of the sword with large buckets of ice, thus causing heat to flow out of the warmer interior of the sword's blade to the ends. If the sword were cooled sufficiently, its diameter might shrink slightly, and just possibly he could pull the sword out.

But remembering the thermodynamics course causes Russell's dream to alter direction, and he then dreams of a related problem that appeared on that course's final exam many years ago. A thin 1-m-long rod (the sword's shaft) whose lateral surface is insulated to prevent heat flow through that surface has its ends suddenly held at the freezing point of water, $0\,°C$ (contact with the buckets of ice). Taking one end of the rod to be at $x = 0$ and the other at $x = L = 1$, the initial temperature (T) distribution was $T(x, t = 0) = 100\, x\,(1 - x)$, a parabolic profile with a maximum temperature of $25°$ at the midpoint $x = \frac{1}{2}$. In the exam, he was asked to determine the temperature distribution $T(x, t)$ for any time $t > 0$. Despite the passage of time, Russell remembers his approach to solving this problem very well. His method of attack was to solve the diffusion equation

$$\frac{\partial T(x,t)}{\partial t} = d\,\frac{\partial^2 T(x,t)}{\partial x^2}, \tag{5.7}$$

by the method of *separation of variables*, i.e., assume that $T(x,t) = S(x)\,F(t)$. Substituting this assumed form into (5.7) and dividing by $T(x, t)$ yields

$$\frac{1}{F(t)}\,\frac{dF(t)}{dt} = \frac{d}{S(x)}\,\frac{d^2 S(x)}{dx^2}. \tag{5.8}$$

Since the lhs of (5.8) involves a function of t alone and the rhs involves a function of x alone, the only way that it can be generally true is for both sides to be equal to a common constant, called the *separation constant*. The separated ODE for $S(x)$ is solved, subject to the boundary conditions, and the complete product solution constructed, subject to the initial condition.

Russell's wild dream is interrupted by the roar of a big jet passing over his house on its way into Sky Harbor Airport. Waking up, and unable to get back to sleep, he decides to make a cup of instant decaf coffee and implement an animated Maple solution of the heat diffusion problem on his computer. To carry out the animation, the plots library package is loaded. Noting that the diffusion coefficient d could be absorbed into either the spatial or time variable, Russell sets $d = 1$ without loss of generality. He also specifies the rod's length, $L = 1$, and enters the heat equation (5.7).

```
>  restart: with(plots): d:=1: L:=1:
>  heateq:=diff(T(x,t),t)=d*diff(T(x,t),x,x);
```

$$heateq := \frac{\partial}{\partial t}\,T(x,\,t) = \frac{\partial^2}{\partial x^2}\,T(x,\,t)$$

A general solution to the heat equation is sought using the PDE solve (`pdsolve`) command, the HINT option explicitly telling Maple to separate variables.

---

[1] King Arthur has not shown up yet.

```
> pdsolve(heateq,HINT=S(x)*F(t));
```

$$(T(x, t) = S(x) F(t)) \;\&\text{where} \; \left[\left\{\frac{d}{dt}F(t) = \_c_1 F(t), \frac{d^2}{dx^2}S(x) = \_c_1 S(x)\right\}\right]$$

The heat equation has been separated into two ODEs, $\_c_1$ being the separation constant. If the INTEGRATE option is also included, the general solution of each ODE will be generated as Russell now illustrates.

```
> pdsolve(heateq,HINT=S(x)*F(t),INTEGRATE);
```

$$(T(x, t) = S(x) F(t)) \;\&\text{where}$$
$$[\{\{F(t) = \_C3\, e^{(-\_c_1\, t)}\}, \{S(x) = \_C1\, e^{(\sqrt{-\_c_1}\, x)} + \_C2\, e^{(-\sqrt{-\_c_1}\, x)}\}\}]$$

There are three unknown constants in the output, namely $\_C1$, $\_C2$, and $\_C3$. Finally, including the build option produces the general product solution, sol.

```
> sol:=pdsolve(heateq,HINT=S(x)*F(t),INTEGRATE,build);
```

$$sol := T(x, t) = \_C3\, e^{(-\_c_1\, t)} \_C1\, e^{(\sqrt{-\_c_1}\, x)} + \frac{\_C3\, e^{(-\_c_1\, t)} \_C2}{e^{(\sqrt{-\_c_1}\, x)}}$$

Russell then substitutes $\_c_1 = -k^2$ on the rhs of sol and applies the simplify command with the symbolic option, which is equivalent to assuming that $k > 0$.

```
> temp:=simplify(subs(_c[1]=-k^2,rhs(sol)),symbolic);
```

$$temp := \_C3\, (\_C1\, e^{(k\,(-t\,k+x\,I))} + \_C2\, e^{(-k\,(t\,k+x\,I))})$$

The resulting form is complex in appearance, so the complex evaluation command is applied to separate $temp$ into real and imaginary terms.

```
> temp:=evalc(temp);
```

$$temp := \_C3\, (\_C1\, e^{(-k^2\, t)} \cos(k\,x) + \_C2\, e^{(-k^2\, t)} \cos(k\,x))$$
$$+ \_C3\, (\_C1\, e^{(-k^2\, t)} \sin(k\,x) - \_C2\, e^{(-k^2\, t)} \sin(k\,x))\, I$$

The temperature has been expressed in terms of cosine, sine, and exponential terms, which are now successively collected.

```
> temp:=collect(%,[cos,sin,exp]);
```

$$temp := \_C3\, (\_C1 + \_C2)\, e^{(-k^2\, t)} \cos(k\,x) + \_C3\, (\_C1 - \_C2)\, e^{(-k^2\, t)} \sin(k\,x)\, I$$

Since $T(x = 0, t) = 0$ for all times and $\cos(k\,x)$ does not vanish at $x = 0$, Russell removes it from $temp$. This is the first boundary condition (bc1).

```
> temp:=remove(has,temp,cos); #bc1
```

$$temp := \_C3\, (\_C1 - \_C2)\, e^{(-k^2\, t)} \sin(k\,x)\, I$$

Similarly, the temperature must also vanish at $x = L = 1$ for all t (the second boundary condition). Therefore $\sin(k\,L) = 0$, so $k = n\,\pi/L$ with $n = 1, 2, 3, \ldots$ The general solution must involve a linear combination of terms involving all possible n values. With the coefficient labeled A_n, the nth term in the *Fourier series* representation of the temperature will be of the following form.

```
> T[n]:=A[n]*subs(k=n*Pi/L,exp(-k^2*t)*sin(k*x)); #bc2
```

$$T_n := A_n\, e^{(-n^2\, \pi^2\, t)} \sin(n\,\pi\,x)$$

The complete series solution is $T = \sum_{n=1}^{\infty} T_n$, with the form of A_n to be determined from the initial condition. At $t = 0$, one must have

$$T(x,0) = \sum_{n=1}^{\infty} A_n \sin(n \pi x) = 100 \, x \, (1 - x).$$

If $T(x,0)$ is multiplied by $\sin(m \pi x)$ and integrated from $x = 0$ to $x = L = 1$, the lhs will integrate to zero for every term in the series except for the term corresponding to $m = n$. This is a result of the independence, or orthogonality, of the sine terms in the series. The resulting equation can then be solved for A_n. Russell will now apply this approach. He evaluates T_n at $t = 0$, and enters the initial temperature profile, $T0$.

> X:=eval(T[n],t=0); T0:=100*x*(1-x);

$$X := A_n \sin(n \pi x) \qquad T0 := 100 \, x \, (1 - x)$$

Since only the nth term in the series survives, Russell forms $T0 - X$, multiplies this by $\sin(n \pi x)$, and integrates from $x = 0$ to 1 in eq. Then eq is set equal to zero and solved for A_n.

> eq:=int((T0-X)*sin(n*Pi*x),x=0..1);
> A[n]:=solve(eq=0,A[n]);

$$A_n := -\frac{200 \, (-2 + n \pi \sin(n \pi) + 2 \cos(n \pi))}{n^2 \pi^2 \, (n \pi - \sin(n \pi) \cos(n \pi))}$$

Assuming that n is an integer, the nth Fourier term T_n is simplified, A_n having been automatically substituted. The double colon in assuming is a "type match" command.

> T[n]:=simplify(T[n]) assuming n::integer;

$$T_n := -\frac{400 \, (-1 + (-1)^n) \, e^{(-n^2 \pi^2 t)} \sin(n \pi x)}{n^3 \pi^3}$$

Thus, the temperature distribution in the rod satisfying the boundary and initial conditions is given by the following infinite Fourier series:

> Temp:=Sum(T[n],n=1..infinity);

$$Temp := \sum_{n=1}^{\infty} \left(-\frac{400 \, (-1 + (-1)^n) \, e^{(-n^2 \pi^2 t)} \sin(n \pi x)}{n^3 \pi^3} \right)$$

The sum of the first five terms gives a good approximation to $T(x,t)$, because the contribution of higher-order terms drops very rapidly with increasing time.

> TT:=sum(T[n],n=1..5);

$$TT := \frac{800 \, e^{(-\pi^2 t)} \sin(\pi x)}{\pi^3} + \frac{800}{27} \frac{e^{(-9 \pi^2 t)} \sin(3 \pi x)}{\pi^3} + \frac{32}{5} \frac{e^{(-25 \pi^2 t)} \sin(5 \pi x)}{\pi^3}$$

Then TT is animated so its spatial and temporal evolution can be clearly seen.

> animate(plot,[TT,x=0..1],t=0..1,frames=50);

Russell is feeling sleepy, so is going back to bed. If you wish to see how the initial parabolic temperature profile decays to zero everywhere inside the rod, execute the work sheet, click on the plot, and then on the play arrow.

PROBLEMS:

Problem 5-11: Hot rod

For the heated rod discussed in the text, at what time is the temperature at the center of the rod equal to one-third of its initial value?

Problem 5-12: The switch

The temperatures at the ends $x = 0$ and $x = 100$ of a rod (insulated on its sides) 100 cm long are held at $0°$ and $100°$, respectively, until steady state is achieved. Then at the instant $t = 0$, the temperatures of the two ends are interchanged. Determine the resultant temperature distribution $T(x,t)$. Animate your solution and discuss its behavior.

5.2.2 Aussie Barbecue

He who does not mind his belly, will hardly mind anything else.
Samuel Johnson, English author and lexicographer (1709–1784)

When coauthor Richard spent a sabbatical leave at the Australian National University, in Canberra, he constructed a barbecue consisting of a scrap of rectangular iron plate placed on a primitive[2] brickwork support. The myriad eucalyptus trees that dot the landscape of the Australian Capital Territory and New South Wales constantly shed branches, which were gathered to be used as free barbecue fuel. With the iron plate slightly tilted to allow the grease from the sizzling lamb chops (accompanied with a cask of wine, the staple of the Australian outback barbecue) to drip off into the fire, the flames on the downhill edge of the plate tended to make that edge considerably hotter than the other edges of the plate. The temperature distribution in the plate was uneven, and vigilance was necessary to prevent the chops from being turned into charcoal. The Aussie barbecue is the inspiration for the following two-dimensional boundary value example, in which the steady-state temperature profile in a thin rectangular plate, with prescribed temperatures on the edges, is found.

Consider a uniform solid rectangular plate stretching between $x = 0$ and $x = L$ and $y = 0$ and $y = h$, where L and h are measured in meters. Since what is important is the temperature differences across the plate, without loss of generality the coldest edge can be set equal to zero. Specifically, we shall imagine the plate to have three "cold" edges whose temperatures are set equal to zero, i.e., $T(0,y) = T(x,0) = T(x,h) = 0$. The "hot" edge $x = L$ will be assumed to have a parabolic temperature distribution given by $T(L,y) = 400\,y\,(h-y)\,°\mathrm{C}$. If, for example, $h = 1$ m, the temperature along the hot edge will vary from zero at the corners to a temperature $100°$ hotter in the middle of that edge. Our goal is to determine the steady-state temperature distribution $T(x,y)$ in the plate.

---

[2]Note: Richard is a theoretician, not an experimentalist.

The answer will require that the two-dimensional form of Laplace's equation,

$$\frac{\partial^2 T(x,y)}{\partial x^2} + \frac{\partial^2 T(x,y)}{\partial y^2} = 0, \tag{5.9}$$

be solved, subject to the four boundary conditions. After loading the plots package, we enter Laplace's equation (LE).

```
>   restart:  with(plots):
>   LE:=diff(T(x,y),x,x)+diff(T(x,y),y,y)=0;
```

$$LE := \left(\frac{\partial^2}{\partial x^2} T(x,\, y)\right) + \left(\frac{\partial^2}{\partial y^2} T(x,\, y)\right) = 0$$

Using the pdsolve command with the HINT=f(x)*g(y),INTEGRATE, and build options, we generate the general product solution of Laplace's equation.

```
>   sol:=pdsolve(LE,HINT=f(x)*g(y),INTEGRATE,build);
```

$$sol := T(x,\, y) = \_C3 \sin(\sqrt{-c_1}\, y)\, \_C1\, e^{(\sqrt{-c_1}\, x)} + \frac{\_C3 \sin(\sqrt{-c_1}\, y)\, \_C2}{e^{(\sqrt{-c_1}\, x)}}$$

$$+ \_C4 \cos(\sqrt{-c_1}\, y)\, \_C1\, e^{(\sqrt{-c_1}\, x)} + \frac{\_C4 \cos(\sqrt{-c_1}\, y)\, \_C2}{e^{(\sqrt{-c_1}\, x)}}$$

The solution involves one separation constant $\_c_1$ and four unknown coefficients. For notational convenience, we substitute $\sqrt{-c_1} = m$ in the rhs of sol.

```
>   T:=subs(sqrt(_c[1])=m,rhs(sol));
```

$$T := \_C3 \sin(m\, y)\, \_C1\, e^{(m\, x)} + \frac{\_C3 \sin(m\, y)\, \_C2}{e^{(m\, x)}} + \_C4 \cos(m\, y)\, \_C1\, e^{(m\, x)}$$

$$+ \frac{\_C4 \cos(m\, y)\, \_C2}{e^{(m\, x)}}$$

The coefficients are determined from the four boundary conditions. To satisfy $T(x,0) = 0$, the $\cos(m\, y)$ terms are removed since they don't vanish at $y = 0$.

```
>   T:=remove(has,T,cos(m*y));  #bc1
```

$$T := \_C3 \sin(m\, y)\, \_C1\, e^{(m\, x)} + \frac{\_C3 \sin(m\, y)\, \_C2}{e^{(m\, x)}}$$

Then T is converted completely to trigonometric form.

```
>   T:=expand(convert(T,trig));
```

$$T := \_C3 \sin(m\, y)\, \_C1 \cosh(m\, x) + \_C3 \sin(m\, y)\, \_C1 \sinh(m\, x)$$

$$+ \frac{\_C3 \sin(m\, y)\, \_C2}{\cosh(m\, x) + \sinh(m\, x)}$$

The second boundary condition is that $T(0,y) = 0$. Since $\cosh(m\, x)$ doesn't vanish at $x = 0$, terms involving $\cosh(m\, x)$ are removed from T.

```
>   T:=remove(has,T,cosh(m*x));  #bc2
```

$$T := \_C3 \sin(m\, y)\, \_C1 \sinh(m\, x)$$

The third boundary condition, $T(x,h) = 0$, requires that $\sin(m\, h) = 0$, so that $m = n\pi/h$ with n a positive integer. This relation is substituted into T, and the awkward coefficient combination replaced with A_n. The resulting form, labeled T_n, is the nth term in the infinite Fourier series solution.

```
>  T[n]:=A[n]*subs(m=n*Pi/h,T)/(_C3*_C1); #bc3
```

$$T_n := A_n \sin\left(\frac{n\pi y}{h}\right) \sinh\left(\frac{n\pi x}{h}\right)$$

The remaining boundary condition must be applied along the "hot" edge at $x = L$. Then T_n is evaluated at this value in S, and the temperature distribution $T0 = 400\, y\,(h - y)$ along the hot edge entered.

```
>  S:=eval(T[n],x=L); T0:=400*y*(h-y);
```

$$S := A_n \sin\left(\frac{n\pi y}{h}\right) \sinh\left(\frac{n\pi L}{h}\right) \qquad T0 := 400\, y\,(h - y)$$

On the edge $x = L$, we have

$$T(L, y) = \sum_{n=1}^{\infty} A_n \sin\left(\frac{n\pi y}{h}\right) \sinh\left(\frac{n\pi L}{h}\right) = T0 = 400\, y\,(h - y).$$

If $T(L, y)$ is multiplied by $\sin(m\pi y/h)$ and integrated from $y = 0$ to h, the lhs will yield zero except for $m = n$. Thus, the coefficient A_n can be determined in a manner similar to the last recipe. The next two command lines carry out this procedure and calculate A_n.

```
>  eq:=int((S-T0)*sin(n*Pi*y/h),y=0..h)=0: #bc4
>  A[n]:=solve(eq,A[n]):
```

Assuming that n is an integer, the nth Fourier term is simplified, A_n having been automatically substituted.

```
>  T[n]:=simplify(T[n]) assuming n::integer;
```

$$T_n := -\frac{1600\, h^2 \left(-1 + (-1)^n\right) \sin\left(\frac{n\pi y}{h}\right) \sinh\left(\frac{n\pi x}{h}\right)}{n^3 \sinh\left(\frac{n\pi L}{h}\right) \pi^3}$$

The complete solution is $T(x, y) = \sum_{n=1}^{\infty} T_n$. To plot the solution, we will take $L = 1$, $h = 1$, and keep terms in the series up to $n = 7$,

```
>  L:=1: h:=1: Temp:=sum(T[n],n=1..7);
```

$$Temp := \frac{3200 \sin(\pi y) \sinh(\pi x)}{\sinh(\pi)\, \pi^3} + \frac{3200}{27} \frac{\sin(3\pi y) \sinh(3\pi x)}{\sinh(3\pi)\, \pi^3}$$
$$+ \frac{128}{5} \frac{\sin(5\pi y) \sinh(5\pi x)}{\sinh(5\pi)\, \pi^3} + \frac{3200}{343} \frac{\sin(7\pi y) \sinh(7\pi x)}{\sinh(7\pi)\, \pi^3}$$

which is sufficiently accurate here. By itself, the series representation doesn't tell us much. To see the steady-state temperature distribution $T(x, y)$ in the plate, a three-dimensional colored contour plot with 20 equally spaced contours is created.

```
>  plot3d(Temp,x=0..L,y=0..h,axes=boxed,style=patchcontour,
   contours=20,orientation=[-132,35],tickmarks=[2,2,2]);
```

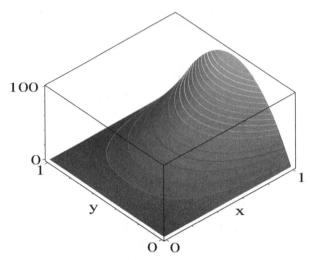

Figure 5.6: Three-dimensional contour plot of the temperature distribution.

The resulting picture is shown in Figure 5.6. From the figure, the reader can get a good feeling for how the temperature distribution drops off from the hot edge. If it is desired to know the temperature at a particular point, a more accurate answer can be obtained by substituting the coordinates of the point into the series solution. For example, the temperature in the middle ($x = L/2$, $y = h/2$) of the plate is evaluated,

>   `middle:=evalf(eval(Temp,{x=L/2,y=h/2}));`

$$middle := 20.53145858$$

and found to be about $20\frac{1}{2}°$ warmer than on the cold edges.

PROBLEMS:
Problem 5-13: Electric potential
An infinitely long hollow conductor has a rectangular cross section with sides L and $2\,L$. One of the longer sides is charged to a potential $V = V_0$ and the other three sides are held at zero potential. By solving Laplace's equation, determine the potential distribution in the interior region. Taking $V_0 = 1$ and $L = 1$, plot the equipotential lines. Calculate the electric field $\vec{E} = -\nabla V$ in the interior region. Plot $\vec{E}$ on the same graph as the equipotential lines.

Problem 5-14: Barbecue plate
Suppose that a rectangular barbecue plate has a finite thickness $c = 0.01$ m in the z-direction and dimensions $a = 0.5$ m and $b = 0.5$ m in the x- and y-directions, respectively. Assuming that the bottom of the plate is uniformly heated and is $100°$ hotter than the other five sides, determine the steady-state temperature distribution $T(x, y, z)$ inside the plate and make a suitable plot.

5.2.3 Benny's Solution

A good scientist is a person with original ideas. A good engineer is a person who makes a design that works with as few original ideas as possible. There are no prima donnas in engineering.
Freeman Dyson, British-born U.S. physicist (1923–)

Greg Arious Nerd is currently teaching the mathematical physics course to a mixture of future engineers and physicists at Erehwon's most famous academic institution, EIT (Erehwon Institute of Technology). The students are being instructed in the use of the Elpam computer algebra system in solving their mathematical physics problems. As a classroom example, Greg selects a somewhat artificial, but pedagogically useful, two-dimensional static potential problem. A circular annulus has an angular potential distribution $\phi(10, \theta) = 15 \cos \theta$ specified on the inner radius $r_1 = 10$ cm and a potential $\phi(20, \theta) = 30 \sin \theta$ on the outer radius $r_2 = 20$ cm. The question to be answered is, "What is the potential distribution in the annular region, and what do the equipotentials look like in this region?"

The following recipe for solving this problem has been submitted by one of Greg's engineering students, Benjamin Beetlebrox III. Although, as a descendent of one of the founding families on Erehwon, he doesn't like his first name shortened, we shall take Freeman Dyson's words to heart and call him Benny.

In addition to the plots package needed for plotting the equipotentials, Benny loads the VectorCalculus package, because it contains the `Laplacian` command, which will enable him to easily generate Laplace's equation for the potential $\phi(r, \theta)$ in polar coordinates. The radial (r) and angular (θ) polar coordinates are related to the Cartesian coordinates (x, y) through the relations $x = r \cos \theta$, $y = r \sin \theta$.

```
>   restart: with(plots): with(VectorCalculus):
```

Curious about the coordinate systems that the Elpam system supports, Benny enters the following command line. On executing this line, a list of the available two- and three-dimensional coordinate systems appears in a Help page.

```
>   ?coords;
```

Through the hyperlinks at the bottom of the Help page, Benny is led to the `coordplot` and `coordplot3d` commands for plotting representative curves and surfaces in two and three dimensions, corresponding to holding each coordinate equal to a constant value. Closing the Help window, Benny uses `coordplot` to plot the lines $r =$ constant and $\theta =$ constant in polar coordinates. The `grid` option is used to control the number of constant values and therefore the number of lines drawn. The default is `grid=[12,12]`. The values of the constants are added to the graph by including `labeling=true`. More detailed explanations and other options may be found on the `coordplot` Help page.

```
>   coordplot(polar,grid=[5,7],labelling=true,
    scaling=constrained);
```

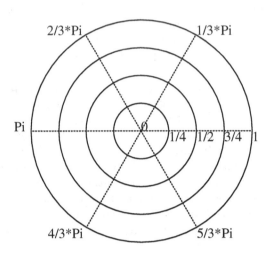

Figure 5.7: Constant r and θ lines in polar coordinates.

The resulting picture is reproduced in Figure 5.7, circles being produced for $r = 0, \frac{1}{4}, \frac{1}{2}, \frac{3}{4}, 1$ and polar lines for $\theta = 0, \pi/3, 2\pi/3, \pi, \ldots$.

Benny now enters Laplace's equation (LE) in polar coordinates.

> LE:=expand(Laplacian(phi(r,theta),'polar'[r,theta]))=0;

$$LE := \frac{\frac{\partial}{\partial r}\,\phi(r,\,\theta)}{r} + \left(\frac{\partial^2}{\partial r^2}\,\phi(r,\,\theta)\right) + \frac{\frac{\partial^2}{\partial\theta^2}\,\phi(r,\,\theta)}{r^2} = 0$$

The pdsolve command with the HINT=f(r)*g(theta), INTEGRATE and build options is used to find the general product solution of Laplace's equation.

> sol:=pdsolve(LE,HINT=f(r)*g(theta),INTEGRATE,build);

$$sol := \phi(r,\,\theta) = \_C3\sin(\sqrt{\_c_1}\,\theta)\,\_C1\,r^{(\sqrt{\_c_1})} + \frac{\_C3\sin(\sqrt{\_c_1}\,\theta)\,\_C2}{r^{(\sqrt{\_c_1})}}$$
$$+\,\_C4\cos(\sqrt{\_c_1}\,\theta)\,\_C1\,r^{(\sqrt{\_c_1})} + \frac{\_C4\cos(\sqrt{\_c_1}\,\theta)\,\_C2}{r^{(\sqrt{\_c_1})}}$$

Benny notes that the solution must reduce to a $\cos\theta$ form on one boundary and a $\sin\theta$ form on the other, so he accordingly sets the separation constant $\sqrt{\_c_1} = 1$ on the rhs of sol.

> phi:=subs(sqrt(_c[1])=1,rhs(sol));

$$\phi := \_C3\sin(\theta)\,\_C1\,r + \frac{\_C3\sin(\theta)\,\_C2}{r} + \_C4\cos(\theta)\,\_C1\,r + \frac{\_C4\cos(\theta)\,\_C2}{r}$$

The terms are then grouped by successively collecting $\cos\theta$, $\sin\theta$, and r terms.

```
> phi:=collect(phi,[cos(theta),sin(theta),r]);
```

$$\phi := \left(\_C4 \_C1\, r + \frac{\_C4 \_C2}{r} \right) \cos(\theta) + \left(\_C3 \_C1\, r + \frac{\_C3 \_C2}{r} \right) \sin(\theta)$$

To simplify the coefficients in ϕ, Benny makes the following substitutions.

```
> phi:=subs(_C3=a/_C1,_C2=b*_C1/a,_C4=c/_C1,phi);
```

$$\phi := \left(c\,r + \frac{c\,b}{a\,r} \right) \cos(\theta) + \left(a\,r + \frac{b}{r} \right) \sin(\theta)$$

```
> phi:=algsubs(c*b=a*d,phi);
```

$$\phi := \cos(\theta) \left(c\,r + \frac{d}{r} \right) + \left(a\,r + \frac{b}{r} \right) \sin(\theta)$$

To determine the four unknown coefficients a, b, c, and d, four boundary conditions are required. On the inner boundary, $r = 10$ cm, the coefficient of the $\cos\theta$ term must equal 15, while the coefficient of the $\sin\theta$ term must equal zero. This yields two boundary conditions, labeled *bc1* and *bc2*.

```
> bc1:=eval(coeff(phi,cos(theta)),r=10)=15;
```

$$bc1 := 10\,c + \frac{d}{10} = 15$$

```
> bc2:=eval(coeff(phi,sin(theta)),r=10)=0;
```

$$bc2 := 10\,a + \frac{b}{10} = 0$$

On the outer boundary, $r = 20$ cm, the coefficient of the $\sin\theta$ term must equal 30, while the coefficient of the $\cos\theta$ term is equal to zero. This yields two more boundary conditions, *bc3* and *bc4*.

```
> bc3:=eval(coeff(phi,sin(theta)),r=20)=30;
```

$$bc3 := 20\,a + \frac{b}{20} = 30$$

```
> bc4:=eval(coeff(phi,cos(theta)),r=20)=0;;
```

$$bc4 := 20\,c + \frac{d}{20} = 0$$

The four boundary condition equations are solved for a, b, c, and d,

```
> coefficients:=solve({bc1,bc2,bc3,bc4},{a,b,c,d});
```

$$coefficients := \left\{ d = 200,\ a = 2,\ c = \frac{-1}{2},\ b = -200 \right\}$$

which are assigned to produce the final solution ϕ to the potential problem.

```
> assign(coefficients): phi:=phi;
```

$$\phi := \cos(\theta) \left(-\frac{r}{2} + \frac{200}{r} \right) + \left(2\,r - \frac{200}{r} \right) \sin(\theta)$$

As requested by Professor Nerd, Benny will now use ϕ to plot the equipotentials in the annular region between $r = 10$ and 20. He first converts the potential into Cartesian coordinates by substituting $r = \sqrt{x^2 + y^2}$, $\cos\theta = x/\sqrt{x^2 + y^2}$, and $\sin\theta = y/\sqrt{x^2 + y^2}$, into ϕ.

```
> phi:=subs({r=sqrt(x^2+y^2),cos(theta)=x/sqrt(x^2+y^2),
           sin(theta)=y/sqrt(x^2+y^2)},phi):
```

He then creates a piecewise potential function Φ equal to ϕ for $100 \le x^2 + y^2 \le 400$ and zero otherwise. (The Φ output is suppressed here.)

```
> Phi:=piecewise(100<=x^2+y^2 and x^2+y^2<=400,phi,0);
```

A functional operator F is formed, which uses the `implicitplot` command, to plot Φ in potential steps of $5\,i$, where i will be allowed to take on integer values.

```
> F:=i->implicitplot({Phi=5*i},x=-20..20,y=-20..20,
         scaling=constrained,grid=[100,100]):
```

The sequence of constant-potential plots is then displayed for $i = -6$ to $+6$, the result being shown in Figure 5.8.

```
> display({seq(F(i),i=-6..6)});
```

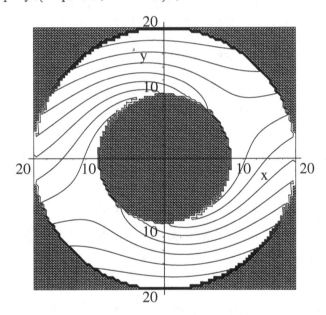

Figure 5.8: Equipotential lines for the circular annulus.

Although, he hasn't bothered to label the equipotential lines, Benny feels that the plot already conveys a better sense of the equipotentials than could be gained by staring at the formula, simple as it is. He has left the labeling of the equipotentials for you to carry out as a problem.

PROBLEMS:
Problem 5-15: Labeling of equipotential lines
Using the `textplot` command, add appropriate potential values to Figure 5.8 so that the equipotential lines are clearly identified.

Problem 5-16: Split potential

Along the circumference of a circle of radius $r = 1$, the potential $\phi(r, \theta)$ has the value $\phi = 1$ for the angular range $0 < \theta < \pi$ and $\phi = 0$ when $\pi < \theta < 2\pi$. Determine $\phi(r, \theta)$ for $r < 1$ and for $r > 1$. Plot the equipotentials inside and outside the circle.

Problem 5-17: Temperature distribution

A plate has the form of an annular region bounded by two concentric circles $r_1 = 1$ m and $r_2 = 2$ m. The temperature in degrees Celsius along r_1 is $T_1 = 75 \sin \theta$ and along r_2 is $T_2 = 60 \cos \theta$. Determine the steady-state temperature at each point of the annular region and plot the isotherms.

Problem 5-18: Alternating surface potential

An infinitely long hollow conducting cylinder of radius $r = a$ is divided into equal quarters, alternate segments being held at the potentials $+V$ and $-V$. Using the two-dimensional form of Laplace's equation, determine the potential inside the cylinder. Taking $a = 1$ and $V = 1$, plot the equipotential surfaces.

Problem 5-19: The bipolar coordinate system

For the two-dimensional bipolar coordinate system, use `?coords` to find out how bipolar and Cartesian coordinates are related. Plot the bipolar coordinate system, using options of your own choice. Suggest an electrostatic boundary value problem in which this coordinate system would be useful.

Problem 5-20: Three-dimensional coordinate plots

Use `coordplot3d` to plot the surfaces corresponding to holding each coordinate equal to a constant value for the following three-dimensional systems:
(a) cylindrical; (b) spherical; (c) paraboloidal; (d) sixsphere.

Problem 5-21: Entering Laplace's equation

Making use of the VectorCalculus package, determine the form of Laplace's equation for the scalar field ψ in the following coordinate systems:

(a) cylindrical, with $\psi = \psi(r, \theta, z)$;

(b) spherical, with $\psi = \psi(r, \theta, \phi)$;

(c) bispherical, with $\psi = \psi(\xi, \eta, \phi)$;

(d) paraboidal, with $\psi = \psi(\eta, \xi, \phi)$;

(e) prolate spheroidal, with $\psi = \psi(u, v, \phi)$;

(f) elliptic, with $\psi = \psi(u, v)$.

Problem 5-22: Solving Laplace's equation

Solve Laplace's equation in the following coordinate systems. In each case, look at what Maple does as each additional option is included, ending up with the complete product solution:

(a) cylindrical, with $\psi = \psi(r, \theta, z)$;

(b) spherical, with $\psi = \psi(r, \theta, \phi)$;

(c) paraboloidal, with $\psi = \psi(\eta, \xi, \phi)$.

5.2.4   Hugo and the Atomic Bomb

If the radiance of a thousand suns were to burst forth at once in the sky, that would be like the splendor of the Mighty One.
Bhagavad Gita, a philosophical dialogue that is a sacred Hindu text, found in the *Mahabharata*, one of the ancient Sanskrit epics (250 BC–250 AD)

Hugo, who was formerly a scientist in country X, has emigrated to the New World in search of a better life. Unfortunately, he has been forced to temporarily drive taxis until he can find a job more suitable to his educational training. While waiting for his next fare, he tries to keep his mind sharp by carrying out model calculations on his laptop computer. At this particular moment, Hugo is working on the problem of the growth of the neutron density in a nuclear chain reaction. Let's eavesdrop on what Hugo is doing.

Hugo knows that if uranium nuclei are bombarded with neutrons, a given nucleus may absorb a neutron, resulting in the splitting of the uranium nucleus into two parts with the release of substantial energy as well as two or three of the neutrons that were already present in the nucleus. This splitting process is called *nuclear fission* and is the first step in a *chain reaction*. Whether the reaction will keep on going depends on how many of the released neutrons are available to initiate another fission process. The factor by which the number of neutrons increases between one step and the next in the chain reaction is called the *multiplication factor*. In a nuclear reactor, the multiplication factor is kept at unity (called the *critical* condition) by using boron or cadmium control rods to "soak up" excess neutrons. In this case, the chain reaction proceeds at a constant rate with a steady output of energy. If the multiplication factor is greater than unity (the *supercritical* condition), the chain reaction leads to a geometrically increasing number of fissions in a very short time interval with the accompanying release of an enormous amount of energy. A nuclear explosion takes place — the basis of the atomic bomb.

Hugo decides that he can learn more about the underlying role that the neutrons play in the chain reaction by modeling the time evolution of some specified initial neutron distribution inside a mass of fissionable material. For calculational purposes, he takes the mass to be cylindrical in shape with a radius $r = a$, the lower face of the cylinder at $z = 0$ and the upper face at $z = h$. For simplicity, Hugo takes the neutron density N (number of neutrons per unit volume) to be independent of the angular coordinate θ, i.e., $N = N(r, z, t)$. In the absence of any production of neutrons by fission, the neutron density would obey the linear diffusion equation. To account for the production of neutrons by fission, Hugo adds a neutron source term βN, where β is a positive rate constant, to the diffusion equation, viz.,

$$\frac{\partial N}{\partial t} = d \nabla^2 N + \beta N. \tag{5.10}$$

In order to use the cylindrical polar form of the Laplacian, he calls up the VectorCalculus package.

```
>  restart: with(plots): with(VectorCalculus):
```
The diffusion equation is entered with the Laplacian in cylindrical coordinates.

```
>  de:=diff(N(r,z,t),t)-beta*N(r,z,t)
      =d*Laplacian(N(r,z,t),'cylindrical'[r,theta,z]);
```

$$de := \frac{\partial}{\partial t} N(r, z, t) - \beta N(r, z, t)$$

$$= \frac{d\left(\left(\frac{\partial}{\partial r} N(r, z, t)\right) + r\left(\frac{\partial^2}{\partial r^2} N(r, z, t)\right) + r\left(\frac{\partial^2}{\partial z^2} N(r, z, t)\right)\right)}{r}$$

The diffusion equation *de* is solved with the **pdsolve** command, the **HINT** option being omitted, because Maple already assumes a general product solution. After the introduction of two separation constants $\_c_1$ and $\_c_2$, the PDE is separated into three ODEs which are solved and the product formed.

```
>  sol:=pdsolve(de,INTEGRATE,build);
```

$$sol := N(r, z, t)$$
$$= e^{(\sqrt{-c_2}\,z)} \_C5\, e^{(t\,d\,-c_1)}\, e^{(t\,d\,-c_2)}\, e^{(t\,\beta)} \_C1\, \text{BesselJ}(0, \sqrt{-\_c_1}\, r) \_C3$$
$$+ \frac{\_C5\, e^{(t\,d\,-c_1)}\, e^{(t\,d\,-c_2)}\, e^{(t\,\beta)} \_C1\, \text{BesselJ}(0, \sqrt{-\_c_1}\, r) \_C4}{e^{(\sqrt{-c_2}\,z)}}$$
$$+ e^{(\sqrt{-c_2}\,z)} \_C5\, e^{(t\,d\,-c_1)}\, e^{(t\,d\,-c_2)}\, e^{(t\,\beta)} \_C2\, \text{BesselY}(0, \sqrt{-\_c_1}\, r) \_C3$$
$$+ \frac{\_C5\, e^{(t\,d\,-c_1)}\, e^{(t\,d\,-c_2)}\, e^{(t\,\beta)} \_C2\, \text{BesselY}(0, \sqrt{-\_c_1}\, r) \_C4}{e^{(\sqrt{-c_2}\,z)}}$$

The solution *sol* involves a linear combination of a zeroth-order Bessel function of the first kind ($J_0(\sqrt{-\_c_1}\, r)$ in math notation) and a zeroth-order Bessel function of the second kind ($Y_0(\sqrt{-\_c_1}\, r)$). The latter Bessel function diverges at $r = 0$ at arbitrary time, so Hugo imposes the first boundary condition, removing terms involving **BesselY** from the rhs of the solution.

```
>  N:=remove(has,rhs(sol),BesselY); #bc1
```

$$N := e^{(\sqrt{-c_2}\,z)} \_C5\, e^{(t\,d\,-c_1)}\, e^{(t\,d\,-c_2)}\, e^{(t\,\beta)} \_C1\, \text{BesselJ}(0, \sqrt{-\_c_1}\, r) \_C3$$
$$+ \frac{\_C5\, e^{(t\,d\,-c_1)}\, e^{(t\,d\,-c_2)}\, e^{(t\,\beta)} \_C1\, \text{BesselJ}(0, \sqrt{-\_c_1}\, r) \_C4}{e^{(\sqrt{-c_2}\,z)}}$$

Hugo simplifies N by substituting $\_c_1 = -\mu^2$ and $\_c_2 = -\nu^2$ and applying the **simplify** command with the **symbolic** option.

```
>  N:=simplify(subs({_c[1]=-mu^2,_c[2]=-nu^2},N),symbolic);
```

$$N := \_C5\, \_C1\, \text{BesselJ}(0, \mu\, r)\, e^{(-t\,d\,\mu^2 - t\,d\,\nu^2 + t\,\beta)} \left(\_C3\, e^{(\nu\,z\,I)} + \_C4\, e^{(-I\,\nu\,z)}\right)$$

The complex form is split into real and imaginary parts with the complex evaluation command.

```
>  N:=evalc(N);
```

$$N := \_C5 \, \_C1 \, \text{BesselJ}(0, \, \mu \, r) \, e^{(-t \, d \, \mu^2 - t \, d \, \nu^2 + t \, \beta)} \, (\_C3 \cos(\nu \, z) + \_C4 \cos(\nu \, z))$$
$$+ \_C5 \, \_C1 \, \text{BesselJ}(0, \, \mu \, r) \, e^{(-t \, d \, \mu^2 - t \, d \, \nu^2 + t \, \beta)} \, (\_C3 \sin(\nu \, z) - \_C4 \sin(\nu \, z)) \, I$$

Hugo assumes that the neutron density at the surface of the cylindrical mass is zero. Thus, since it doesn't go to zero at $z = 0$, the $\cos(\nu \, z)$ term is removed (a second boundary condition) from N and the result factored.

```
>  N:=factor(remove(has,N,cos)); #bc2
```

$$N := \_C5 \, \_C1 \, \text{BesselJ}(0, \, \mu \, r) \, e^{(-t \, (-\beta + d \, \mu^2 + d \, \nu^2))} \sin(\nu \, z) \, (-\_C4 + \_C3) \, I$$

In the next command line, Hugo removes the "ugly" Maple coefficient combination from N by forming the product of the 4th, 5th, and 6th operands of N, the other operands corresponding to the various constants.

```
>  N2:=op(4,N)*op(5,N)*op(6,N);
```

$$N2 := \text{BesselJ}(0, \, \mu \, r) \, e^{(-t \, (-\beta + d \, \mu^2 + d \, \nu^2))} \sin(\nu \, z)$$

Since the neutron density is zero on the cylindrical surface $r = a$ and at the end $z = h$, two more boundary conditions must be imposed. At $r = a$, $J_0(\mu \, a) = 0$, so that $\mu = \lambda_m/a$, where λ_m is the mth zero of J_0. These zeros may be found with the command `BesselJZeros(0,m)`, where $m = 1, 2, \ldots$. At $z = h$, $\sin(\nu \, h) = 0$, so $\nu = n \, \pi/h$, with n a positive integer. This pair of boundary condition relations is substituted into $N2$.

```
>  N2:=subs({mu=BesselJZeros(0,m)/a,nu=n*Pi/h},N2); #bcs
```

$$N2 := \text{BesselJ}\left(0, \, \frac{\text{BesselJZeros}(0, \, m) \, r}{a}\right)$$
$$e^{\left(-t \, \left(-\beta + \frac{d \, \text{BesselJZeros}(0, \, m)^2}{a^2} + \frac{d \, n^2 \, \pi^2}{h^2}\right)\right)} \sin\left(\frac{n \, \pi \, z}{h}\right)$$

By the principle of linear superposition, the complete solution will be

$$N(r, z, t) = \sum_{m=1}^{\infty} \sum_{n=1}^{\infty} C_{m,n} \, N2 \equiv \sum_{m=1}^{\infty} \sum_{n=1}^{\infty} N_{m,n},$$

with the $C_{m,n}$ determined by the initial neutron density $N(r, z, 0) = f$. Evaluating $N2$ at $t = 0$, and labeling the result g,

```
>  g:=eval(N2,t=0);
```

$$g := \text{BesselJ}\left(0, \, \frac{\text{BesselJZeros}(0, \, m) \, r}{a}\right) \sin\left(\frac{n \, \pi \, z}{h}\right)$$

one has

$$N(r, z, 0) = \sum_{m=1}^{\infty} \sum_{n=1}^{\infty} C_{m,n} \, g = f.$$

If we multiply both sides of $N(r, z, 0)$ by $r \, J_0(\lambda_{m'} \, r/a) \sin(n' \, \pi \, z/h)$ and integrate r from 0 to a and z from 0 to h, the lhs will yield zero unless $m = m'$ and $n = n'$. Solving for $C_{m',n'}$, and dropping the primes, the coefficients can be calculated from

$$C_{m,n} = \frac{\int_0^a \int_0^h r \, f \, g \, dr \, dz}{\int_0^a \int_0^h r \, g^2 \, dr \, dz}.$$

For an initial neutron density satisfying the boundary conditions, Hugo takes f to be of the following simple structure:

```
> f:=(1-r^2/a^2)*sin(Pi*z/h);
```

$$f := \left(1 - \frac{r^2}{a^2}\right) \sin\left(\frac{\pi z}{h}\right)$$

The coefficients $C_{m,n}$ are calculated, and then the Fourier term $N_{m,n}$.

```
> C[m,n]:=int(int(r*f*g,z=0..h),r=0..a)
         /int(int(r*g^2,z=0..h),r=0..a):

> N[m,n]:=C[m,n]*N2;
```

For integer $n > 1$, $N_{m,n}$ should be equal to zero, since only an $n = 1$ term occurs in f. Hugo checks that this is the case.

```
> simplify(N[m,n]) assuming n::integer,n>1;
```

$$0$$

To determine $N_{m,1}$ greater care must be exercised, and the limit taken as $n \to 1$. The result then is simplified with respect to the exponentials.

```
> N[m,1]:=simplify(limit(N[m,n],n=1),exp);
```

$$N_{m,1} := 8 \sin\left(\frac{\pi z}{h}\right) \text{BesselJ}\left(0, \frac{\text{BesselJZeros}(0, m) r}{a}\right)$$

$$e^{\left(-\frac{(-\beta a^2 h^2 + d\,\text{BesselJZeros}(0, m)^2 h^2 + d\,\pi^2 a^2) t}{a^2 h^2}\right)}$$

$$\Big/ \text{BesselJ}(1, \text{BesselJZeros}(0, m)) \text{BesselJZeros}(0, m)^3$$

The total neutron density at time t then is $\sum_{m=1}^{\infty} N_{m,1}$. To plot the density, Hugo takes the nominal values $a = 1$, $h = 1$, and $d = 1$. In the time-dependent part of the density, the coefficient of t is $\beta - d(\lambda_m^2/a^2 + \pi^2/h^2)$. The largest possible positive value of this coefficient occurs when $m = 1$, corresponding to the first zero of J_0. In this case $\lambda_1 \approx 2.405$. If

$$\beta < \beta_c = d\left(\frac{\lambda_1^2}{a^2} + \frac{\pi^2}{h^2}\right),$$

all terms in the series representation of the neutron density will decay with time. If β is larger than the critical value, β_c, exponential growth of the neutron density will take place. If the cylinder dimensions a and h are such that $\beta = \beta_c$, then the corresponding mass M_c equals $\rho(\pi a^2) h$, where ρ is the mass density of the fissionable material, is called the "critical mass." According to Ohanian [Oha85], the critical mass for uranium 235 is about 53 kg and for a spherical shape corresponds to a diameter of 18 cm.

For his nominal input parameters, Hugo finds that the critical value of β is slightly less than 15.7. To simulate growth that starts out relatively slowly but then increases rapidly, Hugo takes $\beta = 15.7$. The total neutron density is evaluated with the above parameters, three terms being kept in the series.

```
>  Density:=sum(eval(N[m,1],{a=1,h=1,d=1,beta=15.7}),m=1..3);
```

$Density :=$

$1.108022261 \sin(3.141592654\, z) \, \text{BesselJ}(0.,\, 2.404825558\, r)\, e^{(0.047209632\, t)}$

$- 0.1397775054 \sin(3.141592654\, z) \, \text{BesselJ}(0.,\, 5.520078110\, r)\, e^{(-24.64086674\, t)}$

$+ 0.04547647069 \sin(3.141592654\, z) \, \text{BesselJ}(0.,\, 8.653727913\, r)\, e^{(-69.05661119\, t)}$

The numerical coefficient of t in the first term of the output is slightly positive while the coefficients of t in the other terms are large negative numbers. Thus, the first term in the series will grow with time and all remaining terms exponentially decay. If the neutron density grows fast enough, an uncontrolled chain reaction, i.e., an explosion, will occur.

To create a dynamic 3-dimensional picture of the explosive growth in neutron density (the "nuclear explosion"), Hugo makes use of the animation command.

```
>  animate(plot3d,[Density,r=-1..1,z=0..1],t=0..100,
   frames=25,orientation=[65,40],shading=zhue,
   style=patchnogrid,grid=[25,25],axes=framed);
```

As Hugo contentedly looks at the "explosion" on his computer screen, running in the loop mode and with 200% zoom magnification, he hears a gruff voice.

"Hey Buddy, if you don't mind, I am running late and have to get to the airport."

Interrupted by harsh reality, Hugo puts his computer away, but hopes that one of his job applications will pan out soon.

PROBLEMS:

Problem 5-23: Spherical mass
Carry out a calculation similar to that in the text recipe for a fissionable spherical mass of radius $r = a$ and an initial neutron density $f(r) = 1 - r^2/a^2$. Make an animated plot of the neutron density inside the sphere when the mass slightly exceeds the critical mass.

Problem 5-24: Semicircular plate
A thin semicircular plate of radius $r = 1$ has its edges held at zero temperature and its flat faces insulated. If the initial temperature distribution inside the plate is $T(r,\theta,0) = 100\, r^2 \cos(2\,\theta)$, determine the temperature $T(r,\theta,t)$ for $t > 0$. Taking the diffusion constant $d = 1$, animate the temperature profile.

Problem 5-25: Temperature distribution
A solid has the shape of an infinitely long quarter-cylinder of radius $r = 1$ and diffusion constant d. The flat sides are insulated, so no heat flows through them, while the curved surface is kept at $100°C$. Assuming that the temperature initially varies as the fourth power of the distance from the axis, find the temperature distribution at any point inside the solid for $t > 0$. Choosing nominal parameter values, create an animated plot of the temperature profile.

Problem 5-26: Cylindrical temperature profile
A cylinder of unit radius and unit height has its circular ends $z = 0$ and $z = 1$

kept at $T = 0$ and $T = 1$, respectively, while the circular surface is kept at $T = 1$. Determine the steady-state temperature profile inside the cylinder and plot the contours of equal temperature.

Problem 5-27: Elliptic cross section
An infinitely long bar of elliptic cross section has its curved surface, $x^2 + 4\,y^2 = 1$, kept at $T = 0°$. Determine the temporal evolution of T in the cylinder if initially $T = 100°$ everywhere inside and the diffusion constant $d = 1$. Animate the result.

5.2.5 Hugo Prepares for His Job Interview

I'm notorious for giving a bad interview. I'm an actor and I can't help but feel I'm boring when I'm on as myself.
Rock Hudson, American movie actor (1925–1985)

Hugo has been invited by International Hydrodynamics Inc. to interview for a research position with their company, which specializes in designing streamlined hulls for surface vessels as well as underwater craft. As part of his preparation, Hugo decides to create some files for the interview that demonstrate his skills at simulating liquid flow around a variety of rigid geometrical shapes. As a "warm-up" exercise, he recalls from his undergraduate days a problem involving uniform incompressible fluid flow around a sphere, a problem that can be solved analytically. Hugo remembers that the fluid can be characterized by a velocity potential U that satisfies Laplace's equation. The velocity $\vec{V}$ of a fluid element is then given by $\vec{V} = -\nabla U$.

Since a spherical boundary of, say, radius a is involved, Hugo chooses to use spherical coordinates (r, θ, ϕ) with the origin at the center of the sphere and the physicist's convention that θ is measured from the positive z-axis and ϕ from the positive x-axis. Then, the Cartesian and spherical polar coordinates are connected by the relations $x = r \sin\theta \cos\phi$, $y = r \sin\theta \sin\phi$, and $z = r \cos\theta$. The range of r is from 0 to ∞, θ from 0 to π, and ϕ from 0 to 2π.

At a distance r far from the sphere, the fluid flow is assumed to be uniform and directed along the z-axis. The asymptotic (large r) form of the velocity is $\vec{V} = V_0\,\hat{e}_z$ where V_0 is the undisturbed fluid speed. Thus, since

$$\vec{V} = -\frac{\partial U}{\partial x}\hat{e}_x - \frac{\partial U}{\partial y}\hat{e}_y - \frac{\partial U}{\partial z}\hat{e}_z = V_0\,\hat{e}_z, \tag{5.11}$$

then, on equating vector components and integrating, the asymptotic velocity potential is given (to within an arbitrary constant) by $U = -V_0\,z = -V_0\,r\cos\theta$. This will serve as one of the boundary conditions on the solution. The other boundary condition is at the surface of the sphere. If the surface is idealized to be absolutely rigid, it will acquire no momentum from the fluid. This implies that the normal component of the fluid velocity vector must vanish at $r = a$. Writing the gradient operator in spherical polar coordinates, this means that the velocity potential must satisfy the boundary condition $\partial U/\partial r|_{r=a} = 0$.

To solve Laplace's equation, Hugo loads the VectorCalculus package,

> restart: with(plots): with(VectorCalculus):

and inputs Laplace's equation in spherical coordinates. He notes that the problem has rotational symmetry about the z-axis, the direction of fluid flow, so that there should be no ϕ dependence in the final solution. Accordingly, he takes $U = U(r, \theta)$.

> LE:=expand(Laplacian(U(r,theta),'spherical'[r,theta,phi]))=0;

$$LE := \frac{2\left(\frac{\partial}{\partial r} U(r, \theta)\right)}{r} + \left(\frac{\partial^2}{\partial r^2} U(r, \theta)\right) + \frac{\cos(\theta)\left(\frac{\partial}{\partial \theta} U(r, \theta)\right)}{r^2 \sin(\theta)} + \frac{\frac{\partial^2}{\partial \theta^2} U(r, \theta)}{r^2} = 0$$

With no HINT provided, a general product solution is built with the pdsolve command, the answer involving one separation constant, $\_c_1$, and four arbitrary constants. The result is then expanded.

> sol:=expand(pdsolve(LE,INTEGRATE,build));

$$sol := U(r, \theta) = \frac{\_C1\, r^{(1/2\,\sqrt{1+4\,\_c_1})}\, \_C3\, \text{LegendreP}\left(\frac{1}{2}\sqrt{1 + 4\,\_c_1} - \frac{1}{2}, \cos(\theta)\right)}{\sqrt{r}}$$

$$+ \frac{\_C1\, r^{(1/2\,\sqrt{1+4\,\_c_1})}\, \_C4\, \text{LegendreQ}\left(\frac{1}{2}\sqrt{1 + 4\,\_c_1} - \frac{1}{2}, \cos(\theta)\right)}{\sqrt{r}}$$

$$+ \frac{\_C2\, r^{(-1/2\,\sqrt{1+4\,\_c_1})}\, \_C3\, \text{LegendreP}\left(\frac{1}{2}\sqrt{1 + 4\,\_c_1} - \frac{1}{2}, \cos(\theta)\right)}{\sqrt{r}}$$

$$+ \frac{\_C2\, r^{(-1/2\,\sqrt{1+4\,\_c_1})}\, \_C4\, \text{LegendreQ}\left(\frac{1}{2}\sqrt{1 + 4\,\_c_1} - \frac{1}{2}, \cos(\theta)\right)}{\sqrt{r}}$$

The solution is expressed in terms of Legendre functions of the first kind (LegendreP) and of the second kind (LegendreQ). The former are well-behaved polynomial functions of $\cos \theta$, but the latter diverge at the ends of the angular range and therefore must be removed on physical grounds.

> U:=remove(has,rhs(sol),LegendreQ);

$$U := \frac{\_C1\, r^{(1/2\,\sqrt{1+4\,\_c_1})}\, \_C3\, \text{LegendreP}\left(\frac{1}{2}\sqrt{1 + 4\,\_c_1} - \frac{1}{2}, \cos(\theta)\right)}{\sqrt{r}}$$

$$+ \frac{\_C2\, r^{(-1/2\,\sqrt{1+4\,\_c_1})}\, \_C3\, \text{LegendreP}\left(\frac{1}{2}\sqrt{1 + 4\,\_c_1} - \frac{1}{2}, \cos(\theta)\right)}{\sqrt{r}}$$

In standard math notation, the Legendre polynomials of the first kind would be written as $P_n(\cos\theta)$, where $n = \frac{1}{2}\sqrt{1 + 4\_c_1} - \frac{1}{2}$. To express U in the Maple equivalent of the math notation, Hugo substitutes $\_c_1 = -\frac{1}{4} + (n + \frac{1}{2})^2$, and simplifies the square root that would otherwise appear by including the symbolic option.

```
> U:=simplify(subs(_c[1]=-1/4+(n+1/2)^2,U),symbolic);
```

$$U := \frac{\_C3 \, \text{LegendreP}(n, \cos(\theta)) \, (\_C1 \, r^{(n+1/2)} + \_C2 \, r^{(-n-1/2)})}{\sqrt{r}}$$

To simplify the notation, and without any loss of generality, Hugo sets $\_C3 = 1$, $\_C1 = A$, and $\_C2 = B$, in U. On some executions of the recipe the coefficient $\_C3$ in U is replaced with $\_C4$, so the latter is also set equal to one.

```
> U:=subs({_C3=1,_C4=1,_C1=A,_C2=B},U);
```

$$U := \frac{\text{LegendreP}(n, \cos(\theta)) \, (A \, r^{(n+1/2)} + B \, r^{(-n-1/2)})}{\sqrt{r}}$$

Only two coefficients, A and B, remain to be evaluated, so the two boundary conditions mentioned earlier are used. The radial derivative of U must be zero at $r = a$, which is now entered, and solved for B in terms of A.

```
> bc:=eval(diff(U,r),r=a)=0;
> B:=simplify(solve(bc,B));
```

$$B := \frac{n \, A \, a^{(2n+1)}}{n + 1}$$

Now Hugo recalls that the first few Legendre polynomial functions of the first kind are $P_0(\cos\theta) = 1$, $P_1(\cos\theta) = \cos\theta$, and $P_2(\cos\theta) = \frac{1}{2}(3\cos^2\theta - 1)$, ... Although the general solution will involve a sum over n from 1 to ∞, only the $n = 1$ term has the correct $\cos\theta$ dependence to match the asymptotic form of the velocity potential. The coefficients of all terms for which $n \neq 1$ must be equal to zero. As $r \to \infty$, $U \to A \, P_1(\cos\theta) \, r = -V_0 \, r\cos\theta$, so $A = -V_0$. The potential U is now evaluated with $A = -V_0$ and $n = 1$, and simplified.

```
> U:=simplify(eval(U,{A=-V[0],n=1}));
```

$$U := -\frac{1}{2} \frac{\cos(\theta) \, V_0 \, (2 \, r^3 + a^3)}{r^2}$$

So the velocity potential U is now completely known. In standard texts, the potential and velocity distributions would probably only be sketched, if drawn at all. Hugo knows that this is not good enough for his prospective employers. He decides to create a plot of the region outside the sphere that shows the equipotential lines as well as the velocity vectors $\vec{V} = -\nabla U$. For graphing purposes, he chooses $V_0 = 1$ and $a = 1$, and switches to Cartesian coordinates, taking $x = 0$ so that a two-dimensional plot can be made.

```
> U:=subs({V[0]=1,a=1,cos(theta)=z/sqrt(z^2+y^2),
          r=sqrt(z^2+y^2)},U):
```

Since the region of interest is outside the sphere, Hugo forms a piecewise function $U2$ that is equal to U for $z^2 + y^2 \geq 1$, and zero otherwise.

```
> U2:=piecewise(z^2+y^2>=1,U,0);
```

$$U2 := \begin{cases} -\dfrac{z\,(2\,(z^2 + y^2)^{(3/2)} + 1)}{2\,(z^2 + y^2)^{(3/2)}} & 1 \le z^2 + y^2 \\ 0 & \textit{otherwise} \end{cases}$$

A functional operator **F**, which makes use of the `implicitplot` command, is formed to plot the equipotentials in steps of $0.25\,i$, where i will be taken to be an integer. Instead of the 25×25 default grid, which is too coarse here, Hugo take the grid to be 100×100.

```
>  F:=i->implicitplot(U2=.25*i,z=-2..2,y=-2..2,grid=[100,100]):
```

Next, Hugo wants to form a graph of the velocity vectors. He calculates the velocity in the z-y plane and again forms piecewise functions to give the velocity components outside the sphere.

```
>  Vel:=Gradient(-U,'cartesian'[z,y]);
>  Ve[1]:=piecewise(z^2+y^2>=1,Vel[1],0):
>  Ve[2]:=piecewise(z^2+y^2>=1,Vel[2],0):
```

The `fieldplot` command is used to produce a graph of the velocity vectors.

```
>  gr:=fieldplot([Ve[1],Ve[2]],z=-2..2,y=-2..2,grid=[18,18],
       arrows=MEDIUM,color=blue):
```

The velocity vectors and the equipotential lines corresponding to $i = -8$ to 8 ($U = -2$ to $+2$ in steps of 0.25) are displayed in Figure 5.9.

```
>  display({gr,seq(F(i),i=-8..8)},axes=boxed,
     scaling=constrained);
```

Looking at the figure, Hugo can easily "see" the fluid flow around the sphere, with the velocity vectors perpendicular to the equipotentials and varying in magnitude near the sphere.

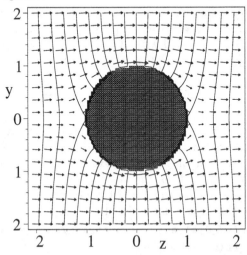

Figure 5.9: Velocity vectors and equipotentials for fluid flow around a sphere.

With this simple example under his belt, Hugo feels confident that if he can come up with some fluid flow examples for different hull shapes, he might make a good impression on his interviewers and be hired. He realizes that for more complicated shapes he will probably have to use Maple's numerical capability.

So let's leave Hugo to finish his job interview preparation, so he doesn't pull a Rock Hudson and give a bad interview.

PROBLEMS:

Problem 5-28: Split potential

On the surface of a hollow sphere of radius a, the electric potential is

$$\Phi(a, \theta) = \begin{cases} +V, & 0 \leq \theta < \pi/2, \\ -V, & \pi/2 < \theta \leq \pi. \end{cases}$$

By solving Laplace's equation in spherical coordinates, determine $\Phi(r, \theta)$ for both the regions $r < a$ and $r > a$. Plot the equipotential lines and electric field vectors for both regions, taking $a = 1$ and $V = 1$.

Problem 5-29: Another split potential

Two thin concentric spherical shells of radius a and $b > a$ are each divided into two hemispheres by the same horizontal plane. The potential on the top hemisphere of the inner shell is V and the potential on the bottom hemisphere is zero, whereas the potential on the top hemisphere of the outer shell is zero and the potential on the lower hemisphere is V. Using Laplace's equation in spherical polar coordinates, determine the potential in the region between the spheres. Choosing your own parameter values, plot the equipotential lines in this region.

Problem 5-30: Temperature of an iron sphere

A uniform solid iron sphere of radius 20 cm is heated to a temperature of $100°C$ throughout. Its surface is to be kept at the constant temperature $0°C$.

(a) Explicitly determine $T(\vec{r}, t)$.

(b) If the heat diffusion coefficient $d = 0.185$ cgs units, find the temperature of the center of the sphere 15 minutes after the cooling has begun.

(c) Plot the constant-temperature profiles at this time.

(d) Animate the temperature profile inside the sphere.

Problem 5-31: Another spherical temperature distribution

The temperature of the surface of a solid sphere of radius a is prescribed to be $T = T_0 (1 - \cos\theta)$. Find the steady-state temperature distribution at any point inside the sphere. Plot the constant-temperature profiles inside the sphere, taking $a = 1$ and $T_0 = 100$.

Chapter 6

Linear PDE Models. Part 2

In the previous chapter, the computer algebra recipes concentrated on mathematical models formulated in terms of the diffusion and Laplace equations. In this continuation of our study of linear PDE models, the wave equation and a variety of models involving semi-infinite and infinite domains are presented.

6.1 Wave Equation Models

The wave equation,

$$\nabla^2 \psi(\vec{r}, t) = \frac{1}{c^2} \frac{\partial^2 \psi(\vec{r}, t)}{\partial t^2}, \tag{6.1}$$

with ∇^2 the Laplace operator, ψ a scalar or vector function of position $\vec{r}$ and time t, and c the wave velocity, applies to many different types of waves, e.g.,

- electromagnetic waves (with ψ a vector);
- sound waves;
- tidal waves in an incompressible fluid;
- elastic waves in a solid;
- vibrations of elastic strings and membranes.

In our first example, Vectoria plucks the humble string.

6.1.1 Vectoria Encounters Simon Legree

"We called him Tortoise because he taught us," said the Mock Turtle angrily. "Really you are very dull!"
Lewis Carroll, *Alice's Adventures in Wonderland*, 1865

It is some time later in the undergraduate career of Vectoria, the physics student who has been featured in several of our earlier stories. Recall that, inspired by the movie *Casablanca*, Vectoria used Maple to check and animate a textbook formula for the transverse motion of an initially horizontal piano string that has been struck. Since then Vectoria has progressed in her studies and is currently enrolled in a mathematical physics course. She has begun to learn how to solve wave equation problems using the separation of variables method.

Professor Simon Legree,[1] who is teaching the course, has a reputation for assigning large numbers of often difficult problems, so Vectoria decides once again to let Maple help her in deriving the solutions. Although she could do the problems by hand, it seems smarter in the long run to develop a computer algebra approach to lighten the workload and to avoid mathematical mistakes. On talking to the professor, she finds out that, surprisingly, Legree not only agrees but suggests that it might be wise to start with a relatively simple problem, before tackling the more difficult ones. Amazed that Legree, despite his hard-nosed reputation, has been so helpful, Vectoria decides to follow his advice and selects a problem involving the transverse motion of a string that has been plucked and released from rest.

Since she doesn't yet know how to include stiffness in the string, she opts to make use of the linear wave equation that neglects stiffness. The horizontal string is fixed at its endpoints, $x = 0$ and $x = L > 0$, and is given an initial transverse displacement $\psi(x, t = 0) = 2\,h\,x/L$ for $0 \leq x \leq L/2$ and $\psi(x, 0) = 2\,h\,(L - x)/L$ for $L/2 \leq x \leq L$. Vectoria recognizes that this is a triangular profile with a maximum displacement h at the center $(x = L/2)$ of the string. Professor Legree has also stressed that his marking assistant has been instructed not to give full credit for a problem solution unless it is accompanied by some sort of meaningful plot as well as some pertinent discussion.

So Vectoria, anticipating that she will animate the vibrational motion of the string, begins her recipe with a call to the plots package, and enters the wave equation *WE* for the tranverse displacement $\psi(x, t)$ of the string at time t.

```
>   restart: with(plots):
>   WE:=diff(psi(x,t),x,x)=(1/c^2)*diff(psi(x,t),t,t);
```

$$WE := \frac{\partial^2}{\partial x^2}\,\psi(x,\,t) = \frac{\frac{\partial^2}{\partial t^2}\,\psi(x,\,t)}{c^2}$$

On applying the `pdsolve` command to *WE* without any options,

```
>   pdsolve(WE);
```

$$\psi(x,\,t) = \_F1(c\,t + x) + \_F2(c\,t - x)$$

Vectoria finds that the solution is given by $\psi(x, t) = \_F1(c\,t + x) + \_F2(c\,t - x)$, where $\_F1$ and $\_F2$ are arbitrary functions. This is the well-known general solution of the wave equation, which is not too useful for solving the present problem with specified boundary and initial conditions. Clearly, a hint should be provided. She could be quite general and enter `HINT=f(x)*g(t)`, but she realizes that sine and cosine functions clearly satisfy the wave equation. Since $\sin(k\,x)$, where k is an undetermined constant, is equal to zero at $x = 0$ and thus satisfies the boundary condition there, she provides the `HINT=sin(k*x)*g(t)` and integrates and builds up the solution.

---

[1]Unfortunately, Mrs. Legree, when naming her newborn son, chose Simon as a first name. She was unaware that Simon Legree was the cruel plantation owner in Harriet Beecher Stowe's novel, *Uncle Tom's Cabin*.

```
>  sol:=pdsolve(WE,HINT=sin(k*x)*g(t),INTEGRATE,build);
```

$$sol := \psi(x,\, t) = \sin(k\,x)\,\_C1\,\sin(c\,k\,t) + \sin(k\,x)\,\_C2\,\cos(c\,k\,t)$$

The string is initially at rest, so the transverse velocity $\dot{\psi}(x, 0)$ is equal to 0. Mentally differentiating the above output, clearly only the $\cos(c\,k\,t)$ term should be retained, since its derivative vanishes at $t = 0$, whereas the derivative of the $\sin(c\,k\,t)$ term does not. So Vectoria selects the term on the rhs of the solution containing the cosine term.

```
>  sol2:=select(has,rhs(sol),cos);
```

$$sol2 := \sin(k\,x)\,\_C2\,\cos(c\,k\,t)$$

The displacement of the string must also vanish at $x = L$ for arbitrary times, so one must have $\sin(k\,L) = 0$, which yields $k = n\,\pi/L$, with n a positive integer. She substitutes this result into *sol2*, and temporarily removes any coefficients by setting $\_C1$ and $\_C2$ equal to 1.

```
>  F[n]:=subs({k=n*Pi/L,_C1=1,_C2=1},sol2);
```

$$F_n := \sin\left(\frac{n\,\pi\,x}{L}\right)\cos\left(\frac{c\,n\,\pi\,t}{L}\right)$$

The general solution satisfying the boundary conditions at $x = 0$ and L, and with zero initial transverse velocity, then is

$$\psi(x,t) = \sum_{n=1}^{\infty} A_n\,F_n = \sum_{n=1}^{\infty} A_n \sin\left(\frac{n\,\pi\,x}{L}\right)\cos\left(\frac{c\,n\,\pi\,t}{L}\right), \qquad (6.2)$$

where the coefficients A_n have to be determined from the initial string profile. This profile is given by the piecewise function f.

```
>  f:=piecewise(x<L/2,2*h*x/L,x>L/2,2*h*(L-x)/L);
```

$$f := \begin{cases} \dfrac{2\,h\,x}{L} & x < \dfrac{L}{2} \\[2mm] \dfrac{2\,h\,(L-x)}{L} & \dfrac{L}{2} < x \end{cases}$$

Evaluating equation (6.2) at $t = 0$ and equating to f, Vectoria sees that the orthogonality of the independent sine functions for different values of n yields

$$A_n = \int_0^L f \sin\left(n\,\pi\,x/L\right) dx \,\Big/ \int_0^L \sin^2(n\,\pi\,x/L)\,dx,$$

which Vectoria now calculates. In order for the piecewise integral in the numerator to actually be carried out, it is necessary to assume that $L > 0$.

```
> A[n]:=int(f*sin(n*Pi*x/L),x=0..L)/int(sin(n*Pi*x/L)^2,x=0..L)
       assuming L>0:
```

The nth Fourier term in the infinite series representing the string displacement is then $\psi_n = A_n\,F_n$, the result being simplified by assuming that n is an integer.

```
>  psi[n]:=simplify(A[n]*F[n]) assuming n::integer;
```

$$\psi_n := \frac{8\,h\sin\left(\dfrac{n\,\pi}{2}\right)\sin\left(\dfrac{n\,\pi\,x}{L}\right)\cos\left(\dfrac{c\,n\,\pi\,t}{L}\right)}{n^2\,\pi^2}$$

The transverse displacement of the string at arbitrary time t is $\psi(x, t) = \sum_{n=1}^{\infty} \psi_n$. For animation purposes, Vectoria takes $L = 10$, $h = 1$, and $c = 1$.

> L:=10: h:=1: c:=1:

To obtain a good approximation to the initial string profile, Vectoria keeps terms in the series up to $n = 25$, only the partial output being shown here.

> psi:=sum(psi[n],n=1..25);

$$
\psi := \frac{8\sin\left(\frac{\pi x}{10}\right)\cos\left(\frac{\pi t}{10}\right)}{\pi^2} - \frac{8}{9}\frac{\sin\left(\frac{3\pi x}{10}\right)\cos\left(\frac{3\pi t}{10}\right)}{\pi^2}
$$

$$
+ \frac{8}{25}\frac{\sin\left(\frac{\pi x}{2}\right)\cos\left(\frac{\pi t}{2}\right)}{\pi^2} + \cdots\cdots + \frac{8}{441}\frac{\sin\left(\frac{21\pi x}{10}\right)\cos\left(\frac{21\pi t}{10}\right)}{\pi^2}
$$

$$
- \frac{8}{529}\frac{\sin\left(\frac{23\pi x}{10}\right)\cos\left(\frac{23\pi t}{10}\right)}{\pi^2} + \frac{8}{625}\frac{\sin\left(\frac{5\pi x}{2}\right)\cos\left(\frac{5\pi t}{2}\right)}{\pi^2}
$$

The displacement ψ is animated over the time interval $t = 0$ to 20, with 100 equally spaced time frames.

> animate(plot,[psi,x=0..L],t=0..20,frames=100);

On running the animation command, Vectoria is at first surprised by the behavior of the string, but on thinking about it decides that it makes sense. After executing the recipe, does the behavior make sense to you? What should Vectoria write in her explanation of the wave motion?

PROBLEMS:

Problem 6-1: Another plucked string

A light homogeneous horizontal string of length L fixed at its ends initially has a parabolic shape with height h in the middle. If it is released from rest, solve the wave equation to determine its subsequent transverse displacement at arbitrary time t. Animate the vibrations for parameters of your own choice and choose enough frames to produce a smooth animation.

Problem 6-2: Struck piano string

If a horizontal piano string fixed at $x = 0$ and $x = L$ is struck in such a way that its initial displacement $\psi(x, 0)$ is zero and its initial transverse velocity is

$$
\dot{\psi}(x, 0) = \begin{cases} 4vx/L, & 0 < x < L/4, \\ (4v/L)(L/2 - x), & L/4 < x < L/2, \\ 0, & L/2 < x < L, \end{cases}
$$

solve the wave equation to determine the transverse displacement of the string for all times t. Taking $L = 20$ cm and $v = 5$ cm/s, animate the solution. Choose enough frames to produce a smooth animation. Discuss the results.

6.1.2 Homer's Jiggle Test

The fundamental cause of trouble in the world today is that the stupid are cocksure while the intelligent are full of doubt.
Bertrand Russell, British mathematician and philosopher (1872–1970)

Having successfully used Maple to solve the plucked string problem, Vectoria now tackles the first problem on Professor Legree's wave equation assignment. She is asked to determine the transverse displacement $\psi(x, y, t)$ of a light, horizontal, rectangular elastic membrane having sides of length a and $2a$ that is held fixed along all four edges. Specifically, the boundary conditions are $\psi(0, y, t) = \psi(a, y, t) = 0$ for $0 \leq y \leq 2a$ and $\psi(x, 0, t) = \psi(x, 2a, t) = 0$ for $0 \leq x \leq a$. The membrane is given an initial displacement

$$\psi(x, y, 0) \equiv f = \frac{4 h x^2 (a - x) y^3 (2a - y)}{a^7}$$

and is released from rest. Professor Legree also requests that the membrane displacement be animated for $a = 1$, $h = 1$, and wave velocity $c = 1$.

Vectoria realizes that this membrane problem is mathematically equivalent to two fixed-ends string problems in the x and y directions. So her "shortcut" approach to separating variables in the last recipe can be easily extended.

After loading the plots package, she enters the wave equation *WE* in two dimensions.

```
>   restart: with(plots):
>   WE:=diff(psi(x,y,t),x,x)+diff(psi(x,y,t),y,y)
        =(1/c^2)*diff(psi(x,y,t),t,t);
```

$$WE := \left(\frac{\partial^2}{\partial x^2} \psi(x, y, t)\right) + \left(\frac{\partial^2}{\partial y^2} \psi(x, y, t)\right) = \frac{\frac{\partial^2}{\partial t^2} \psi(x, y, t)}{c^2}$$

The pdsolve command is applied to *WE* with HINT=sin(p*x)*sin(q*y)*T(t), where p and q are constants. The assumed form satisfies the boundary conditions along the fixed edges $x = 0$ and $y = 0$.

```
>   sol:=pdsolve(WE,HINT=sin(p*x)*sin(q*y)*T(t),INTEGRATE,build);
```

$$sol := \psi(x, y, t) = \sin(p\,x) \sin(q\,y) \_C1 \sin(c\,\sqrt{p^2 + q^2}\,t)$$
$$+ \sin(p\,x) \sin(q\,y) \_C2 \cos(c\,\sqrt{p^2 + q^2}\,t)$$

The initial transverse velocity is zero, so once again only the term involving $\cos(c\,\sqrt{p^2 + q^2}\,t)$ is kept.

```
>   sol2:=select(has,rhs(sol),cos);
```

$$sol2 := \sin(p\,x) \sin(q\,y) \_C2 \cos(c\,\sqrt{p^2 + q^2}\,t)$$

At $x = a$, $\sin(p\,a) = 0$, so $p = m\,\pi/a$, where m is a positive integer. Similarly, at $y = 2a$, $\sin(q\,2a) = 0$, so $q = n\,\pi/(2a)$, n being a positive integer. Vectoria substitutes the p and q relations into *sol2*, and temporarily removes all coefficients by setting $\_C1 = \_C2 = 1$.

```
>  F[m,n]:=subs({p=m*Pi/a,q=n*Pi/(2*a),_C1=1,_C2=1},sol2);;
```

$$F_{m,n} := \sin\left(\frac{m\,\pi\,x}{a}\right) \sin\left(\frac{n\,\pi\,y}{2\,a}\right) \cos\left(c\sqrt{\frac{m^2\,\pi^2}{a^2} + \frac{n^2\,\pi^2}{4\,a^2}}\,t\right)$$

Since she will need it for determining the coefficients, Vectoria evaluates $F_{m,n}$ at $t = 0$, labeling the result $g_{m,n}$.

```
>  g[m,n]:=eval(F[m,n],t=0);
```

$$g_{m,n} := \sin\left(\frac{m\,\pi\,x}{a}\right) \sin\left(\frac{n\,\pi\,y}{2\,a}\right)$$

The initial profile of the rectangular membrane is entered.

```
>  f:=4*h*x^2*(a-x)*y^3*(2*a-y)/a^7;
```

$$f := \frac{4\,h\,x^2\,(a - x)\,y^3\,(2\,a - y)}{a^7}$$

Now, the general form of the transverse displacement at time t will be given by the following linear superposition,

$$\psi(x, y, t) = \sum_{m=1}^{\infty} \sum_{n=1}^{\infty} A_{m,n}\,F_{m,n} \equiv \sum_{m=1}^{\infty} \sum_{n=1}^{\infty} \psi_{m,n}, \tag{6.3}$$

with the coefficients $A_{m,n}$ determined by the initial profile f. Setting $t = 0$ in (6.3) yields

$$\sum_{m=1}^{\infty} \sum_{n=1}^{\infty} A_{m,n}\,g_{m,n} = f.$$

Multiplying both sides by $\sin(m'\pi\,x/a)\sin(n'\pi\,y/(2\,a))$, integrating over x from 0 to a and over y from 0 to $2\,a$, and taking the double sum will yield zero unless $m = m'$ and $n = n'$. Removing the primes and solving for $A_{m,n}$ yields

$$A_{m,n} = \int_0^{2a} \int_0^a f\,g_{m,n}\,dx\,dy \Big/ \int_0^{2a} \int_0^a g_{m,n}^2\,dx\,dy.$$

Using this relation, we calculate $A_{m,n}$ in the next command line.

```
>  A[m,n]:=int(int(f*g[m,n],x=0..a),y=0..2*a)/int(int(g[m,n]^2,
          x=0..a),y=0..2*a):
```

Then, the general Fourier term $\psi_{m,n} = A_{m,n}\,F_{m,n}$ is determined, the result being simplified assuming that m and n are integers and $a > 0$.

```
>  psi[m,n]:=simplify(A[m,n]*F[m,n]) assuming m::integer,
          n::integer,a>0;
```

$$\psi_{m,n} := 3072\,h\,\left(4 + 8\,(-1)^m + \pi^2\,n^2\,(-1)^n + 2\,(-1)^{(m+n)}\,\pi^2\,n^2\right.$$

$$\left. + 8\,(-1)^{(1+m+n)} + 4\,(-1)^{(1+n)}\right) \sin\left(\frac{m\,\pi\,x}{a}\right) \sin\left(\frac{n\,\pi\,y}{2\,a}\right)$$

$$\cos\left(\frac{c\,\pi\,\sqrt{4\,m^2 + n^2}\,t}{2\,a}\right) \Big/ (m^3\,\pi^8\,n^5)$$

Given the mathematical form of f, Vectoria is not surprised at how complicated the coefficient turns out to be in $\psi_{m,n}$.

The parameter values suggested by Professor Legree are entered,

> a:=1: h:=1: c:=1:

and the membrane displacement approximated by the finite sum

$$\psi = \sum_{m=1}^{15} \sum_{n=1}^{15} \psi_{m,n}.$$

Vectoria claims that she has kept $15^2 = 225$ terms so as to get an accurate animation. We will let you be the judge of this, but we have overruled her desire to show the lengthy result and advised her to put a colon on the command line.

> psi:=sum(sum(psi[m,n],m=1..15),n=1..15):

Finally, a 3-dimensional zhue colored animation of ψ is produced over the time interval $t = 0$ to 5, with 40 frames.

> animate(plot3d,[psi,x=0..a,y=0..2*a],t=0..5,
>   frames=40,axes=frame,shading=zhue,orientation=[-130,50],
>   tickmarks=[2,2,3]);

On executing the animation, Vectoria is reminded of a rerun of an episode of a popular cartoon series that she watched recently with her fiancé, Mike. The main character, Homer Simpson, who has a rather fat stomach, went to the doctor for a medical checkup. The doctor supposedly measured the fat content of Homer's stomach by giving it a jiggle and studying the motion of the resulting wave form. The initial shape of the membrane in this animation and its subsequent movement somewhat resembles that of Homer's stomach in the doctor's jiggle test. Vectoria wonders whether Professor Legree was partially motivated in setting up this problem by watching the same cartoon episode. Perhaps, underneath his stern exterior, Professor Legree has a sense of humor.

PROBLEMS:

Problem 6-3: Vibrations of a square membrane

A square membrane whose sides are of unit length is given an initial transverse displacement $\psi(x, y, 0) = x\,y\,(1 - x)\,(1 - y)$ and then released. Determine the displacement $\psi(x, y, t)$ of the membrane for $t > 0$ and animate the solution for nominal values of the parameters.

Problem 6-4: Free edges

Consider a rectangular membrane having sides of length a between $x = 0$ and $x = a$ and sides of length $2\,a$ between $y = 0$ and $y = 2\,a$. The edges at $x = 0$ and $x = a$ are fixed, while those at $y = 0$ and $y = 2\,a$ are "free." At a free edge, the slope is zero. Explicitly determine $\psi(x, y, t)$ if the membrane is initially at rest and has the initial shape

$$\psi(x, y, 0) = \begin{cases} 2\,x\,h/a, & 0 \le x \le a/2, \\ (2\,h/a)\,(a - x), & a/2 \le x \le a. \end{cases}$$

Animate the solution for $a = 1$, $h = 1$, and $c = 1$.

6.1.3   Vectoria's Second Problem

Oh yes, there is a vast difference between the savage and the civilized man, but it is never apparent to their wives until after breakfast.
Helen Rowland, American journalist, *A Guide to Men*, 1922

The second problem on Professor Legree's assignment is to determine the transverse displacement $\psi(r, \theta, t)$ of a light, horizontal, circular membrane or drumhead of radius a fixed on its perimeter, given the initial conditions

$$\psi(r, \theta, 0) \equiv f = 20\, r \left(1 - \frac{r^2}{a^2} \right) \sin(2\,\theta), \quad \dot\psi(r, \theta, 0) = 0.$$

The motion of the drumhead is then to be animated for $a \doteq 1$ and wave velocity $c = 1$, and its behavior discussed.

After loading the plots and VectorCalculus packages, Vectoria uses the Laplacian command to enter the wave equation in polar coordinates.

> restart: with(plots): with(VectorCalculus):

> WE:=expand(Laplacian(psi(r,theta,t),'polar'[r,theta])
    =(1/c^2)*diff(psi(r,theta,t),t,t));

$$WE := \frac{\dfrac{\partial}{\partial r}\psi(r,\theta,t)}{r} + \left(\frac{\partial^2}{\partial r^2}\psi(r,\theta,t) \right) + \frac{\dfrac{\partial^2}{\partial \theta^2}\psi(r,\theta,t)}{r^2} = \frac{\dfrac{\partial^2}{\partial t^2}\psi(r,\theta,t)}{c^2}$$

Based on her experience with the first problem on the assignment, the initial condition $\dot\psi(r, \theta, 0) = 0$ can be satisfied if she assumes a general product solution of the form $F(r)\, G(\theta)\, \cos(c\,k\,t)$, where the functions F and G and the constant k remain to be determined. The assumed form is provided as a HINT in the pdsolve command.

> sol:=pdsolve(WE,HINT=F(r)*G(theta)*cos(c*k*t),INTEGRATE,
    build);

$$\begin{aligned}
sol := \psi(r,\theta,t) = {}& \cos(c\,k\,t)\_C3 \sin(\sqrt{\_c_1}\,\theta)\_C1\ \mathrm{BesselJ}(\sqrt{\_c_1},\,k\,r) \\
& + \cos(c\,k\,t)\_C3 \sin(\sqrt{\_c_1}\,\theta)\_C2\ \mathrm{BesselY}(\sqrt{\_c_1},\,k\,r) \\
& + \cos(c\,k\,t)\_C4 \cos(\sqrt{\_c_1}\,\theta)\_C1\ \mathrm{BesselJ}(\sqrt{\_c_1},\,k\,r) \\
& + \cos(c\,k\,t)\_C4 \cos(\sqrt{\_c_1}\,\theta)\_C2\ \mathrm{BesselY}(\sqrt{\_c_1},\,k\,r)
\end{aligned}$$

The Bessel functions of the second kind (Y) diverge at $r = 0$, so are removed.

> sol2:=remove(has,rhs(sol),BesselY);

$$\begin{aligned}
sol2 := {}& \cos(c\,k\,t)\_C3 \sin(\sqrt{\_c_1}\,\theta)\_C1\ \mathrm{BesselJ}(\sqrt{\_c_1},\,k\,r) \\
& + \cos(c\,k\,t)\_C4 \cos(\sqrt{\_c_1}\,\theta)\_C1\ \mathrm{BesselJ}(\sqrt{\_c_1},\,k\,r)
\end{aligned}$$

Vectoria simplifies the separation constant by substituting $\sqrt{\_c_1} = p$ into *sol2*.

> sol3:=subs(sqrt(_c[1])=p,sol2);

$$\begin{aligned}
sol3 := {}& \cos(c\,k\,t)\_C3 \sin(p\,\theta)\_C1\ \mathrm{BesselJ}(p, k\,r) \\
& + \cos(c\,k\,t)\_C4 \cos(p\,\theta)\_C1\ \mathrm{BesselJ}(p, k\,r)
\end{aligned}$$

Recognizing that the sines and cosines are independent functions and to be consistent with the $\sin(2\theta)$ term in the initial displacement, Vectoria removes the $\cos(p\theta)$ term from *sol3* and sets $p = 2$ in the resulting expression.

```
>  sol4:=eval(remove(has,sol3,cos(p*theta)),p=2);
```

$$sol4 := \cos(c\,k\,t)\,\_C3\,\sin(2\,\theta)\,\_C1\,\text{BesselJ}(2,\,k\,r)$$

The drumhead is fixed on its perimeter for all t, i.e., $\psi(a,\theta,t) = 0$, so that the Bessel function $J_2(k\,a)$ equals 0. The allowed k values are the zeros of $J_2(x)$. The mth zero, k_m, is entered,

```
>  k[m]:=BesselJZeros(2,m)/a;
```

$$k_m := \frac{\text{BesselJZeros}(2,\,m)}{a}$$

and substituted into *sol4* with all Maple constants removed.

```
>  F[2,m]:=subs(k=k[m],BesselJ(2,k*r)*sin(2*theta)*cos(c*k*t));
```

$$F_{2,\,m} := \text{BesselJ}\left(2,\,\frac{\text{BesselJZeros}(2,\,m)\,r}{a}\right)\sin(2\,\theta)\cos\left(\frac{c\,\text{BesselJZeros}(2,\,m)\,t}{a}\right)$$

The displacement of the drumhead at time t will be given by

$$\psi(r,\theta,t) = \sum_{m=1}^{\infty} A_{2,m}\,F_{2,m} \equiv \sum_{m=1}^{\infty} \psi_{2,m},$$

with the coefficients $A_{2,m}$ determined by the initial profile f, which is entered.

```
>  f:=20*r*(1-r^2/a^2)*sin(2*theta);
```

$$f := 20\,r\left(1 - \frac{r^2}{a^2}\right)\sin(2\,\theta)$$

Evaluating $F_{2,m}$ at $t = 0$ yields $g_{2,m}$.

```
>  g[2,m]:=eval(F[2,m],t=0);
```

$$g_{2,\,m} := \text{BesselJ}\left(2,\,\frac{\text{BesselJZeros}(2,\,m)\,r}{a}\right)\sin(2\,\theta)$$

The mth coefficient $A_{2,m}$ is given by

$$A_{2,m} = \int_0^a \int_0^{2\pi} r\,f\,g_{2,m}\,d\theta\,dr \,\Big/\, \int_0^a \int_0^{2\pi} r\,(g_{2,m})^2\,d\theta\,dr,$$

which is now calculated.

```
>  A[2,m]:=int(int(r*f*g[2,m],theta=0..2*Pi),r=0..a)
           /int(int(r*g[2,m]^2,theta=0..2*Pi),r=0..a):
```

Then, the mth Fourier term $\psi_{2,m} = A_{2,m}\,F_{2,m}$ is determined, the lengthy and quite formidable result being suppressed here in the text.

```
>  psi[2,m]:=A[2,m]*F[2,m];
```

For animation purposes $\psi_{2,m}$ is evaluated at $a = 1$, $c = 1$,

```
>  psi[2,m]:=eval(psi[2,m],{a=1,c=1}):
```

and converted to Cartesian coordinates by noting that $\sin(2\theta) = 2\sin\theta\cos\theta$ and substituting $r = \sqrt{x^2 + y^2}$ and $\sin(2\theta) = 2\,x\,y/(x^2 + y^2)$.

```
>  psi[2,m]:=subs({r=sqrt(x^2+y^2),sin(2*theta)=2*x*y
              /(x^2+y^2)},psi[2,m]):
```

Vectoria will animate the sum $\sum_{m=1}^{5} \psi_{2,m}$, which is sufficient to give a good animation. Labeling this result as ψ, she numerically evaluates the terms and displays them.

```
>  psi:=evalf(sum(psi[2,m],m=1..5));
```

$$\psi := \frac{33.78863892 \, \text{BesselJ}(2., 5.135622302 \, \sqrt{x^2 + y^2}) \, x \, y \cos(5.135622302 \, t)}{x^2 + y^2}$$

$$+ \frac{1.727434530 \, \text{BesselJ}(2., 8.417244140 \, \sqrt{x^2 + y^2}) \, x \, y \cos(8.417244140 \, t)}{x^2 + y^2}$$

$$+ \frac{4.730070500 \, \text{BesselJ}(2., 11.61984117 \, \sqrt{x^2 + y^2}) \, x \, y \cos(11.61984117 \, t)}{x^2 + y^2}$$

$$+ \frac{0.8068048260 \, \text{BesselJ}(2., 14.79595178 \, \sqrt{x^2 + y^2}) \, x \, y \cos(14.79595178 \, t)}{x^2 + y^2}$$

$$+ \frac{1.787787351 \, \text{BesselJ}(2., 17.95981949 \, \sqrt{x^2 + y^2}) \, x \, y \cos(17.95981949 \, t)}{x^2 + y^2}$$

Since the displacement ψ applies only to the drumhead, not the region outside, Vectoria forms a piecewise function that is equal to ψ for $x^2 + y^2 \leq 1$ and equal to zero otherwise.

```
>  psi2:=piecewise(x^2+y^2<=1,psi,0):
```

The piecewise function is animated, a PATCHCONTOUR style being chosen so as to best display any nodal lines in the animation. Along a nodal line, the membrane remains stationary for all times. Vectoria has taken only 15 frames in her animation, but if you have a fast computer you could take more frames.

```
>  animate(plot3d,[psi2,x=-1..1,y=-1..1],t=0..2,frames=15,
    axes=framed,shading=zhue,style=PATCHCONTOUR,
    orientation=[45,30]);
```

On executing the animation, Vectoria observes that in the initial frame of the animation the membrane displays two up peaks and two down peaks, located radially at $r = a/\sqrt{3}$ and separated into four quadrants by perpendicular intersecting nodal lines. These nodal lines correspond to $\sin(2\theta) = 0$, or $\theta = 0$, $\pi/2$, π, and $3\pi/2$. The up and down peaks correspond to plus and minus signs as θ ranges from 0 to 2π in $\sin(2\theta)$. For $\theta = 0$ to $\pi/2$ and π to $3\pi/2$, $\sin(2\theta)$ is positive; for $\theta = \pi/2$ to π and $3\pi/2$ to 2π, it is negative.

As time progresses the up peaks become down peaks and vice versa, the "flipping" of peaks occurring periodically. Having written down her observations, she decides that this is enough for the time being, since she has three more problems to do on Professor Legree's assignment. She can add more discussion, if necessary, when she does a final review of all her answers. She has a date with Mike and would like to finish the assignment before he calls on her.

PROBLEMS:

Problem 6-5: Circular drumhead

A circular drumhead of radius $r = a$ and fixed on its perimeter is displaced a distance h at its center at time $t = 0$ so that it has a conical shape. Determine the displacement of the membrane for $t > 0$. Taking $a = h = 1$ and $c = 1$, animate the motion of the membrane.

Problem 6-6: A different initial shape

A circular drumhead of radius a fixed on its outer edge has an initial displacement $\psi(r, \theta, 0) = (1 - r^2/a^2) \sin(4\theta)$ and its initial velocity is zero. Determine the subsequent displacement $\psi(r, \theta, t)$, and animate the solution for parameters of your own choosing.

6.1.4   Sound of Music?

Music is spiritual. The music business is not.
Van Morrison, Irish rock musician, *Times*, London, 6 July 1990

The third problem on Vectoria's assignment, which Professor Legree has worded as follows, might be considered to be noteworthy.

Some musically inclined people like to sing in the shower stall when taking a shower. Suppose that a large "shower stall" is empty without the water running and consists of a completely enclosed hollow vertical metal cylinder of radius a and height h with (approximately) rigid walls and all fixtures removed. The speed of sound for the air inside the cylinder is c. By solving the scalar Helmholtz equation for the spatial part of the velocity potential, determine the allowed normal modes inside the cylinder. For rigid walls, the normal component of the fluid velocity (or the normal derivative of the potential) must vanish at each wall. Taking $a = 1.83$ m, $h = 3.04$ m, and $c = 344$ m/s (assuming dry air), determine the three lowest eigenfrequencies. By either consulting a musically inclined friend or a music reference book, or going to the Internet, find the closest musical notes on the equal-tempered scale to these eigenfrequencies and relate these notes to those found on a piano keyboard.

Vectoria knows that the velocity potential ϕ satisfies the scalar wave equation, with c the speed of sound. The velocity of a fluid element is given by $\vec{v} = -\nabla\phi$. A normal mode of frequency ω is obtained by assuming a solution of the form $\phi(\vec{r}, t) = S(\vec{r}) \cos(\omega t)$. The wave equation then reduces to the *scalar Helmholtz equation*,

$$\nabla^2 S + k^2 S = 0, \quad \text{with} \ \ k = \omega/c. \tag{6.4}$$

Professor Legree in his lectures uses the acronym SHE for the scalar Helmholtz equation, and jokingly refers to this equation as "she who must be obeyed". Evidently, this refers to a standard phrase uttered by the main character, a British barrister, about his wife in a PBS television series (*Rumpole of the Bailey*) that Legree saw some years ago.

Vectoria begins her solution by loading the VectorCalculus package,

```
> restart: with(VectorCalculus):
```
and entering SHE in cylindrical coordinates. The z direction is taken along the cylinder axis, r is the radial coordinate measured perpendicular to the z-axis, and θ is the angular coordinate measured about this axis.

```
> SHE:=expand(Laplacian(S(r,theta,z),'cylindrical'[r,theta,z])
    +k^2*S(r,theta,z))=0;
```

$$SHE := \frac{\frac{\partial}{\partial r} S(r, \theta, z)}{r} + \left(\frac{\partial^2}{\partial r^2} S(r, \theta, z) \right) + \frac{\frac{\partial^2}{\partial \theta^2} S(r, \theta, z)}{r^2} + \left(\frac{\partial^2}{\partial z^2} S(r, \theta, z) \right)$$

$$+ k^2 S(r, \theta, z) = 0$$

SHE is solved with the **pdsolve** command, a general product solution of the form $S(r, \theta, z) = R(r) \, \Theta(\theta) \, Z(z)$ being assumed.

```
> S:=rhs(pdsolve(SHE,HINT=R(r)*Theta(theta)*Z(z),INTEGRATE,
    build));
```

$$S := e^{(\sqrt{-c_3}\,z)} e^{(\sqrt{-c_2}\,\theta)} \_C5 \_C3 \_C1 \,\%2 + e^{(\sqrt{-c_3}\,z)} e^{(\sqrt{-c_2}\,\theta)} \_C5 \_C3 \_C2 \,\%1$$
$$+ \frac{e^{(\sqrt{-c_3}\,z)} \_C5 \_C4 \_C1 \,\%2}{e^{(\sqrt{-c_2}\,\theta)}} + \frac{e^{(\sqrt{-c_3}\,z)} \_C5 \_C4 \_C2 \,\%1}{e^{(\sqrt{-c_2}\,\theta)}}$$
$$+ \frac{e^{(\sqrt{-c_2}\,\theta)} \_C6 \_C3 \_C1 \,\%2}{e^{(\sqrt{-c_3}\,z)}} + \frac{e^{(\sqrt{-c_2}\,\theta)} \_C6 \_C3 \_C2 \,\%1}{e^{(\sqrt{-c_3}\,z)}}$$
$$+ \frac{\_C6 \_C4 \_C1 \,\%2}{e^{(\sqrt{-c_3}\,z)} e^{(\sqrt{-c_2}\,\theta)}} + \frac{\_C6 \_C4 \_C2 \,\%1}{e^{(\sqrt{-c_3}\,z)} e^{(\sqrt{-c_2}\,\theta)}}$$
$$\%1 := \text{BesselY}(\sqrt{-\_c_2}, \sqrt{-c_3 + k^2}\, r)$$
$$\%2 := \text{BesselJ}(\sqrt{-\_c_2}, \sqrt{-c_3 + k^2}\, r)$$

The Bessel functions of the second kind diverge at $r = 0$ and must be removed.

```
> S2:=remove(has,S,BesselY);
```

$$S2 := e^{(\sqrt{-c_3}\,z)} e^{(\sqrt{-c_2}\,\theta)} \_C5 \_C3 \_C1 \,\%1 + \frac{e^{(\sqrt{-c_3}\,z)} \_C5 \_C4 \_C1 \,\%1}{e^{(\sqrt{-c_2}\,\theta)}}$$
$$+ \frac{e^{(\sqrt{-c_2}\,\theta)} \_C6 \_C3 \_C1 \,\%1}{e^{(\sqrt{-c_3}\,z)}} + \frac{\_C6 \_C4 \_C1 \,\%1}{e^{(\sqrt{-c_3}\,z)} e^{(\sqrt{-c_2}\,\theta)}}$$
$$\%1 := \text{BesselJ}(\sqrt{-\_c_2}, \sqrt{-c_3 + k^2}\, r)$$

To simplify the notation for matching the boundary conditions, the separation constants $\_c_2$ and $\_c_3$ are replaced with $-m^2$ and $-q^2$ in S2. Vectoria also substitutes $k^2 = \alpha^2 + q^2$, and simplifies the result with the **symbolic** option.

```
> S3:=simplify(subs({_c[2]=-m^2,_c[3]=-q^2,k^2=alpha^2+q^2},
    S2),symbolic);
```

$$S3 := \_C1 \, \text{BesselJ}(m, \alpha\, r)(\_C5 \_C3 \, e^{((q\,z + m\,\theta)\,I)} + \_C5 \_C4 \, e^{((q\,z - m\,\theta)\,I)}$$
$$+ \_C6 \_C3 \, e^{(-I\,(q\,z - m\,\theta))} + \_C6 \_C4 \, e^{(-I\,(q\,z + m\,\theta))})$$

Then *S3* is converted to trig form and the result expanded.

> S4:=expand(convert(S3,trig));

$$S4 := \_C1 \,\text{BesselJ}(m,\, \alpha\, r)\, \_C5\, \_C3 \cos(q\, z)\cos(m\,\theta)$$
$$- \_C1 \,\text{BesselJ}(m,\, \alpha\, r)\, \_C5\, \_C3 \sin(q\, z)\sin(m\,\theta)$$
$$+ \cdots \cdots \cdots \cdots \cdots \cdots \cdots$$

Approximating the metallic walls of the cylinder as being rigid, the normal component of the fluid velocity must vanish at the walls. Since the velocity is equal to minus the gradient of the potential, this implies that the normal derivative of the potential must vanish at the walls. In the z direction, one must have $dZ/dz = 0$ at $z = 0$, so only terms involving $\cos(q\, z)$ are selected.

> S5:=select(has,S4,cos(q*z));

$$S5 := \_C1 \,\text{BesselJ}(m,\, \alpha\, r)\, \_C5\, \_C3 \cos(q\, z)\cos(m\,\theta) + \cdots$$

To satisfy the derivative boundary condition on the top face of the cylinder at $z = h$, one must have $\sin(q\, h) = 0$, or $q = n\,\pi/h$, with $n = 0, 1, 2, \ldots$.

For the angular part to remain single-valued as θ increases by $2\,\pi$, one must have $\cos(m\,(\theta + 2\,\pi)) = \cos(\theta)$, which is satisfied if $m = 0, 1, 2, \ldots$. Negative integer values of m need not be considered because the minus sign can be absorbed in the arbitrary constants. Similar remarks apply to $\sin(m\,\theta)$, which also appears in *S5*. So the allowed Bessel functions are J_0, J_1, J_2, etc.

To satisfy the derivative condition on the cylindrical wall, one must have $(d/dr)J_m(\alpha_{m,s}\, r)|_{r=a} = 0$, where s labels the zeros of the derivative of the mth-order Bessel function. For later convenience, Vectoria sets $\alpha_{m,s} = \pi\, p_{m,s}/a$, the values of $p_{m,s}$ still to be determined. The allowed values of q and α are subsituted into *S5* and the result factored.

> S6:=factor(subs({q=n*Pi/h,alpha=Pi*p[m,s]/a},S5));

$$S6 := (-\cos(m\,\theta)\, \_C3\, I - \cos(m\,\theta)\, \_C4\, I + \sin(m\,\theta)\, \_C3 - \sin(m\,\theta)\, \_C4)$$
$$\cos\left(\frac{n\,\pi\, z}{h}\right)\text{BesselJ}\left(m,\, \frac{\pi\, p_{m,s}\, r}{a}\right)\, \_C1\,(\_C5 + \_C6)\, I$$

Replacing the awkward coefficient combinations with $A_{n,m,s}$ and $B_{n,m,s}$ gives normal modes of the following form. For each m value, except $m = 0$, there are actually two modes corresponding to $\cos(m\,\theta)$ and $\sin(m\,\theta)$.

> NM:=(A[n,m,s]*cos(m*theta)+B[n,m,s]*sin(m*theta))
>   *select(has,S6,{cos(n*Pi*z/h),BesselJ});

$$NM := (A_{n,m,s}\cos(m\,\theta) + B_{n,m,s}\sin(m\,\theta))\cos\left(\frac{n\,\pi\, z}{h}\right)\text{BesselJ}\left(m,\, \frac{\pi\, p_{m,s}\, r}{a}\right)$$

Noting that the boundary condition $(d/dr)(J_m(\pi\, p_{m,s}\, r/a)|_{r=a} = 0$ may be rewritten as $(d/dp)(J_m(\pi\, p)|_{p} = p_{m,s} = 0$, Vectoria creates a functional operator f to apply the derivative boundary condition at the cylinder wall.

> f:=m->diff(BesselJ(m,Pi*p),p)=0:

Then using f, she employs a do loop to numerically determine $p_{m,s}$ for $m = 0$ and 1 and $s = 0$ to $s = 4$.

```
>   for m from 0 to 1 do
>   sol[m]:=seq(p[m,s]=fsolve(f(m),p,s..s+1),s=0..4);
>   assign(sol[m]):
>   end do;
```

$$sol_0 := p_{0,0} = 0., \ p_{0,1} = 1.219669891, \ p_{0,2} = 2.233130594,$$
$$p_{0,3} = 3.238315484, p_{0,4} = 4.241062864$$

$$sol_1 := p_{1,0} = 0.5860669999, \ p_{1,1} = 1.697050942, \ p_{1,2} = 2.717193891,$$
$$p_{1,3} = 3.726137088, \ p_{1,4} = 4.731227206$$

Recalling that $\omega = c\,k = c\,\sqrt{\alpha^2 + q^2}$, the eigenfrequencies are given by

$$\omega_{n,m,s} = \pi\,c\,\sqrt{\left(\frac{p_{m,s}}{a}\right)^2 + \left(\frac{n}{h}\right)^2} \quad \text{radians per second,}$$

or, on using $\omega = 2\,\pi\,\nu$,

$$\nu_{n,m,s} = \frac{c}{2}\,\sqrt{\left(\frac{p_{m,s}}{a}\right)^2 + \left(\frac{n}{h}\right)^2} \quad \text{hertz.}$$

A functional operator F is formed to calculate the $\nu_{n,m,s}$.

```
>   F:=(n,m,s)->(c/2)*sqrt((p[m,s]/a)^2+(n/h)^2):
```

The given values of the cylinder radius a, cylinder height h, and speed of sound c are entered,

```
>   a:=1.83: h:=3.04: c:=344:
```

and the three lowest allowed frequencies calculated using F.

```
>   nu[0,1,0]:=F(0,1,0);
```

$$\nu_{0,1,0} := 55.08389289$$

```
>   nu[1,0,0]:=F(1,0,0);
```

$$\nu_{1,0,0} := 56.57894736$$

```
>   nu[1,1,0]:=F(1,1,0);
```

$$\nu_{1,1,0} := 78.96462843$$

Consulting *The Acoustical Foundations of Music*, by John Backus [Bac69], Vectoria determines that the closest musical notes are A_1 (55.000 Hz) for the first two frequencies and $D_2{}^{\#}$ (77.782 Hz) for the third. Not having much of a feeling for these numbers, Vectoria goes to the Internet and finds that the lowest-frequency note on a piano is 27.500 Hz and the highest frequency is 4186.0 Hz. So the allowed frequencies in this "shower stall" example are at the low-frequency end of the piano keyboard.

PROBLEMS:

Problem 6-7: Acoustical waveguide

A sound wave of frequency ω is generated at one end ($z = 0$) of a very long straight cylindrical pipe of radius a having rigid walls.

(a) Determine and discuss in detail the allowed modes of propagation and the *cutoff frequency* for wave propagation. The cutoff frequency is the minimum frequency for propagation of a specific mode.

(b) If the speed of sound in air is 1100 feet per second and if the frequency of the sound wave is 500 Hz, show that only a plane wave will be propagated if $a < 7.73$ inches.

Problem 6-8: Vibrating cylinder
An infinitely long circular cylinder of radius $r = a$ is surrounded by an ideal compressible fluid. The cylinder's surface is vibrating with a radial velocity $V_0 \cos(\omega t)$. The fluid velocity is given by $\vec{v} = -\nabla\phi(r, t)$, where ϕ satisfies the wave equation in cylindrical coordinates and r is measured from the cylinder axis. Assuming that the cylinder's surface is rigid, the fluid velocity must equal the velocity of the vibrating surface. Noting that far from the surface the waves in the fluid must be outgoing from the cylinder, analytically determine $\vec{v}(r, t)$ in the fluid and animate the solution for nominal values of the parameters. Note that the Bessel function of the second kind must be kept, since the origin of the cylindrical coordinates lies inside the cylinder and outside the fluid.

6.2 Semi-infinite and Infinite Domains

To solve diffusion and wave equation boundary value problems involving semi-infinite or infinite domains, a standard approach is to use integral transform methods to solve the governing PDE for some specified initial condition. This approach is illustrated in the following recipes, where use is made of Maple's integral transform library package.

6.2.1 Vectoria's Fourth Problem

It is nothing short of a miracle that modern methods of instruction have not yet entirely strangled the holy curiosity of inquiry.
Albert Einstein, Nobel laureate in physics (1879–1955)

The fourth problem on Professor Legree's assignment involves a thin insulated semi-infinite $(0 \le x \le \infty)$ rod that has the end $x = 0$ held at the constant temperature $T(0, t) = T0 = 100°C$ and whose interior $(x > 0)$ is at zero degrees at time $t = 0$. Vectoria is asked to determine the temperature distribution $T(x, t)$ in the interior of the rod for $t \ge 0$ and to animate $T(x, t)$ for a heat diffusion constant $d = 100$ cm^2/s.

Vectoria has learned that semi-infinite domain problems such as this one, where the function is specified on one boundary, may be solved using the *Fourier sine transform* method. The Fourier sine transform (FST) of a function $f(x)$

that vanishes at $x = \infty$ is defined as

$$FST(f(x)) \equiv F(s) = \sqrt{\frac{2}{\pi}} \int_0^\infty f(x) \sin(s\,x)\,dx, \tag{6.5}$$

and the inverse Fourier sine transform of $F(s)$ by

$$f(x) = \sqrt{\frac{2}{\pi}} \int_0^\infty F(s) \sin(s\,x)\,ds. \tag{6.6}$$

For the present problem, the temperature distribution will satisfy the one-dimensional heat diffusion equation (with d the heat diffusion coefficient),

$$\frac{\partial T(x,t)}{\partial t} = d\,\frac{\partial^2 T(x,t)}{\partial x^2}. \tag{6.7}$$

Taking the *FST* of (6.7) with respect to x and integrating the rhs twice by parts, and assuming $f'(x \to \infty) \to 0$, will convert the PDE into a first-order ODE for the transformed quantity $F(s,t)$. The ODE will depend on the boundary condition $T(0,t)$. The ODE is then solved for $F(s,t)$ making use of the Fourier sine transform of the initial condition. Finally, on performing the inverse transform with respect to s, the temperature distribution $T(x,t)$ is found.

Once again, Vectoria disdains doing the problem "the old-fashioned way" favored by some instructors, but instead intends to let the computer assist her. To this end, she loads the plots and integral transform packages,

> `restart: with(plots): with(inttrans):`

and enters the diffusion equation DE,

> `DE:=diff(T(x,t),t)=d*diff(T(x,t),x,x);`

$$DE := \frac{\partial}{\partial t} T(x,\,t) = d\left(\frac{\partial^2}{\partial x^2} T(x,\,t)\right)$$

and the boundary condition $T(0,t) = T0$.

> `T(0,t):=T0:`

To keep the recipe general for the moment, Vectoria has not yet specified the value of *T0*. She then takes the *FST* of *DE* with respect to x, the integration by parts and boundary condition substitution being automatically done.

> `FST:=fouriersin(DE,x,s);`

$$FST := \frac{\partial}{\partial t}\,\text{fouriersin}(\,T(x,t),x,s) = \frac{d\,s\left(\sqrt{2}\,\,T0 - s\,\text{fouriersin}(\,T(x,t),x,s)\,\sqrt{\pi}\,\right)}{\sqrt{\pi}}$$

To simplify the notation and facilitate solving the above ODE, she replaces fouriersin($T(x,t),x,s$) with $F(t)$, temporarily suppressing the argument s.

> `FST:=subs(fouriersin(T(x,t),x,s)=F(t),FST);`

$$FST := \frac{d}{dt} F(t) = \frac{d\,s\left(\sqrt{2}\,\,T0 - s\,F(t)\,\sqrt{\pi}\,\right)}{\sqrt{\pi}}$$

The Fourier sine transform of the initial condition is zero, so Vectoria solves the first-order linear differential equation *FST* for $F(t)$, subject to $F(0) = 0$.

> `eq:=dsolve({FST,F(0)=0},F(t));`

$$eq := F(t) = \frac{\sqrt{2}\, T0}{s\,\sqrt{\pi}} - \frac{e^{(-d\,s^2\,t)}\,\sqrt{2}\, T0}{s\,\sqrt{\pi}}$$

The Fourier (inverse) sine transform of the rhs of *eq* with respect to *s* is taken,

> `T:=fouriersin(rhs(eq),s,x);`

$$T := T0 - T0\,\mathrm{erf}\left(\frac{x}{2\,\sqrt{d}\,t}\right)$$

yielding the temperature T expressed in terms of the error function. Now T is evaluated with the given parameter values $(d=100,\ T0=100)$ and simplified.

> `T:=simplify(eval(T,{d=100,T0=100}));`

$$T := 100 - 100\,\mathrm{erf}\left(\frac{x}{20\,\sqrt{t}}\right)$$

Choosing the spatial range to be $x = 0$ to 400 cm, Vectoria animates T over the time interval $t = 0$ to 150 seconds, 50 frames being taken.

> `animate(plot,[T,x=0..400],t=0..150,frames=50,thickness=2);`

On running the animation, Vectoria is pleased with the "sweetness" of the whole computer algebra derivation.

Looking at her watch, she realizes that Mike will be picking her up at 5:00 p.m., so she had better finish the last problem on Professor Legree's assignment before he shows up. She will enjoy her dinner at Giraffes on the waterfront more if she has finished her work.

PROBLEMS:

Problem 6-9: Fourier sine transform
Calculate the Fourier sine transform of each $f(x)$ below and plot the answers:

(a) $f(x) = e^{-3\,|x|}$; (b) $f(x) = \cos(2\,x)$; (c) $f(x) = \sin(x)^2$;
(d) $f(x) = x/(x^2 + 1)$.

Problem 6-10: A different $T(x,0)$
Use the Fourier sine transform approach to solve the heat conduction problem for $d = 10$ cm^2/s in a semi-infinite rod $(0 \le x \le \infty)$ that has the boundary condition $T(0,t) = 0$ and initial interior temperature distribution $T(x,0) = 10\,x/(x^2 + 1)$. Animate the plot.

6.2.2 Assignment Complete!

It is not knowledge, but the act of learning, not possession, but the act of getting there, which grants the greatest enjoyment.
Carl Friedrich Gauss, German mathematician (1777–1855)

The last problem on Vectoria's assignment is similar to the fourth one, but involves a different boundary condition and initial condition. It is now supposed that the semi-infinite rod $(0 \le x \le \infty)$ has an initial temperature distribution

$T(x, 0) = A \delta(x - a)$, i.e., is a *Dirac delta function*[2] of amplitude A located at $x = a > 0$. The end $x = 0$ is perfectly insulated so that no heat can flow across this end. For no heat flow to occur, the boundary condition is $\partial T/\partial x|_{x=0} = 0$, i.e., the gradient of the temperature is zero. The evolution of the temperature profile is to be animated over the spatial range $x = 0$ to 20 for the time interval $t = 1$ to 200, with $a = 5$, $A = 5$, and a heat diffusion constant $d = 1$.

In her mathematical physics course, Vectoria has learned that when a gradient boundary condition is involved for a semi-infinite domain problem the *Fourier cosine transform* should be used. The Fourier cosine transform $F(s)$ of a function $f(x)$ for which both $f(x)$ and $f'(x) \to 0$ as $x \to \infty$ and its inverse transform are defined as

$$F(s) = \sqrt{\frac{2}{\pi}} \int_0^\infty f(x) \cos(s\,x)\,dx, \quad f(x) = \sqrt{\frac{2}{\pi}} \int_0^\infty F(s) \cos(s\,x)\,ds.$$

Guided by her easy conquest of the previous problem, Vectoria realizes that she will be done by the time Mike arrives to pick her up.

She again loads the plots and integral transform library packages, and enters the 1-dimensional heat diffusion equation.

```
>  restart: with(plots): with(inttrans):
>  DE:=diff(T(x,t),t)=d*diff(T(x,t),x,x);
```

$$DE := \frac{\partial}{\partial t} T(x,\,t) = d \left(\frac{\partial^2}{\partial x^2} T(x,\,t) \right)$$

The boundary condition bc and the initial condition ic are specified. The differential command $D[1](T)(0,t)$ is used to enter $\partial T(x,t)/\partial x|_{x=0}$, while the Dirac delta function is entered with the Dirac command.

```
>  bc:=D[1](T)(0,t)=0; ic:=A*Dirac(x-a);
```

$$bc := D_1(T)(0,\,t) = 0 \qquad ic := A \operatorname{Dirac}(-x + a)$$

The Fourier cosine transforms of the initial condition (assuming that $a > 0$) and the diffusion equation are calculated. The boundary condition is substituted into the latter result.

```
>  F0:=fouriercos(ic,x,s) assuming a>0;
```

$$F0 := \frac{A\sqrt{2}\cos(s\,a)}{\sqrt{\pi}}$$

```
>  FCT:=subs(bc,fouriercos(DE,x,s));
```

$$FCT := \frac{\partial}{\partial t} \operatorname{fouriercos}(T(x,\,t),\,x,\,s) = -d\,s^2 \operatorname{fouriercos}(T(x,\,t),\,x,\,s)$$

As in the previous recipe, fouriercos($T(x,\,t)$, x, s) is replaced with $F(t)$ in FCT.

```
>  FCT:=subs(fouriercos(T(x,t),x,s)=F(t),FCT);
```

$$FCT := \frac{d}{dt} F(t) = -d\,s^2 F(t)$$

---

[2] The Dirac delta function $\delta(x - a)$ is an infinitely tall spike located at $x = a$ and is equal to zero for $x \neq a$. One of its most important properties is the *sifting property*, viz., for a smooth function $f(x)$ (not another delta function) and $\epsilon > 0$, $\int_{a-\epsilon}^{a+\epsilon} f(x) \delta(x - a)\,dx = f(a)$.

Then, the ordinary differential equation FCT is analytically solved for $F(t)$, subject to the initial condition $F(0) = F0$,

> `eq:=dsolve({FCT,F(0)=F0},F(t));`

$$eq := F(t) = \frac{A\sqrt{2}\cos(s\,a)\,e^{(-d\,s^2\,t)}}{\sqrt{\pi}}$$

and the inverse transform performed on the rhs of *eq*.

> `T:=fouriercos(rhs(eq),s,x);`

$$T := \frac{A\sqrt{\dfrac{\pi}{t\,d}}\,e^{\left(-\frac{a^2+x^2}{4\,t\,d}\right)}\cosh\left(\dfrac{a\,x}{2\,t\,d}\right)}{\pi}$$

Looking at the analytic result for T, Vectoria is pleased that she has spent time learning how to use the Maple computer algebra system. Once she has set up a template for a certain type of problem, then any other problem of that type is generally trivial to tackle. In the real world, certain integrals and other steps may not be carried out analytically, but traditionally in the world of academia, problems are assigned by instructors for which analytic answers exist.

To finish off the problem and complete the assignment, Vectoria evaluates T with the given parameter values ($d = 1$, $a = 5$, $A = 5$),

> `T:=eval(T,{d=1,a=5,A=5});`

$$T := \frac{5\sqrt{\dfrac{\pi}{t}}\,e^{\left(-\frac{25+x^2}{4\,t}\right)}\cosh\left(\dfrac{5\,x}{2\,t}\right)}{\pi}$$

and animates the solution over the suggested time interval $t = 1$ to 200. Clearly, the solution cannot be plotted at $t = 0$, because the initial profile has been assumed to be a Dirac delta function. The number of points is controlled to product a smooth curve.

> `animate(plot,[T,x=0..20],t=1..200,numpoints=200,`
> `frames=50,thickness=2);`

As she watches the animation on the computer screen, Mike arrives to whisk her off on their date.

PROBLEMS:
Problem 6-11: Fourier cosine transform
Calculate the Fourier cosine transform of each $f(x)$ below and plot the answers where possible:

(a) $f(x) = e^{-3\,|x|}$; **(b)** $f(x) = \cos(2\,x)$; **(c)** $f(x) = \sin(x)^2$;
(d) $f(x) = x/(x^2 + 1)$.

Problem 6-12: Different initial condition
Modify the text recipe to find the temperature distribution inside a semi-infinite rod ($0 \le x \le \infty$) that is insulated at $x = 0$ and has the initial temperature distribution $T(x > 0, 0) = 25\,x^2/(x^2 + 25)$ and $d = 1$. Animate $T(x, t)$.

6.2.3   Radioactive Contamination

The unexamined life is not worth living.
Socrates, Greek philosopher (470–399 BC)

When coauthor Richard was a student, he worked one summer as a chemistry lab assistant at a uranium mine on Great Bear Lake, which straddles the Arctic Circle in Northern Canada. Because it was a summer job, there was a period of several weeks when the sun never set, so evening baseball games were never called off because of darkness. The baseball diamond was located on the leveled mine tailings, the only relatively flat spot in the small northern min-

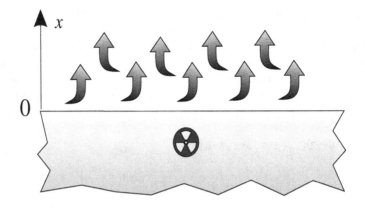

Figure 6.1: The radioactive disposal site.

ing town of Port Radium, which is perched on almost treeless primordial rock formations. The author has always wondered what long-term health problems eventually arose among the permanent workers because of working in the mine and playing baseball on the radioactive mine tailings. Motivated by these reminiscences, we could not resist including a related example involving radioactive contamination.

A radioactive gas is diffusing at a steady rate into the atmosphere from a leveled contaminated disposal site. The ground and the atmosphere will be taken to be semi-infinite media with $X = 0$ at the boundary as shown in Figure 6.1. The concentration $C(X, T)$ of radioactive gas in the atmosphere obeys the concentration equation (a modified diffusion equation)

$$\frac{\partial C(X, T)}{\partial T} = d\,\frac{\partial^2 C(X, T)}{\partial X^2} - \lambda\,C(X, T), \tag{6.8}$$

where d is the diffusion constant and λ is the decay rate of the radioactive gas. The boundary condition at $X = 0$ is given by *Fick's law*,

$$-d\,\frac{\partial C(0, T)}{\partial X} = K, \tag{6.9}$$

where K is a positive constant with units in $kg/(m^2 \cdot s)$. The coefficients can be eliminated and the two equations cast into dimensionless form by introducing the new variables $t \equiv \lambda T$, $x \equiv \sqrt{\lambda/d}\, X$, and $c = (\sqrt{\lambda d}/K)\, C$. Then the concentration equation becomes

$$\frac{\partial c(x,t)}{\partial t} = \frac{\partial^2 c(x,t)}{\partial x^2} - c(x,t), \qquad (6.10)$$

with $\partial c/\partial x = -1$ as the boundary condition at $x = 0$.

Assuming that initially $c = 0$ for $x \geq 0$, we want to determine the distribution of radioactive gas in the atmosphere for times $t \geq 0$ and animate the solution. The method of attack will be to make use of the Laplace transform.

The Laplace transform of a function $f(t)$ is defined as

$$L(f(t)) \equiv F(s) = \int_0^\infty f(t)\, e^{-st}\, dt. \qquad (6.11)$$

Integrating by parts and assuming that $e^{-st} f(t) \to 0$ as $t \to \infty$, then

$$L\left(\frac{df}{dt}\right) = s\, F(s) - f(0), \text{ and } L\left(\frac{d^2 f}{dt^2}\right) = s^2\, F(s) - s\, f(0) - \frac{df(0)}{dt}. \qquad (6.12)$$

To solve the concentration equation (6.10), subject to the boundary and initial conditions, the Laplace transform can be applied to the time part of the equation, the resultant second-order ODE in x solved, and the inverse Laplace transform performed to regain the time dependence.

This program is now carried out using the `laplace` command contained in the integral transform library package.

```
>  restart: with(plots): with(inttrans):
```
The dimensionless concentration equation is entered,
```
>  CE:=diff(c(x,t),t)=diff(c(x,t),x,x)-c(x,t);
```

$$CE := \tfrac{\partial}{\partial t}\, c(x, t) = \left(\tfrac{\partial^2}{\partial x^2}\, c(x, t)\right) - c(x, t)$$

and the initial concentration specified.
```
>  c(x,0):=0:
```
The Laplace transform of CE is taken with respect to the time variable, and the function laplace($c(x, t), t, s$) then replaced with $f(x)$ in LT.
```
>  LT:=laplace(CE,t,s);
>  LT:=subs(laplace(c(x,t),t,s)=f(x),LT);
```

$$LT := s\, f(x) = \left(\tfrac{d^2}{dx^2} f(x)\right) - f(x)$$

This second-order ODE is solved for $f(x)$, the following DEtools command line being used to obtain exponential solutions, which are convenient here.
```
>  sol:=DEtools[expsols](LT,f(x));
```

$$sol := [e^{(\sqrt{s+1}\, x)}, e^{(-\sqrt{s+1}\, x)}]$$

As $x \to \infty$, the concentration must go to zero, so the negative exponent solution is selected and multiplied by an arbitrary coefficient B to yield f.

```
>   f:=B*sol[2];
```

$$f := B e^{(-\sqrt{s+1}\,x)}$$

The Laplace-transformed boundary condition bc takes the following form:

```
>   bc:=eval(diff(f,x),x=0)=laplace(-1,t,s);
```

$$bc := -B\sqrt{s+1} = -\frac{1}{s}$$

Then bc is solved for B, and the result BB substituted into f.

```
>   BB:=solve(bc,B); f:=subs(B=BB,f);
```

$$BB := \frac{1}{\sqrt{s+1}\,s} \qquad f := \frac{e^{(-\sqrt{s+1}\,x)}}{\sqrt{s+1}\,s}$$

To perform the inverse Laplace transform, the substitution $s = y - 1$ is first made in f. Looking back at the definition of the Laplace transform, it is then necessary to multiply this result by e^{-t}, producing $f2$.

```
>   f2:=exp(-t)*subs(s=y-1,f);
```

$$f2 := \frac{e^{(-t)}\,e^{(-\sqrt{y}\,x)}}{\sqrt{y}\,(y-1)}$$

The inverse Laplace transform of $f2$ with respect to y is carried out assuming that $x > 0$. This yields the analytic form c for the concentration, expressed in terms of the *complementary error function* ($\mathrm{erfc}(z) = 1 - \mathrm{erf}(z)$).

```
>   c:=invlaplace(f2,y,t) assuming x>0;
```

$$c := -\frac{1}{2}\,\mathrm{erfc}\left(\frac{x+2t}{2\sqrt{t}}\right)e^{x} + \frac{1}{2}\,\mathrm{erfc}\left(-\frac{-x+2t}{2\sqrt{t}}\right)e^{(-x)}$$

The concentration is animated over the spatial region $x = 0$ to 5 and time interval $t = 0$ to 3.

```
>   animate(plot,[c,x=0..5],t=0..3,frames=100,thickness=2);
```

On running the code and observing the animated concentration profile in the region $x > 0$, you will see the radioactive gas diffusing into the atmosphere, with the concentration attaining a steady-state profile that decreases (approximately) exponentially from the contaminated surface as the distance x increases from zero. Steady state occurs when there is a balance between the rate at which radioactive gas atoms are diffusing into the atmosphere and the rate at which they are decaying.

PROBLEMS:
Problem 6-13: Laplace transform
Calculate the Laplace transforms of the following functions, simplifying the answer where necessary, and identifying any special functions that occur:

(a) $f(t) = e^{-a\sqrt{t}}$ with $a > 0$; **(b)** $f(t) = t\,\cos(a\,t)$; **(c)** $f(t) = \arctan(t)$;

(d) $f(t) = t^n \ln(t)$ with $n > 0$; **(e)** $f(t) = \tanh(t)$; **(f)** $f(t) = \tanh^{-1}(t)$;

(g) $f(t) = \dfrac{\sin(3\sqrt{t})}{t^{1/4}}$; **(h)** $f(t) = J_0(t) J_1(t)$.

Problem 6-14: Inverse Laplace transform

Calculate the inverse Laplace transforms of the following functions, identifying any special functions that occur in the answer:

(a) $F(s) = \dfrac{1}{s^2}$; **(b)** $F(s) = \dfrac{a}{s^2 + a^2}$; **(c)** $F(s) = \dfrac{s^2}{(s^2 + a^2)^{3/2}}$;

(d) $F(s) = \dfrac{1}{\sqrt{s^2 + a^2}}$; **(e)** $F(s) = e^{-as}$; **(f)** $F(s) = \dfrac{\sin(as)}{s}$;

Problem 6-15: Heat flow in a semi-infinite rod

Consider a semi-infinite rod spanning the range $x = 0$ to $x = \infty$. The initial temperature of the rod is zero. For $t > 0$, the temperature at $x = 0$ is $T(x, 0) = T_0$. By Laplace transforming the temporal part of the diffusion equation, determine the temperature distribution inside the rod for $t > 0$. Animate the temperature profile for nominal values of the parameters.

Problem 6-16: Heat flow in a bar of varying cross section

The heat flow along an insulated semi-infinite bar whose cross section varies exponentially is described by

$$\frac{\partial}{\partial x}\left(e^{\alpha x}\frac{\partial T}{\partial x}\right) = e^{\alpha x}\frac{\partial T}{\partial t}.$$

If $T(x, 0) = 0$ for $x > 0$, $T(0, t) = 1$, and $T(\infty, t) = 0$, use the Laplace transform approach to show that for $t > 0$, the temperature distribution in the bar is

$$T(x, t) = \frac{e^{-\alpha x}}{2}\left(\operatorname{erfc}\left(\frac{1}{2}\left(x\,t^{-1/2} - \alpha\,t^{1/2}\right)\right) + e^{\alpha x}\operatorname{erfc}\left(\frac{1}{2}\left(x\,t^{-1/2} + \alpha\,t^{1/2}\right)\right)\right).$$

Taking $\alpha = 1$, animate $T(x, t)$ over the range $x = 0$ to 5 for $t = 0$ to 10.

Problem 6-17: Convolution theorem

An important property of the Laplace transform is the *convolution theorem*. If $f_1(t)$ and $f_2(t)$ are two functions, their *convolution* is defined to be

$$C(T) = \int_0^T f_1(T - t)\, f_2(t)\, dt.$$

If $F(s)$, $F_1(s)$, and $F_2(s)$ are the Laplace transforms of $C(T)$, $f_1(t)$, and $f_2(t)$, respectively, the convolution theorem states that

$$F(s) = F_1(s)\, F_2(s).$$

Using the integral transform package, take the Laplace transform of $C(t)$ and confirm the convolution theorem.

6.2.4 "Play It, Sam" Revisited

How can you tell Al Gore from a roomful of Secret Service agents?
He's the stiff one.

Al Gore, former U.S. Vice President, joking about his reputation for stiffness,
New York Times, 14 September 1996

As pointed out in the subsection, **Play It, Sam**, the small transverse vibrations of a light, horizontal, stretched, elastic string are well modeled by the
linear wave equation. However, a piano "string," the subject of that section,
is not actually a flexible elastic string, but rather a wire, possessing a degree
of *stiffness*. To understand the concept of stiffness, imagine holding a string
at one end between your fingers. The unsupported end of the string will flop
vertically downward at the juncture with your fingers. The internal forces of
the string are unable to balance the shear force exerted on the string by Earth's
gravitational pull. For a wire, the unsupported end will sag, but not flop vertically downward at the juncture point. The wire is said to have a degree of
stiffness.

For a string under tension, the stiffness is negligible compared to the tension,
but for a piano wire stiffness should be included in the equation of motion. As
shown in Morse [Mor48], the transverse displacement $\psi(x,t)$ of a horizontal
wire is governed by the following fourth-order PDE:

$$\frac{\partial^2 \psi}{\partial x^2} - \frac{1}{2\,\alpha^2} \frac{\partial^4 \psi}{\partial x^4} = \frac{1}{c^2} \frac{\partial^2 \psi}{\partial t^2}. \tag{6.13}$$

The parameter α is a measure of the ratio of tension to stiffness. If $\alpha \to \infty$,
tension predominates and the usual string wave equation results. For intermediate values of α, the full wire equation must be solved. If, on the other hand,
stiffness is all important, then α is very small and the fourth spatial derivative
term dominates over the second spatial derivative. In this limiting case, the
equation relevant to a vibrating "bar" results. As a wire is made thicker and
thicker it becomes a bar. Some common examples of bars are the steel girders
supporting bridges and used in high-rise construction and railway tracks.

Labeling $a^2 \equiv c^2/(2\,\alpha^2)$, the transverse vibrations of a bar are governed by

$$\frac{\partial^2 \psi}{\partial t^2} + a^2 \frac{\partial^4 \psi}{\partial x^4} = 0. \tag{6.14}$$

Because of the fourth spatial derivative, the vibrations of a bar are substantially
different from those of a string.

As a simple example, but one with a complicated answer, suppose that an
infinitely long bar is initially at rest $(\partial \psi(x,0)/\partial t = 0)$ and is given a transverse
displacement

$$\psi(x,0) = A\,e^{-b^2 x^2},$$

with A the amplitude. We want to analytically determine the displacement of
the bar for times $t > 0$ and animate the solution. Our approach will be to make
use of the "full" *Fourier transform* and its inverse.

Given a function $f(x)$ that (along with all of its derivatives) vanishes as $x \to \pm\infty$, the Fourier transform of $f(x)$ and its inverse are defined as

$$F(k) = \int_{-\infty}^{\infty} f(x)\, e^{-I\,k\,x}\, dx, \quad f(x) = \frac{1}{2\pi} \int_{-\infty}^{\infty} F(k)\, e^{I\,k\,x}\, dk. \quad (6.15)$$

If the Fourier transform of $f(x)$ (denoted by $\mathcal{F}[f(x)]$) is $F(k)$, then

$$\mathcal{F}\left[\frac{d^n f(x)}{dx^n}\right] = (I\,k)^n\, F(k).$$

Fourier transforming the spatial part of the bar equation will involve $n = 4$.

Now let's use the "full" Fourier transform to solve the vibrating bar problem, first loading the integral transform package, which contains the necessary **fourier** and **invfourier** (inverse Fourier) commands.

```
>  restart: with(plots): with(inttrans):
```

The parameters a in the bar equation and b in the initial profile are assumed to be positive, as is the time.

```
>  assume(a>0,b>0,t>0):
```

The partial differential equation of motion for the bar is entered, using the shortcut x$4 to enter the fourth spatial derivative.

```
>  pde:=diff(psi(x,t),t,t)+a^2*diff(psi(x,t),x$4)=0;
```

$$pde := \left(\frac{\partial^2}{\partial t^2}\psi(x,\,t)\right) + a^2\left(\frac{\partial^4}{\partial x^4}\psi(x,\,t)\right) = 0$$

The Fourier transform of the spatial part of *pde* is performed, and the function $fourier(\psi(x,\,t),\,x,\,k)$ replaced with $F(t)$.

```
>  FT:=fourier(pde,x,k);
>  FT:=subs(fourier(psi(x,t),x,k)=F(t),FT);
```

$$FT := a^2\, k^4\, F(t) + \left(\frac{d^2}{dt^2}F(t)\right) = 0$$

Since a fourth spatial derivative was involved in the transform, a term $(I\,k)^4 = k^4$ has resulted in the output.

The Fourier transform of the initial profile is calculated,

```
>  F0:=fourier(A*exp(-b^2*x^2),x,k);
```

$$F0 := A\,e^{\left(-\frac{k^2}{4\,b^2}\right)}\sqrt{\frac{\pi}{b^2}}$$

and the second-order differential equation FT solved for $F(t)$, subject to the initial conditions $F(0) = F0$, $\dot{F}(0) = 0$.

```
>  sol:=dsolve({FT,F(0)=F0,D(F)(0)=0},F(t));
```

$$sol := F(t) = \frac{A\,e^{\left(-\frac{k^2}{4\,b^2}\right)}\sqrt{\pi}\,\cos(a\,k^2\,t)}{b}$$

To facilitate the calculation of the inverse Fourier transform, the right-hand side of *sol* is converted to exponential form and simplified.

```
> F2:=simplify(convert(rhs(sol),exp));
```

$$F2 := \frac{1}{2}\frac{A\sqrt{\pi}\left(e^{\left(\frac{k^2(-1+4Iatb^2)}{4b^2}\right)}+e^{\left(-\frac{k^2(1+4Iatb^2)}{4b^2}\right)}\right)}{b}$$

We temporarily set $-1+4Iatb^2 = -B$ and $1+4Iatb^2 = C$ in $F2$.

```
> F3:=subs({-1+4*I*a*t*b^2=-B,1+4*I*a*t*b^2=C},F2);
```

$$F3 := \frac{1}{2}\frac{A\sqrt{\pi}\left(e^{\left(-\frac{k^2 B}{4b^2}\right)}+e^{\left(-\frac{k^2 C}{4b^2}\right)}\right)}{b}$$

To perform the inverse Fourier transform of $F3$, it is necessary to make assumptions about B and C. Clearly the real part of each is positive and assuming $Re(A) > 0$ and $Re(B) > 0$ will work here. A slightly simpler form results if we assume that both B and C are positive, even though they are really complex. The inverse Fourier transform of $F3$ then yields the solution ψ.

```
> psi:=invfourier(F3,k,x) assuming B>0,C>0;
```

$$\psi := \frac{1}{2}\frac{A\left(e^{\left(-\frac{x^2 b^2}{B}\right)}\sqrt{C}+e^{\left(-\frac{x^2 b^2}{C}\right)}\sqrt{B}\right)}{\sqrt{B}\sqrt{C}}$$

The original forms of B and C are substituted back into the displacement ψ, the lengthy complex output being suppressed.

```
> psi:=subs({B=1-4*I*a*t*b^2,C=1+4*I*a*t*b^2},psi):
```

Since the initial profile was real, ψ should be real as well. The complex evaluation command, `evalc`, is applied and the result simplified.

```
> psi:=simplify(evalc(psi));
```

$$\psi := \frac{A e^{\left(-\frac{b^2 x^2}{\%1}\right)}\left(\cos\left(\frac{4b^4 x^2 a t}{\%1}\right)\sqrt{2\sqrt{\%1}+2}+\sin\left(\frac{4b^4 x^2 a t}{\%1}\right)\sqrt{2\sqrt{\%1}-2}\right)}{2\sqrt{\%1}}$$

$$\%1 := 1 + 16 a^2 t^2 b^4$$

As expected, the displacement ψ is completely real. Noting the subexpression $\%1$ (an artifact of exporting the Maple output into the text), the mathematical form of ψ is indeed quite complicated, as predicted earlier.

Taking the nominal values $A = 1$, $a = 0.1$, and $b = 1$ in ψ,

```
> psi:=eval(psi,{A=1,a=0.1,b=1}):
```

the vibrations of the bar are animated over the spatial range $x = -200$ to 200 and time interval $t = 0$ to 400.

```
> animate(plot,[psi,x=-200..200],t=0..400,frames=150,
    numpoints=500,thickness=2);
```

On running the animation, you will observe that the initial Gaussian profile of the bar begins to decrease in the vicinity of $x = 0$, generating oscillations that rapidly disperse away from the origin in both directions. The bar does not overshoot the horizontal equilibrium position near the origin, even for longer times. You should experiment with other values of the parameters.

PROBLEMS:

Problem 6-18: Fourier transform
Calculate the Fourier transforms of the following functions and plot the results:

(a) $f(x) = e^{-3|x|}$; (b)$f(x) = \cos(2x)$; (c) $f(x) = x/(x^2 + 1)$.

Problem 6-19: Inverse Fourier transform
Calculate the inverse Fourier transforms of the following functions, simplifying where necessary, and plot the results:

(a) $f(k) = \cos(\pi k/2)/(1 - k^2)$;

(b) $f(k) = -2Ik/(\pi(k^2 + 2))$;

(c) $f(k) = (1/\sqrt{2\pi})(\sin k/k)^2$.

Problem 6-20: Bandwidth theorem
An approximately monochromatic plane wave packet in one dimension has the instantaneous form $u(x,0) = f(x)e^{Ik_0 x}$, with $f(x)$ the envelope function and k_0 the central wave number. Consider the following functions:

(a) $f(x) = 2e^{-3|x|/2}$;

(b) $f(x) = 4e^{-x^2/4}$;

(c) $f(x) = 5$ for $|x| < 1$ and 0 otherwise.

For each function, perform the following:

- Calculate the wave number spectrum $|A(k)|^2$ where $A(k)$ is the Fourier transform of $f(x)$.

- Plot the intensities $|u(x,0)|^2$ and $|A(k)|^2$.

- Explicitly evaluate the root mean square deviations from the means, Δx and Δk, defined with respect to the above intensities.

- Show that in each case the *bandwidth theorem* (the optical analogue of the *uncertainty principle*) $\Delta x \Delta k \geq \frac{1}{2}$ is satisfied.

Problem 6-21: Temperature distribution in an infinite rod
By Fourier transforming the spatial part of the diffusion equation, determine the temperature distribution $T(x,t)$ in an infinite rod for $t > 0$ when $T(x,0) = T_0$ for $|x| < x_0$, and zero otherwise. Animate $T(x,t)$ for nominal values of the parameters.

Problem 6-22: Another temperature distribution
By Fourier transforming the spatial part of the diffusion equation, determine the temperature distribution $T(x,t)$ in an infinite rod for $t > 0$ when $T(x,0) = T_0 e^{-\alpha^2 x^2}$. Animate $T(x,t)$ for nominal values of the parameters.

6.3 Numerical Simulation of PDEs

Simulated disorder postulates perfect discipline; simulated fear postulates courage; simulated weakness postulates strength.
Sun Tzu, Chinese general, *The Art of War*, c. 490 BC

To solve some linear and almost all nonlinear PDE models, subject to specified initial and/or boundary conditions, one must resort to numerical means and simulate the behavior of the model equation by replacing the spatial and time derivatives with finite difference approximations and introducing an accurate numerical representation of any functional forms. In this section, we will illus-

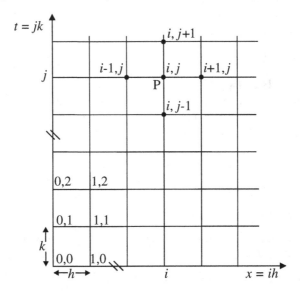

Figure 6.2: Subdividing the x-t plane with a rectangular computational mesh.

trate how fixed-step explicit schemes are created for a one-spatial-dimensional PDE whose dependent variable is some amplitude $\psi(x,t)$, x being the spatial coordinate and t the time. The discussion presented here is meant only to give you the flavor of a vast and important topic.

The general approach is to extend the ideas introduced at the end of Chapter 2 for numerically solving ODEs. Referring to Figure 6.2, the x-t plane can be subdivided for computational purposes into a rectangular grid or mesh, with each rectangle having sides of length h and k in the x- and t-directions, respectively. The coordinates of a representative intersection, or mesh, point P are taken to be $x = i\,h$, $y = j\,k$ with $i, j = 0, 1, 2, \ldots$. In a nonadaptive scheme, since h and k are fixed, the integers i and j may be used to label the mesh points, the point P being indicated by (i, j). The value of ψ at P is written as $\psi_P = \psi(x = i\,h, y = j\,k) \equiv \psi_{i,j}$. A similar subscript notation may be used for the spatial and time derivatives.

For example, in the forward difference approximation, the first spatial and time derivatives at P are written as

$$\left(\frac{\partial \psi}{\partial x}\right)_P = \frac{(\psi_{i+1,j} - \psi_{i,j})}{h}, \quad \left(\frac{\partial \psi}{\partial t}\right)_P = \frac{(\psi_{i,j+1} - \psi_{i,j})}{k}, \tag{6.16}$$

while the "standard" CDAs for the second derivatives at P are

$$\left(\frac{\partial^2 \psi}{\partial x^2}\right)_P = \frac{(\psi_{i+1,j} - 2\psi_{i,j} + \psi_{i-1,j})}{h^2}, \tag{6.17}$$

$$\left(\frac{\partial^2 \psi}{\partial t^2}\right)_P = \frac{(\psi_{i,j+1} - 2\psi_{i,j} + \psi_{i,j-1})}{k^2}. \tag{6.18}$$

Although rectangular meshes are most commonly employed, diamond-shaped grids can also prove to be useful in certain numerical schemes used to model wave equations.

6.3.1 Freeing Excalibur the Numerical Way

"Are five nights warmer than one night, then?"
Alice ventured to ask. "Five times as warm of course."
"But they could be five times as cold, by the same rule—"
"Just so!" cried the Red Queen.
Lewis Carroll (Charles Lutwidge Dodgson), English writer (1832–1898)

Recall that Russell, the aerospace engineer, was having a wild dream about freeing the sword Excalibur from its stony tomb by cooling its ends with buckets of ice water when his dream took a sudden detour and he recalled the following related problem from his undergraduate thermodynamics course.

A thin 1-meter-long rod (the shaft of the sword), whose lateral surface is insulated to prevent heat flow through that surface, has its ends suddenly held at the freezing point of water, $0°$C, by placing them in contact with buckets of ice. Taking one end of the rod to be at $x = 0$ and the other at $x = 1$, the initial temperature distribution was $T(x, t = 0) = 100\, x\, (1 - x)$, a parabolic profile with a maximum temperature of $25°$ at the midpoint $x = \frac{1}{2}$.

Russell was able to determine the analytic solution to this problem, using the separation of variables technique with the aid of Maple. Since it is now necessary in his work to numerically solve a system of nonlinear PDEs, and he hasn't done any numerical work for a while, he decides to tackle the Excalibur problem first, using an explicit finite-difference scheme.

In the linear diffusion equation (with diffusion coefficient $d = 1$), Russell replaces the time derivative with the forward-difference approximation and the second spatial derivative with the standard CDA, so that

$$\frac{\partial T}{\partial t} = \frac{\partial^2 T}{\partial x^2} \Rightarrow \frac{(T_{i,j+1} - T_{i,j})}{k} = \frac{(T_{i+1,j} - 2T_{i,j} + T_{i-1,j})}{h^2}. \tag{6.19}$$

If he sets $r \equiv k/h^2$ and $c \equiv 1 - 2r$, the explicit scheme is

$$T_{i,j+1} = r\,T_{i-1,j} + c\,T_{i,j} + r\,T_{i+1,j}. \qquad (6.20)$$

The unknown value $T_{i,j+1}$ on time step $j+1$ is explicitly determined from the three known values $T_{i-1,j}$, $T_{i,j}$, and $T_{i+1,j}$ on the previous (jth) time step. One starts with the bottom row ($j = 0$) in the numerical mesh shown in Figure 6.2, which corresponds to the initial temperature profile here, and calculates T at each internal mesh point of the first ($j = 1$) time row. The end mesh points of each time row are held at zero temperature in accordance with the boundary conditions. Once all the $T_{i,1}$ values are known on time step 1, the $T_{i,2}$ on time step 2 are calculated in a similar manner, and so on.

Since he intends to use a matrix multiplication approach, Russell makes a call to the LinearAlgebra library package. He divides the spatial range $x = 0$ to $x = 1$ into 12 equal parts, so that there are $M = 11$ internal mesh points on each time row. The value of M is easily increased if necessary.

>   `restart: with(LinearAlgebra): M:=11: begin:=time():`

To display the $M \times M = 11 \times 11$ square matrix that will perform the operation on the rhs of (6.20), the following interface command is entered. The argument `rtablesize=infinity` allows a matrix with $M > 10$ to be explicitly displayed, instead of being represented by a placeholder.

>   `interface(rtablesize=infinity):`

Russell introduces an $M \times M$ *tridiagonal matrix*[3] A using the `BandMatrix` command. In the argument, the list `[r,c,r]` gives the coefficients on the rhs of the explicit scheme (6.20), the number 1 indicates that there is one subdiagonal on either side of the main diagonal, and `M` specifies the size of the matrix A.

>   `A:=BandMatrix([r,c,r],1,M,M);`

$$A := \begin{bmatrix}
c & r & 0 & 0 & 0 & 0 & 0 & 0 & 0 & 0 & 0 \\
r & c & r & 0 & 0 & 0 & 0 & 0 & 0 & 0 & 0 \\
0 & r & c & r & 0 & 0 & 0 & 0 & 0 & 0 & 0 \\
0 & 0 & r & c & r & 0 & 0 & 0 & 0 & 0 & 0 \\
0 & 0 & 0 & r & c & r & 0 & 0 & 0 & 0 & 0 \\
0 & 0 & 0 & 0 & r & c & r & 0 & 0 & 0 & 0 \\
0 & 0 & 0 & 0 & 0 & r & c & r & 0 & 0 & 0 \\
0 & 0 & 0 & 0 & 0 & 0 & r & c & r & 0 & 0 \\
0 & 0 & 0 & 0 & 0 & 0 & 0 & r & c & r & 0 \\
0 & 0 & 0 & 0 & 0 & 0 & 0 & 0 & r & c & r \\
0 & 0 & 0 & 0 & 0 & 0 & 0 & 0 & 0 & r & c
\end{bmatrix}$$

Consulting his old numerical analysis text, Russell finds that the fixed-step explicit scheme for the diffusion equation becomes numerically unstable if the ratio r is greater than $\frac{1}{2}$. Wanting to carry out a reasonably accurate calculation, he takes $r = 0.05$. Inputting the spatial step size $h = 1/(M + 1)$, he

---

[3] A tridiagonal matrix has nonzero matrix elements only on the central diagonal and two adjacent diagonals.

evaluates the time step size $k = r h^2$, as well as the parameter c. Wanting to animate his numerical solution, he decides to create $N = 50$ plots, with each plot separated in time by $s = 20$ steps.

```
>  r:=0.05: h:=1.0/(M+1); k:=r*h^2; c:=1-2*r; N:=50: s:=20:
```
$$h := 0.08333333333 \qquad k := 0.0003472222222 \qquad c := 0.90$$

An operator f will be used to input the initial parabolic temperature profile.

```
>  f:=x->evalf(100*x*(1-x)):
```

With the help of f, the initial ($j = 0$ time row) temperatures at the M internal mesh points are calculated, and expressed as a column vector[4] T_0 so that matrix multiplication can be performed.

```
> T[0]:=<<seq(f(evalf(i/(M+1))),i=1..M)>>; #input temperatures
```

$$T_0 := \begin{bmatrix} 7.638888889 \\ 13.88888889 \\ 18.75000000 \\ 22.22222222 \\ 24.30555556 \\ 25.00000000 \\ 24.30555556 \\ 22.22222222 \\ 18.75000000 \\ 13.88888889 \\ 7.638888886 \end{bmatrix}$$

A graphing operator is formed to plot the temperature profile on the jth step, the plotting points being connected by straight lines. The zero temperatures at the endpoints are included.

```
>  gr:=j->plot([[0,0],seq([i/(M+1),T[j][i,1]],i=1..M),[1,0]],
            labels=["x","T"]):
```

In the following do loop, the temperature profile is calculated every $s = 20$ steps.

```
>  for n from 1 to N do
>  T[n]:=A^s . T[n-1];
>  end do:
```

The CPU time on a 3-GHz personal computer to perform the calculation

```
>  CPUtime:=(time()-begin)*seconds;
```

$$CPUtime := 0.350 \; seconds$$

is a fraction of a second, hardly worth recording. The plots[display] command[5] with the insequence=true option is employed to produce an animated sequence of the temperature profile in the rod.

```
>  plots[display]([seq(gr(j),j=0..N)],insequence=true);
```

---

[4]Entered with the short-hand notation << >>.
[5]A shortcut to first entering with(plots) and then display.

To see Russell's animated diffusion equation solution, you will have to execute the program and use the animation tool bar.

PROBLEMS:

Problem 6-23: Numerical Instability
In the text recipe, confirm that the solution becomes numerically unstable for $r > 0.5$. Numerical instability is signaled by the appearance of increasingly wild oscillations in the solution as time increases.

Problem 6-24: Comparison with exact solution
Explore the change in the percentage error in the numerical mesh values at the end of the run compared with the exact values as M is increased.

Problem 6-25: A different profile
Modify the text recipe to produce an animated numerical solution for the initial temperature profile $T(x, 0) = 25 \sin(\pi x)$. Compare the numerical solution in the center of the rod with the exact solution as a function of time. At what time is the temperature in the middle equal to one-quarter of the initial value?

6.3.2   Enjoy the Klein–Gordon Vibes

The world is never quiet, even its silence eternally resounds with the same notes, in vibrations which escape our ears.
Albert Camus, French-Algerian philosopher, writer (1913–1960)

After tackling the Excalibur heat-diffusion example, Russell decides to numerically investigate the small transverse oscillations of a light stretched horizontal string embedded in a stretched vertical elastic membrane. In the absence of the membrane, the instantaneous displacement $\psi(x, t)$ of the string satisfies the one-dimensional wave equation. The effect of the membrane is to add an additional Hooke's law restoring force, proportional to ψ, on the string, which tends to speed up the vibrations. The relevant transverse wave equation, called the *Klein–Gordon equation* (KGE), then is

$$\frac{\partial^2 \psi}{\partial x^2} - \frac{\partial^2 \psi}{\partial t^2} = a \, \psi, \qquad (6.21)$$

where a is the elastic coefficient of the membrane and the wave speed has been set equal to unity.

If the string is of unit length and fixed at both ends, the boundary conditions are $\psi(0, t) = \psi(1, t) = 0$ for $t \geq 0$. Supposing that the string has the initial transverse profile $f \equiv \psi(x, 0) = x \, (1 - x)^5$ and velocity $g \equiv \dot{\psi}(x, 0) = x^3 \, (1 - x)$, Russell wishes to numerically solve the KGE and animate the oscillations.

At the internal mesh points, he uses the standard CDAs for the second derivatives and approximates the inhomogeneous term with $a \, \psi_{i,j}$. The numerical algorithm for the KGE then is

$$\frac{(\psi_{i+1,j} - 2\psi_{i,j} + \psi_{i-1,j})}{h^2} - \frac{(\psi_{i,j+1} - 2\psi_{i,j} + \psi_{i,j-1})}{k^2} = a\,\psi_{i,j}, \qquad (6.22)$$

or, setting $r \equiv k^2/h^2$ and $c \equiv 2 - 2\,r - k^2\,a$, and rearranging,

$$\psi_{i,j+1} = r\,\psi_{i-1,j} + c\,\psi_{i,j} + r\,\psi_{i+1,j} - \psi_{i,j-1}, \qquad (6.23)$$

with $j = 1, 2, 3, \ldots$. The mesh points involved in this explicit scheme are schematically depicted in Figure 6.3.

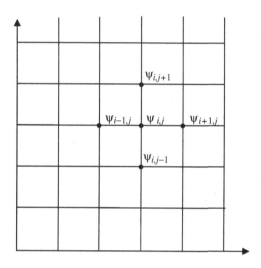

Figure 6.3: Relevant mesh points for numerically solving the KGE.

In this algorithm, the unknown ψ value on time step $(j+1)$ depends on its values on the previous two time steps, j and $(j-1)$. The second time row, corresponding to taking $j = 1$, will be the first to be calculated. The ψ values are known along the zeroth time row from the initial condition $\psi(x_i, 0) = f(x_i)$. To apply the scheme, the ψ values along the first time row must also be known. These may be determined from the initial transverse velocity. Using the forward-difference approximation, the condition $\dot{\psi}(x, 0) = g(x)$ yields[6]

$$\frac{\psi_{i,1} - \psi_{i,0}}{k} = g(x_i), \quad \text{or} \quad \psi_{i,1} \equiv G = \psi_{i,0} + k\,g(x_i). \qquad (6.24)$$

Now Russell programs the explicit scheme (6.23), first loading the Linear-Algebra package so a matrix approach can be used again.

```
>   restart: with(LinearAlgebra): begin:=time():
```

---

[6]In the problem set, Russell will show us a better approximation than this one.

He divides the range $x = 0$ to 1 into 50 equal spatial intervals, so that the number of internal mesh points is $M = 49$. The size of the time step is taken to be $k = 0.005$, the number of time steps is $N = 200$, and $a = 100$.

```
>  M:=49: k:=0.005: N:=200: a:=100:
```

The values of the spatial step size h, the ratio r, and c are calculated.

```
>  h:=1.0/(M+1); r:=k^2/h^2; c:=2-2*r-k^2*a;
```

$$h := 0.02000000000 \quad r := 0.06250000000 \quad c := 1.872500000$$

Except for the inclusion of the last term on the rhs (and a different definition of r), the numerical scheme (6.23) is identical with that for the Excalibur example. So, Russell forms a tridiagonal matrix A as before. Because of its large size, the full matrix is not explicitly displayed, but given by the following placeholder.

```
>  A:= BandMatrix([r,c,r],1,M);
```

$$A := \begin{bmatrix} 49 \text{ x } 49 \text{ Matrix} \\ \text{Data Type : anything} \\ \text{Storage : band}[1,1] \\ \text{Shape : band}[1,1] \\ \text{Order : Fortran\_order} \end{bmatrix}$$

Operators are formed to calculate $f(x)$, $g(x)$, and $G(x)$.

```
>  f:=x->evalf(x*(1-x)^5):
```

```
>  g:=x->evalf(x^3*(1-x)):
```

```
>  G:=x->evalf(f(x)+k*g(x)):
```

Using the operator f, the initial (zeroth time row) displacements are calculated at the M internal mesh points, and put into a column vector format.

```
>  v[0]:=<<seq(f(evalf(i/(M+1))),i=1..M)>>:
```

Similarly, G is used to calculate the displacements at internal mesh points on the first time row and represented as a column vector.

```
>  v[1]:=<<seq(G(evalf(i/(M+1))),i=1..M)>>:
```

A graphing operator is formed to plot the displacement profile on the jth step.

```
>  gr:=j->plot([[0,0],seq([i/(M+1),v[j][i,1]],i=1..M),[1,0]],
             labels=["x","psi"]):
```

The following loop implements the numerical scheme, calculating the displacements at the internal mesh points from the second time row to the Nth row.

```
>  for j from 1 to N do
```

```
>  v[j+1]:=A . v[j]-v[j-1]:
```

```
>  end do:
```

The CPU time is a fraction of a second.

```
>  CPUtime:=(time()-begin)*seconds;
```

$$CPUtime := 0.540 \ seconds$$

Russell animates the transverse oscillations by using the **display** command with the **insequence=true** option.

```
> plots[display]([seq(gr(j),j=0..N)],insequence =true);
```
Once again, you should execute the worksheet to see the oscillations of the string. Feel free to change the parameter values or initial conditions.

Although the KGE can be solved analytically, since it is a linear PDE, Russell's numerical approach allows him to tackle nonlinear wave equations that cannot be solved analytically. For example, the nonlinear KGE

$$\frac{\partial^2 \psi}{\partial x^2} - \frac{\partial^2 \psi}{\partial t^2} = a\,\psi + b\,\psi^3, \qquad (6.25)$$

with a and b positive, might be used to model the symmetric transverse oscillations of the string when the membrane is stretched sufficiently that a nonlinear correction to Hooke's law should be included. The modification of Russell's numerical scheme is left as a problem for the interested reader.

PROBLEMS:

Problem 6-26: $a = 0$
For $a = 0$, the KGE reduces to the linear wave equation. Run the code for $a = 0$ and then explore how the results are affected by increasing a.

Problem 6-27: Nonzero initial velocity
Modify the text recipe for the KGE to handle the initial conditions $f(x) = 0$, $g(x) = \sin(\pi\,x)$. You may have to adjust the viewing box.

Problem 6-28: Nonlinear KGE
Modify the text recipe to numerically simulate the nonlinear KGE (6.25). Take the same parameters as in text recipe and explore what happens when increasing positive values of b are considered.

Problem 6-29: Better first-row approximation
A better approximation to the first time row to use for solving the KGE is

$$\psi_{i,1} = f(x_i) + k\,g(x_i) + \frac{r}{2}(f(x_{i+1}) - 2\,f(x_i) + f(x_{i-1})) + O(k^3).$$

Execute the text recipe with this improved approximation.

6.3.3 Vectoria's Secret

None are so fond of secrets as those who do not mean to keep them.
C. C. Colton, English writer (1780–1832)

Vectoria's secret is out! She and Mike have announced the date of their wedding, which will take place in July in a lupine-dappled alpine meadow near Mount Baker in the North Cascades. Rumor has it that Mike will throw any guest whose cell phone rings into one of the nearby icy Chain Lakes. Since she will be too busy making wedding plans, we will not be seeing Vectoria in the rest of this text. So, we have asked her to favor us with one last recipe for solving the following problem.

In a region of space, the potential $V(x, y)$ satisfies the *Poisson equation*,

$$\frac{\partial^2 V}{\partial x^2} + \frac{\partial^2 V}{\partial y^2} = x\, e^y, \quad 0 \le x \le 2,\ 0 \le y \le 1,$$

with the four boundary conditions $V(0, y) = 0$, $V(2, y) = 2\, e^y$, $V(x, 0) = x$, and $V(x, 1) = e\, x$. Using a central difference approximation for each second derivative at a mesh point (i, j) and evaluating the rhs of Poisson's equation at this point, derive a finite difference scheme for solving the equation. Dividing the x-interval into 30 steps and the y interval into 15 steps, determine V at each mesh point and plot the numerical solution. Here is Vectoria's solution.

She begins by entering the boundaries and dividing them into the suggested intervals in the x- and y-directions.

```
>   restart: with(plots): begin:=time():
```

In the x-direction, the boundaries are at $x = a = 0$ and $x = b = 2.0$. The interval $b - a$ is divided into $m = 30$ steps, and the x step size $h = (b-a)/m$ calculated.

```
>   a:=0: b:=2.0: m:=30: h:=(b-a)/m;
```

$$h := 0.06666666667$$

In the y-direction, the boundaries are at $y = c = 0$ and $y = d = 1.0$. The interval $d - c$ is divided into $n = 15$ steps, and the y step size $k = (d - c)/n$ calculated.

```
>   c:=0: d:=1.0: n:=15: k:=(d-c)/n;
```

$$k := 0.06666666667$$

Using central difference approximations for the second derivatives, evaluating the inhomogeneous term at (i, j), setting $r = (h/k)^2$, and rearranging, Poisson's equation becomes

$$2\,(1 + r)\, V_{i, j} - V_{i+1, j} - V_{i-1, j} - r\,(V_{i, j+1} + V_{i, j-1}) + h^2\, x_i\, e^{y_j} = 0 \quad (6.26)$$

The ratio r is now evaluated.

```
>   r:=(h/k)^2;
```

$$r := 1.000000000$$

The grid coordinates in the x- and y-directions are generated.

```
>   Xcoords:=seq(x[i]=i*h,i=0..m):
>   Ycoords:=seq(y[j]=j*k,j=0..n):
```

The potential V at the grid points along the four bounding edges are generated in the following four boundary conditions.

```
>   bc1:=seq(V[i,0]=i*h,i=0..m):
>   bc2:=seq(V[i,n]=evalf(exp(1))*i*h,i=0..m):
>   bc3:=seq(V[0,j]=0,j=0..n):
>   bc4:=seq(V[m,j]=2*evalf(exp(j*k)),j=0..n):
```

Vectoria then assigns *Xcoords*, *Ycoords*, and the four boundary conditions.

```
>   assign(Xcoords,Ycoords,bc1,bc2,bc3,bc4):
```

An operator f is formed to calculate equation (6.26) at the grid point (i, j).

```
> f:=(i,j)->2*(1+r)*V[i,j]-V[i+1,j]-V[i-1,j]-r*(V[i,j+1]
       +V[i,j-1])+h^2*x[i]*exp(y[j])=0;
```

$$f := (i, j) \rightarrow 2\,(1 + r)\,V_{i,j} - V_{i+1,j} - V_{i-1,j} - r\,(V_{i,j+1} + V_{i,j-1}) + h^2\,x_i\,e^{y_j}$$

The mesh equations are determined for all the internal grid points using two nested sequences running from $i = 1$ to $m - 1$ and from $j = 1$ to $n - 1$.

```
> eqs:={seq(seq(f(i,j),i=1..m-1),j=1..n-1)}:
```

The unknown potentials at the internal grid points are entered as the variables.

```
> vars:={seq(seq(V[i,j],i=1..m-1),j=1..n-1)}:
```

The mesh equations are solved (to 6 digits) for the variables, and the solution is assigned.

```
> sol:=evalf(fsolve(eqs,vars),6): assign(sol):
```

Three-dimensional plotting points are formed for all the grid points, including those on the boundaries.

```
> pts:=seq(seq([x[i],y[j],V[i,j]],i=0..m),j=0..n):
```

Using the `pointplot3d` command, the points representing the potential at each grid point are plotted as size-6 blue circles, the result being shown in Figure 6.4.

```
> pointplot3d([pts],symbol=circle,symbolsize=6,color=blue,
    axes=boxed,orientation=[135,60],tickmarks=[3,3,3],
    labels=["x","y","V"]);
```

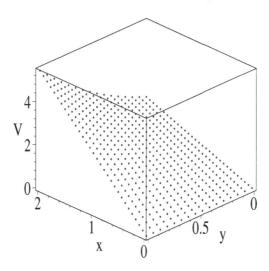

Figure 6.4: Numerical solution of the Poisson equation.

The CPU time to execute the entire recipe is about 6 seconds.

```
> cpu:=time()-begin;
```

$$cpu := 5.978$$

PROBLEMS:
Problem 6-30: Steady-state temperature distribution
The steady-state temperature distribution $T(x, y)$ in a thin square metal plate
0.5 m on a side satisfies Laplace's equation, $\nabla^2 T(x, y) = 0$. The boundary
conditions on the edges of the plate are

$$T(0, y) = 0, \quad T(x, 0) = 0, \quad T(x, 0.5) = 200\, x, \quad T(0.5, y) = 200\, y.$$

Using the standard CDA for the second derivatives, and choosing a suitable
mesh spacing, numerically determine the temperature distribution in the plate
and make a 3-dimensional plot.

Problem 6-31: Potential distribution
A square inner conductor 3 cm on a side is held at a potential of 100 V. A
second square conductor, concentric with the first and 9 cm long on each of its
inner sides, is held at 0 V. The potential $\Phi(x, y)$ in the region between the two
conductors satisfies Laplace's equation, $\nabla^2 \Phi(x, y) = 0$.

(a) Taking the mesh spacing in both directions to be 1 cm, make a mesh
diagram showing all the interior mesh points for which Φ is to be found.

(b) Using CDAs for the second derivatives, write out the mesh equations for
the interior points. Make use of symmetry arguments to show that only
seven interior points need to be used in the calculation of Φ.

(c) Solve the mesh equations and determine Φ at each interior point.

(d) Plot Φ in the region stretching from the inner to the outer conductor.

Part III

THE DESSERTS

The way a child discovers the world constantly
replicates the way science began. You start to
notice what's around you, and you get very curious
about how things work. How things interrelate.
It's as simple as seeing a bug that intrigues you.
You want to know where it goes at night;
who its friends are; what it eats.

David Cronenberg, Canadian filmmaker (1943–)

It's food too fine for angels; yet come, take
And eat thy fill! It's Heaven's sugar cake.

Edward Taylor, English poet (1664–1729)

Chapter 7

The Hunt for Solitons

There is no better ... door by which you can enter into the study of natural philosophy than by considering the ... physical phenomena of a candle.
Michael Faraday, English physicist (1791–1867)

Nonlinear PDEs display a rich spectrum of solutions that in most cases must be obtained by numerical means. However, there exist special analytic solutions to some nonlinear PDEs of physical interest, the best known being *soliton* solutions of nonlinear wave equations. A soliton is a stable *solitary wave*, which is a localized pulse solution that can propagate at some characteristic velocity without changing shape despite the "tug of war" between "competing terms" in the governing equation of motion.

A simple physical example [EJMR81] of a solitary wave is provided by the flame of an ordinary lit candle. There exists a dynamic balance between the diffusion of the heat from the flame into the wax and the nonlinear energy release as the wax vaporizes. The candle flame advances into the wax at a velocity that just maintains the balance. To check whether a solitary wave is stable, i.e., is a soliton, one can subject the solitary wave to some type of perturbation and see whether its integrity is preserved. For example, the candle flame may flicker because of an ambient air current, but it tends to preserve its shape as the candle burns, so the flame displays soliton-like behavior.

Of course, there exist many different possible stability criteria that could be invoked to decide whether a solitary wave is a soliton. Historically, however, mathematicians have decided that in order for a solitary wave to be deemed worthy of the name soliton, it must survive a collision with another solitary-wave solution of the same PDE completely unchanged in shape. There are two main approaches to applying this collisional stability criterion, either numerically or analytically. The numerical simulation approach will be briefly illustrated in the last section. Analytic methods are considerably more complicated to implement (see, e.g., [EM00]) and we will be content here only to quote some of the results in the form of *two-soliton solutions*.

Given a nonlinear PDE, how do we know that it even has the possibility of having soliton solutions? This chapter is about the hunt for solitary waves (possible solitons), using graphical and analytic approaches, and the analytic confirmation that some of these solitary waves are indeed solitons.

First, we should have some idea of what solitary waves look like, and what well-known nonlinear PDEs of physical interest are known to have them. To keep the discussion simple, let's restrict our attention to wave motion in one spatial dimension. Three well-known nonlinear PDEs that describe different types of wave motion are the *Korteweg–de Vries equation* (KdVE), the *sine–Gordon equation* (SGE), and the *nonlinear Schrödinger equation* (NLSE):

- $\dfrac{\partial \psi}{\partial t} + \alpha\,\psi\,\dfrac{\partial \psi}{\partial x} + \dfrac{\partial^3 \psi}{\partial x^3} = 0,$ KdVE;

- $\dfrac{\partial^2 \psi}{\partial x^2} = \dfrac{1}{c^2}\dfrac{\partial^2 \psi}{\partial t^2} + \sin\psi,$ SGE;

- $i\,\dfrac{\partial \psi}{\partial x} \pm \dfrac{1}{2}\dfrac{\partial^2 \psi}{\partial t^2} + |\psi|^2\,\psi = 0,$ NLSE.

Here x is the spatial coordinate, t is the time, α is a numerical scale parameter, $i = \sqrt{-1}$, and $\psi(x,t)$ is the amplitude. All three equations turn out to be very important in nonlinear dynamics because, under suitable approximations, they arise in many different contexts.

The KdVE has been used to describe [SCM73] water waves in shallow canals, magnetohydrodynamic waves in plasmas, longitudinal dispersive waves in elastic rods, pressure waves in liquid–gas bubble mixtures, and so on.

The SGE is applicable [BEMS71] to the propagation of magnetic spins in ferromagnets, magnetic flux in Josephson junctions, crystal dislocations, ultrashort optical pulses, etc.

Undoubtedly, the most important application of the NLSE [Has90] is to the propagation of optical pulses in glass fibers whose refractive index n is of the form $n = n_0 + n_1\,I$, with n_0 and n_1 positive constants and I the light intensity. The light intensity is proportional to $|\psi|^2$, where ψ is the complex electric field amplitude that satisfies the NLSE. The light intensity is experimentally measured rather than the electric field amplitude.

The derivation of these three nonlinear wave equations is beyond the scope of this text. The reader who is interested in such matters should consult the references cited above. As will be demonstrated, all three nonlinear PDEs support collisionally stable solitary-wave solutions, i.e., solitons.

What do solitary waves (solitons) look like? Figure 7.1 shows a sketch of the two commonly occurring types. For the peaked variety (called *nontopological* solitary waves by mathematicians), there exists a localized maximum with the pulse dropping to zero amplitude at $\pm\infty$. Nontopological solitary waves also exist whose pulse amplitude displays a localized dip to zero with the pulse increasing to a constant nonzero amplitude at $\pm\infty$. In terms of the intensity, the NLSE supports both types of nontopological solitary waves. Those optical solitary waves with peaks are called *bright solitary waves*, while those with the

dip are referred to as *black solitary waves*. The bright ones occur for the plus
sign in the NLSE, the black ones for the minus sign. The Korteweg–de Vries
(KdV) equation also possesses nontopological solitary-wave solutions.

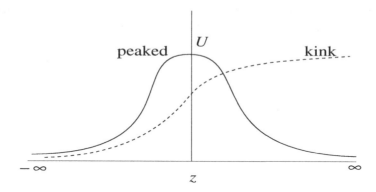

Figure 7.1: Qualitative shapes of two common types of solitary waves.

For the *kink* type (referred to as *topological* solitary waves) in Figure 7.1, the
pulse amplitude U changes from one constant value (e.g., zero in the figure) at
$-\infty$ to a larger constant value at $+\infty$. The region in which the change takes
place is usually quite localized. *Antikink solitary waves* can also exist, for which
the amplitude changes from a constant value at $-\infty$ to a lower constant value
at $+\infty$. The SGE displays both kink and antikink solutions.

Given a nonlinear wave equation, how are these solitary-wave solutions
found? We know that the one-dimensional linear wave equation has a general solution of the structure $\psi(x,t) = f(x-ct) + g(x+ct)$, where f and g
are arbitrary functions. The function $f(x-ct)$ describes a waveform traveling
with speed c in the positive x-direction, while the form $g(x+ct)$ describes a
wave traveling in the negative x-direction. Let us now confine our attention to
waves traveling in the positive x-direction, the discussion for waves traveling
in the opposite direction being similar. The linear wave equation can support
localized solutions $\psi(x,t) = f(z = x - ct)$ such as the peaked solitary waves
shown in Figure 7.1. These solutions will translate unchanged in shape along
the positive x-axis. For nonlinear wave equations, we can look for similar types
of localized solutions. That is to say, we seek solutions of the mathematical form
$\psi(x,t) = \psi(z = x - ct)$ that have profiles qualitatively similar to those shown
in Figure 7.1. Note that this assumed form reduces the number of independent
variables from two (x and t) to one (z), thus reducing the PDE to an ODE.

Borrowing concepts from Chapter 1, we can make a phase-plane portrait for
the nonlinear ODE. As will be shown in the following section, the solitary-wave
solutions will correspond to *separatrix* solutions, a separatrix being a trajectory
in the phase plane that divides the plane into regions with qualitatively different
behaviors. The graphical method of hunting for solitons is quite important,
because it allows for their possible existence to be established, even if analytic
forms do not exist.

7.1 The Graphical Hunt for Solitons

7.1.1 Of Kinks and Antikinks

If you are idle, be not solitary; if you are solitary, be not idle.
Samuel Johnson, letter, 27 October 1779, to James Boswell

In this first example, a phase-plane portrait will reveal that the SGE has both
kink and antikink solitary waves. To produce this portrait, the DEtools library
package must be loaded. In order to implement the assumption $\psi(x,t) = \psi(z = x - ct)$, the dchange command will be employed to make a change of variables,
requiring us also to load PDEtools. The SGE is then entered,

> restart: with(DEtools): with(PDEtools):

> SGE:=diff(psi(x,t),x,x)-diff(psi(x,t),t,t)=sin(psi(x,t));

$$SGE := \left(\frac{\partial^2}{\partial x^2} \psi(x,\,t) \right) - \left(\frac{\partial^2}{\partial t^2} \psi(x,\,t) \right) = \sin(\psi(x,\,t))$$

and a variable transformation tr introduced with the "old" variables x, t, and
$\psi(x,t)$ related to the "new" variables z, τ, and $U(z)$ by $x = z + c\tau$, $t = \tau$, and
$\psi(x,t) = U(z)$, where c is an arbitrary velocity parameter.

> tr:={x=z+c*tau,t=tau,psi(x,t)=U(z)};

$$tr := \{x = z + c\tau,\, t = \tau,\, \psi(x,\,t) = U(z)\}$$

The variable change (dchange) command applies the transformation to SGE.

> de1:=dchange(tr,SGE,[z,tau,U(z)]);

$$de1 := \left(\frac{d^2}{dz^2} U(z) \right) - c^2 \left(\frac{d^2}{dz^2} U(z) \right) = \sin(U(z))$$

The SGE has been reduced to an ODE $de1$ with z as the independent "spatial"
variable. It can be simplified by collecting the second derivatives, $d^2U(z)/dz^2$.

> de2:=collect(de1,diff(U(z),z,z));

$$de2 := (1 - c^2) \left(\frac{d^2}{dz^2} U(z) \right) = \sin(U(z))$$

The nature of the phase-plane portrait will depend on the parameter c. For
$c > 1$ (so $1 - c^2 < 0$), $de2$ is just the undamped plane-pendulum ODE, whose
portrait was "painted" in Chapter 1. For $c < 1$, noting that $\sin U = -\sin(U + \pi)$,
the portrait is the same as for $c > 1$, except that all fixed points are shifted by
π. For $c = 1$, $\sin(U(z)) = 0$ and $U = 0$, or an integer multiple of π, for all z.

The second-order ODE $de2$ is now rewritten as two first-order equations by
setting $dU/dz = Y$ in $de3$, and substituting $de3$ into $de2$ to produce $de4$.

> de3:=diff(U(z),z)=Y(z); de4:=subs(de3,de2);

$$de3 := \frac{d}{dz} U(z) = Y(z)$$

$$de4 := (1 - c^2) \left(\frac{d}{dz} Y(z) \right) = \sin(U(z))$$

Four phase-plane trajectories will be drawn corresponding to the following initial conditions, which are chosen to be close to the saddle points that occur for $c < 1$ at $(U = 0, \pm 2\pi, Y = 0)$ in the Y vs. U phase-plane portrait.

```
>  ic:=[[U(0)=0.01,Y(0)=0],[U(0)=-0.01,Y(0)=0],
        [U(0)=-6.27,Y(0)=0],[U(0)=6.27,Y(0)=0]]:
```

An operator F is formed to apply the `phaseportrait` command to the ODE system *de3* and *de4* for, say, $c = 0.5$ for specified scene parameters A and B. Choosing $A = U$ and $B = Y$ will generate the phase-plane portrait, while $A = z$ and $B = U$ will produce a plot of $U(z)$. By trial and error, the z range is taken to be from 0 to 11. This will produce approximate separatrix trajectories that leave from one saddle point and end up at another saddle point. The tangent arrows are colored green and are taken (using `arrows=MEDIUM`) to be "two-headed." The four trajectories are given different colors, so that they are easily distinguished on the computer screen.

```
>  F:=(A,B)->phaseportrait([de3,eval(de4,c=0.5)],[U(z),Y(z)],
        z=0..11,ic,scene=[A,B],U=-6.5..6.5,Y=-6.5..6.5,
        stepsize=0.05,dirgrid=[30,30],color=green,linecolor=
        [blue,red,black,magenta],arrows=MEDIUM):
```

Entering `F(U,Y)` produces the phase-plane portrait shown in Figure 7.2.

```
>  F(U,Y);
```

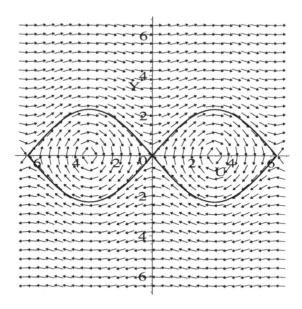

Figure 7.2: Separatixes correspond to kink and antikink solitons.

The solid curves are the separatrixes, two trajectories between the saddle points $(-2\pi, 0)$ and $(0,0)$, and two between $(0,0)$ and $(+2\pi, 0)$. Inside the separatrixes, the solutions are periodic as the trajectories cycle around vortex points at $(U = \pm\pi, Y = 0)$ as z increases. The solutions outside the separatrixes are also oscillatory, resembling "over-the-top" motion for the undamped pendulum.

What do the separatrix solutions look like? Qualitatively, the answer is quite simple. A trajectory starting at the saddle point $(0,0)$ at $z = -\infty$ asymptotically approaches the right saddle point $(2\pi,0)$ as $z \to +\infty$. A similar trajectory connects the left saddle point at $(-2\pi,0)$ to the one at the origin as z varies from $-\infty$ to $+\infty$. These are examples of kink solitary waves, whose profiles $U(z)$ may be seen by entering F(z,U), thus producing Figure 7.3. The curves connecting $U = 2\pi$ to $U = 0$, and $U = 0$ to -2π, are the antikink solutions.

```
>   F(z,U);
```

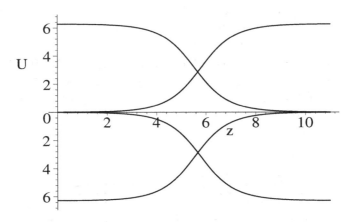

Figure 7.3: Profiles of kink and antikink solitary waves.

An important physical example of a kink is a so-called *Bloch wall* between two magnetic domains in a ferromagnet as schematically depicted in Figure 7.4. The magnetic spins rotate from, say, spin down in one domain to spin up in

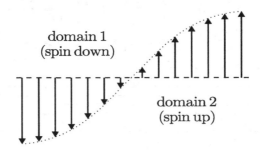

Figure 7.4: Bloch wall between two ferromagnetic domains.

the adjacent domain. The narrow transition region between down and up spins is called a Bloch wall in honor of the theoretical physicist and Nobel laureate

Felix Bloch. Under the influence of an applied magnetic field, the Bloch wall (kink soliton) can propagate according to the SGE without changing in shape.

PROBLEMS:

Problem 7-1: Variation in velocity
Explore how the sine–Gordon solitary waves vary in shape as the velocity c is altered.

Problem 7-2: Cosine–Gordon equation
If the sine term is replaced with a cosine in the SGE, how would the solitary-wave solutions be affected? Confirm your reasoning by running the text recipe with a cosine present, instead of the sine term. You will have to alter the initial conditions to obtain the new separatrixes.

Problem 7-3: Is there or isn't there?
Suppose that the nonlinear term in the SGE is replaced with $\sin^2 \psi$. Using the phase-plane portrait approach, determine whether there is a solitary-wave solution to this modified SGE.

7.1.2 In Search of Bright Solitons

We're all of us sentenced to solitary confinement inside our own skins, for life!
Tennessee Williams, American dramatist (1914–1983)

Our second example illustrates the existence of a bright solitary-wave solution to the NLSE for the situation that the equation has the plus sign.

The DEtools library package is loaded,

> `restart: with(DEtools):`

and the NLSE entered.

> `NLSE:=I*diff(E(x,t),x)+(1/2)*diff(E(x,t),t,t)`
> `          +abs(E(x,t))^2*E(x,t)=0;`

$$NLSE := \left(\frac{\partial}{\partial x} E(x,\,t) \right) I + \frac{1}{2} \left(\frac{\partial^2}{\partial t^2} E(x,\,t) \right) + |E(x,\,t)|^2\, E(x,\,t) = 0$$

Because of the complex nature of the equation, a slightly different assumption is made here than in the sine–Gordon example. In this case, a solitary-wave solution of *NLSE* of the form $E(x,t) = U(t)\, e^{i\,b\,x}$ is sought, where the parameter b, the coordinate x, and $U(t)$ are taken as positive.

> `ode:=eval(NLSE,E(x,t)=U(t)*exp(I*b*x))`
> `          assuming b>0,x>0,U(t)>0;`

$$ode := -U(t)\, b\, e^{(b\,x\,I)} + \frac{1}{2} \left(\frac{d^2}{dt^2}\, U(t) \right) e^{(b\,x\,I)} + U(t)^3\, e^{(b\,x\,I)} = 0$$

This assumption has reduced the NLSE to an ODE, which is simplified by dividing by $e^{i\,b\,x}$ and multiplying by 2.

```
> ode2:=simplify(2*ode/exp(I*b*x));
```

$$ode2 := -2\,U(t)\,b + \left(\frac{d^2}{dt^2}U(t)\right) + 2\,U(t)^3 = 0$$

The second-order ODE is cast into two first-order ODEs by setting $dU/dt = Y$ in *ode3*, and substituting this expression into *ode2*.

```
> ode3:=diff(U(t),t)=Y(t); ode4:=subs(ode3,ode2);
```

$$ode3 := \frac{d}{dt}U(t) = Y(t)$$

$$ode4 := -2\,U(t)\,b + \left(\frac{d}{dt}Y(t)\right) + 2\,U(t)^3 = 0$$

Two phase-plane trajectories will be plotted for initial conditions very close to the origin. The origin will be revealed to be a saddle point.

```
> ic:=[[U(0)=.01,Y(0)=0],[U(0)=-.01,Y(0)=0]]:
```

Taking $b = 1$, an operator F is formed to apply the **phaseportrait** command to *ode3* and *ode4* for specified scene parameters A and B.

```
> F:=(A,B)->phaseportrait([ode3,eval(ode4,b=1)],[U(t),Y(t)],
          t=0..9,ic,scene=[A,B],U=-2..2,Y=-2..2,stepsize=0.05,
          dirgrid=[20,20],color=red,linecolor=blue,arrows=MEDIUM):
```

The phase-plane portrait results on entering F(U,Y),

```
> F(U,Y);
```

the result being shown in Figure 7.5.

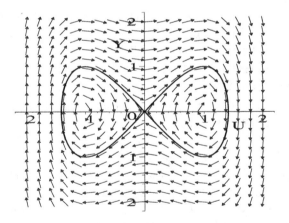

Figure 7.5: Separatrixes correspond to bright solitary waves.

The arrows indicate the direction of increasing t, while the solid curves to the left and right of the origin are the two separatrixes. There are vortex points at $(U = \pm 1, Y = 0)$ and a saddle point at the origin. The separatrix line to the right of the origin starts at $U = 0$ for $t \to -\infty$, grows to a maximum positive value at intermediate t, and returns to zero as $t \to +\infty$. Recalling that the intensity is proportional to $|\psi|^2 = U^2$, we see that the separatrix to the left of the origin will produce exactly the same solitary-wave profile *for the intensity*, which is normally what is measured experimentally. Because this solitary-wave profile is collisionally stable, it is referred to as a bright soliton. The amplitudes $U(t)$ corresponding to the two separatrixes are obtained by entering F(t,U),

>   F(t,U);

the result being shown in Figure 7.6.

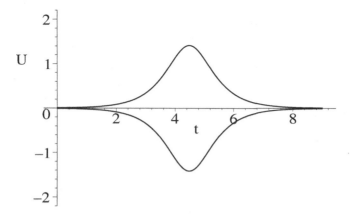

Figure 7.6: Bright solitary-wave amplitudes.

The reader might have been surprised that a different assumed solution was used for the NLSE than in the previous section to derive the bright soliton. This is because in the underlying derivation of the NLSE, which is beyond the scope of this text, a coordinate transformation to a frame moving at the speed of light has already been made. The bright soliton is stationary (has zero velocity) in this moving frame. In the laboratory frame, the bright-soliton solution can be interpreted as follows. At a given point in the medium, the bright-soliton intensity will be essentially zero until the pulse arrives, then grow to a maximum value, and decrease back toward zero as the pulse passes by.

Bright-soliton solutions have been observed [Has90] experimentally in glass fibers, the time interval over which the light intensity is appreciable at a given point in the fiber being of the order of a few picoseconds (1 picosecond (ps) = 10^{-12} seconds).

Because of their narrow widths and their stability, solitons are envisioned by telecommunications engineers as high-bit-rate carriers of digitized information along optical fibers, each soliton representing a "one" and a blank space between adjacent solitons representing a "zero."

Black solitons may be found for the minus-sign case in the NLSE, this being left for you to do as a problem. A few other interesting problems are also presented for your intellectual amusement.

PROBLEMS:

Problem 7-4: Black solitons

Modify the text recipe to determine the solitary-wave solutions for the minus-sign case in the NLSE. Remembering that the physically observed intensity is proportional to $|\psi|^2$, confirm that these solutions are black solitary waves. These solitary waves are collisionally stable, so are black solitons.

Problem 7-5: Saturable refractive index

The NLSE for a dielectric with a saturable refractive index takes the form

$$i\frac{\partial \psi}{\partial x} + \frac{1}{2}\frac{\partial^2 \psi}{\partial t^2} + \frac{|\psi|^2}{1 + a\,|\psi|^2}\,\psi = 0,$$

where $i = \sqrt{-1}$ and a is a positive parameter. Taking $a = 0.5$ and assuming a solution of the form $\psi(x,t) = U(t)\,e^{i\,b\,x}$ with $b = 1$, use the phase-plane portrait to demonstrate graphically that a solitary-wave solution exists. An analytic form is not known for this solitary wave.

Problem 7-6: Burgers' equation

Burger's equation

$$\frac{\partial \psi}{\partial t} + \psi\frac{\partial \psi}{\partial x} = \sigma\frac{\partial^2 \psi}{\partial x^2},$$

with σ a positive parameter, is an example of a *nonlinear diffusion equation*. Graphically show that an antikink solitary-wave solution exists to Burgers' equation for a representative value of the diffusion coefficient σ.

Problem 7-7: Boussinesq's equation

The Boussinesq wave equation, which was first derived in an attempt to describe shallow-water waves (ψ is the surface displacement) propagating in both directions, is

$$\frac{\partial^2 \psi}{\partial x^2} - \frac{\partial^2 \psi}{\partial t^2} + 6\frac{\partial^2 (\psi^2)}{\partial x^2} + \frac{\partial^4 \psi}{\partial x^4} = 0.$$

Using the phase-portrait option, show that a bright solitary-wave solution exists for this equation.

7.1.3   Can Three Solitons Live Together?

... for nothing on earth is solitary ...
Ralph Waldo Emerson, American essayist and philosopher (1803–1882)

An interesting theoretical problem [ER79] in nonlinear optics is the resonant interaction of three collinear waves consisting of two electromagnetic waves labeled with subscripts 1 and 2 and a sound wave with subscript 3. The wave velocities are, respectively, v_1, v_2, and $v_3 \ll v_1, v_2$. The real amplitudes ϕ_1, ϕ_2, and ϕ_3 satisfy the following set of nonlinear PDEs:

$$\frac{\partial \phi_1}{\partial t} + v_1 \frac{\partial \phi_1}{\partial x} = -\beta_1 \, \phi_2 \, \phi_3,$$

$$\frac{\partial \phi_2}{\partial t} + v_2 \frac{\partial \phi_2}{\partial x} = \beta_2 \, \phi_1 \, \phi_3, \qquad (7.1)$$

$$\frac{\partial \phi_3}{\partial t} + v_3 \frac{\partial \phi_3}{\partial x} = -\beta_3 \, \phi_1 \, \phi_2.$$

Here, x is the direction of wave propagation, t is the time, and the coupling parameters β_1, β_2, and β_3 are real and positive. Our goal is to show graphically that there exists a set of three solitary waves, one for each equation, which will propagate along together at a common velocity c.

The DEtools and PDEtools library packages are loaded,

```
>   restart: with(DEtools): with(PDEtools):
```

and the parameter γ, which will shortly be introduced, is unprotected from its Maple assignment as Euler's constant. To generate the $N = 3$ PDEs

```
>   unprotect(gamma): N:=3:
```

with a do loop, let's set $\phi_4 = \phi_1$ and $\phi_5 = \phi_2$.

```
>   phi[4](x,t):=phi[1](x,t): phi[5](x,t):=phi[2](x,t):
```

The following do loop then generates the three relevant PDEs.

```
>   for j from 1 to N do
>   pde[j]:=diff(phi[j](x,t),t)+v[j]*diff(phi[j](x,t),x)
            =(-1)^j*beta[j]*phi[j+1](x,t)*phi[j+2](x,t);
>   end do;
```

$$pde_1 := \left(\tfrac{\partial}{\partial t} \phi_1(x,\, t) \right) + v_1 \left(\tfrac{\partial}{\partial x} \phi_1(x,\, t) \right) = -\beta_1 \, \phi_2(x,\, t) \, \phi_3(x,\, t)$$

$$pde_2 := \left(\tfrac{\partial}{\partial t} \phi_2(x,\, t) \right) + v_2 \left(\tfrac{\partial}{\partial x} \phi_2(x,\, t) \right) = \beta_2 \, \phi_3(x,\, t) \, \phi_1(x,\, t)$$

$$pde_3 := \left(\tfrac{\partial}{\partial t} \phi_3(x,\, t) \right) + v_3 \left(\tfrac{\partial}{\partial x} \phi_3(x,\, t) \right) = -\beta_3 \, \phi_1(x,\, t) \, \phi_2(x,\, t)$$

Solitary-wave solutions are sought that are functions of the single "new" independent "spatial" variable $z = x - ct$, c being an arbitrary velocity for the moment. The relevant variable transformation is entered, with $t = \tau$ and new amplitudes $U_1(z)$, $U_2(z)$, and $U_3(z)$.

```
>   tr:={x=z+c*tau,t=tau,phi[1](x,t)=U[1](z),
        phi[2](x,t)=U[2](z),phi[3](x,t)=U[3](z)};
```

$$tr := \{ x = z + c\tau, \, t = \tau, \, \phi_1(x,\, t) = U_1(z), \, \phi_2(x,\, t) = U_2(z), \, \phi_3(x,\, t) = U_3(z) \}$$

and the dchange command applied to each PDE in the following do loop.

```
>   for j from 1 to N do
>   ode[j]:=dchange(tr,pde[j],[z,tau,U[1](z),U[2](z),U[3](z)]);
```

To simplify the output, the substitution $\beta_j = \gamma_j (v_j - c)$ is made in the jth ODE and each equation divided by $(v_j - c)$ and simplified.

```
>   ode[j]:=subs(beta[j]=gamma[j]*(v[j]-c),ode[j]):
```

```
> ode[j]:=simplify(ode[j]/(v[j]-c)):
> end do:
```

On completion of the do loop, the system *sys* of three resulting nonlinear ODEs is put into a list and displayed.

```
> sys:=[ode[1],ode[2],ode[3]];
```

$$sys := \left[\frac{d}{dz} U_1(z) = -\gamma_1 U_2(z) U_3(z), \frac{d}{dz} U_2(z) = \gamma_2 U_3(z) U_1(z), \right.$$

$$\left. \frac{d}{dz} U_3(z) = -\gamma_3 U_1(z) U_2(z) \right]$$

To graphically find three coexisting solitary-wave solutions, the γ parameters are all set equal to the value 1.

```
> gamma[1]:=1: gamma[2]:=1: gamma[3]:=1:
```

Since the β parameters were all positive, choosing all three γ values to be positive implies that c is less than the smallest v value, i.e., smaller than v_3. Typically, $v_1 \approx v_2 \approx 10^8$ m/s and the speed v_3 of sound is about 10^3 m/s. So, the electromagnetic solitary waves are very unusual, since they would be traveling some five orders of magnitude more slowly than they normally do.

Our graphical procedure will produce a three-dimensional viewing box, so based on trial and error, it is convenient to choose the following two orientations

```
> Orient[1]:=[-90,90]: Orient[2]:=[-90,0]:
```

for the viewing box in the following do loop.

```
> for i from 1 to 2 do
```

The `DEplot3d` plotting command will produce solution trajectories but no tangent field. The independent variable z is allowed to vary from $z = -5$ to $z = 20$. Solitary waves are sought that are kinks (antikinks) or peaked solutions. Peaked solutions would start with $U = 0$ at $z = -\infty$, while kinks would start at some constant value at this limit. One can't literally start at zero, for example, because then it would take forever in terms of z to generate a solution. In the following, U_1 and U_3 start out with the value 0.01 (close to zero) and peaked solutions are sought for these amplitudes. On the other hand, U_2 will start with the value -0.99 (close to -1) and a kink solution sought for this amplitude. The scene option is taken to be `scene=[z,U[1](z),U[2](z)]`, so that with the appropriate orientation, the behavior of U_1 and U_2 as a function of z can be observed.[1]

```
> DEplot3d(sys,[U[1](z),U[2](z),U[3](z)],z=-5..20,
  [[U[1](0)=0.01,U[2](0)=-0.99,U[3](0)=0.01]],
  scene=[z,U[1](z),U[2](z)],U[1]=0..1.2,U[2]=-1.2..1.2,
  stepsize=0.05,orientation=Orient[i],linecolor=blue);
> end do;
```

On completion of the do loop, the two plots are displayed in Figure 7.7.

---

[1] To see U_3 plotted versus z, you can take `scene = [z,U[2](z),U[3](z)]` with an orientation [-90,90].

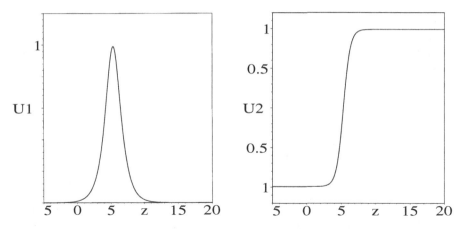

Figure 7.7: Solitary-wave profiles for U_1 (left) and U_2 (right).

On the left is shown the peaked solitary wave for electromagnetic wave number one (U_1) and on the right the coexisting kink solitary-wave solution for the second electromagnetic wave (U_2). The sound-wave amplitude U_3 (not shown) displays a peaked solitary-wave profile as well.

David Kaup [Kau76], [KRB79] has extensively investigated the three-wave problem and analytically established that the three solitary waves are solitons. However, although many years have elapsed since these solitons were predicted, there is still no experimental evidence that such solitons can actually be produced in the laboratory.

PROBLEMS:

Problem 7-8: Solitary sound-wave profile

Modify the recipe in the text to explicitly graph the solitary sound wave profile as a function of z.

Problem 7-9: Variation with γ

Discuss how the three solitary-wave profiles vary in shape as the values of γ_1, γ_2, and γ_3 are altered in the text recipe. For example, try $\gamma_1 = 1$, $\gamma_2 = 2$, and $\gamma_3 = 3$. Support your discussion with the profile plots in each case. Note whether any of the profiles is still a solitary wave or is a wave train.

7.2 Analytic Soliton Solutions

In Chapters 5 and 6, the focus was on solving a wide variety of linear diffusion, Laplace, and wave equation models. The coverage was somewhat lengthy not only because of the physical importance of these PDEs but also because such linear PDEs could be solved analytically, the detailed solutions being readily derived with Maple. The mathematical solution of nonlinear diffusion and wave

equation models is much more difficult, and analytic approaches have consisted
mainly in finding special solutions, some of which are of physical importance.
Many of these approaches are quite complicated in nature, but the seeking of
analytic solitary-wave profiles is relatively easy, provided that analytic solutions
exist. This will now be illustrated in the next two examples.

7.2.1   Follow That Wave!

It's just a job ... waves pound the sand. I beat people up.
Muhammad Ali, American boxer, *New York Times*, 6 April 1977

Probably the first reported observation of soliton behavior recorded in the scientific literature was made by the Scottish engineer and naval architect John
Scott Russell [Rus44]. In the less formal style of scientific reporting of the day,
he wrote:

*I was observing the motion of a boat which was rapidly drawn along a nar-
row channel by a pair of horses, when the boat suddenly stopped — not so the
mass of water in the channel which it had put in motion; it accumulated round
the prow of the vessel in a state of violent agitation, then suddenly leaving it
behind, rolled forward with great velocity, assuming the form of a large solitary
elevation, a rounded smooth and well-defined heap of water, which continued its
course along the channel apparently without change of form or diminution of
speed. I followed it on horseback, and overtook it still rolling on at a rate of
some eight or nine miles an hour, preserving its original figure some thirty feet
long and a foot to a foot and a half in height. Its height gradually diminished,
and after a chase of one or two miles I lost it in the windings of the channel.
Such, in the month of August 1834, was my first chance interview with that
singular and beautiful phenomenon ....*

The "narrow channel" referred to by Russell still exists, being the Union Canal
linking Edinburgh with Glasgow. Actually, Russell was not observing the
"rapidly drawn boat" by accident, but was actually carrying out a series of experiments to determine the force–velocity characteristics of differently shaped
boat hulls in order to determine design parameters for conversion from horse
power to steam power. His solitary-wave observations were followed by ex-
tensive wave-tank experiments in which he established the major properties of
hydrodynamic solitary waves. [EJMR81]

The detailed mathematical explanation of Russell's solitary wave had to wait
50 years until 1895, when the relevant nonlinear Korteweg–de Vries equation,

$$\frac{\partial \psi}{\partial t} + \alpha \, \psi \, \frac{\partial \psi}{\partial x} + \frac{\partial^3 \psi}{\partial x^3} = 0, \tag{7.2}$$

was derived by the Dutch mathematicians Diederik Korteweg and Gustav de
Vries. In the KdV equation, ψ is the transverse displacement of the horizontal
water surface, x the spatial coordinate in the direction of wave propagation, t

the time, and α a numerical factor that can be either scaled out of the equation or, alternatively, assigned a convenient numerical value. The KdV equation doesn't look like a wave equation, having a first time derivative and a third spatial derivative term. However, it does describe the unidirectional propagation of lossless shallow-water waves in a rectangular canal quite well. How a peaked solitary wave solution to the KdV equation occurs is rather interesting.

If the nonlinear term, $\alpha\,\psi\,(\partial\psi/\partial x)$, is neglected, the remaining two terms produce a dispersive (spreading) effect. This is easily understood as follows. A localized propagating pulse can be built up out of a Fourier sum of terms of the plane wave structure $e^{i(k\,x+\omega\,t)}$, where k is the wave number and ω is the frequency. Neglecting the nonlinear term, the *dispersion relation* $\omega = k^3$ results on substituting the plane wave solution. Solving for k, the phase velocity then is $v = \omega/k = \omega^{2/3}$. Therefore, high-frequency Fourier components travel faster than low-frequency components, i.e., dispersion occurs.

On the other hand, if the third-derivative term is ignored, it can be analytically shown [EM00] that the remaining two terms generate a "shock wave" effect. A shock wave is characterized by a progressive steepening of the leading edge of a propagating localized pulse. The solitary-wave solution corresponds to the situation in which the dispersive and shock wave contributions cancel exactly.

To analytically determine the solitary-wave solution, let's follow the procedure of the previous section and assume that $\psi(x,t) = \psi(z = x - ct)$, where $c > 0$ is the velocity, whose value is not yet unspecified. The effect of this simplifying assumption will be to reduce the nonlinear PDE (7.2) to a nonlinear ODE. To carry out this reduction explicitly, the PDEtools library is loaded,

```
>   restart: with(plots): with(PDEtools):
```

and the KdV equation entered.

```
>   KdVE:=diff(psi(x,t),t)+alpha*psi(x,t)*diff(psi(x,t),x)
         +diff(psi(x,t),x,x,x)=0;
```

$$KdVE := \left(\frac{\partial}{\partial t}\,\psi(x,\,t)\right) + \alpha\,\psi(x,\,t)\,\left(\frac{\partial}{\partial x}\,\psi(x,\,t)\right) + \left(\frac{\partial^3}{\partial x^3}\,\psi(x,\,t)\right) = 0$$

The transformation from the "old" independent variables x and t to the new variables z and τ is entered with $x = z + c\tau$ and $t = \tau$. The dependent variable $\psi(x,t)$ is rewritten as $U(z)$.

```
>   tr:={x=z+c*tau,t=tau,psi(x,t)=U(z)};
```

$$tr := \{x = z + c\tau,\, t = \tau,\, \psi(x,\,t) = U(z)\}$$

Then, applying the **dchange** command to *KdVE* with the transformation *tr*,

```
>   ode1:=dchange(tr,KdVE,[z,tau,U(z)]);
```

$$ode1 := -\left(\frac{d}{dz}\,U(z)\right)\,c + \alpha\,U(z)\,\left(\frac{d}{dz}\,U(z)\right) + \left(\frac{d^3}{dz^3}\,U(z)\right) = 0$$

yields the third-order nonlinear ODE shown in *ode1*. The left-hand side of this equation is easily integrated, yielding *ode2*.

> ode2:=int(lhs(ode1),z)=0;

$$ode2 := -c\,U(z) + \frac{1}{2}\,\alpha\,U(z)^2 + \left(\frac{d^2}{dz^2}\,U(z)\right) = 0$$

On the right-hand side of *ode2*, we have set the integration constant equal to zero by assuming that we are seeking a peaked solitary wave similar to that schematically depicted earlier in Figure 7.1. For such a solitary wave, $U(z)$ and all of its derivatives must vanish as $z \to \infty$.

To integrate the second-order nonlinear ODE *ode2*, we can proceed as follows. Letting $V = dU(z)/dz$, then,

$$\frac{d^2U(z)}{dz^2} = \frac{dV}{dz} = \frac{dV}{dU}\frac{dU}{dz} = V\frac{dV}{dU}.$$

Along with $U(z) = U$, this result is substituted into *ode2*.

> ode3:=subs({diff(U(z),z,z)=V(U)*diff(V(U),U),U(z)=U},ode2);

$$ode3 := -c\,U + \frac{\alpha\,U^2}{2} + V(U)\left(\frac{d}{dU}\,V(U)\right) = 0$$

Since $U \to 0$ and $V(U) \equiv dU/dz \to 0$ as $|z| \to \infty$, we can apply the dsolve command to *ode3* subject to the condition $V(U = 0) = 0$,

> sol:=dsolve({ode3,V(0)=0},V(U));

$$sol := V(U) = \frac{\sqrt{-3\,\alpha\,U + 9\,c}\,U}{3}, \quad V(U) = -\frac{\sqrt{-3\,\alpha\,U + 9\,c}\,U}{3}$$

yielding positive and negative square root solutions. The positive answer is selected and simplified.

> sol1:=simplify(sol[1],symbolic);

$$sol1 := V(U) = \frac{\sqrt{-3\,\alpha\,U + 9\,c}\,U}{3}$$

Now, remember that $V \equiv dU/dz$, so from the above output we have

$$z = \int \frac{3\,dU}{U\sqrt{9\,c - 3\,\alpha\,U}}.$$

This integration is now carried out,

> eq:=z=int(1/rhs(sol1),U);

$$eq := z = -\frac{2\,\text{arctanh}\left(\dfrac{\sqrt{-3\,\alpha\,U + 9\,c}}{3\,\sqrt{c}}\right)}{\sqrt{c}}$$

and *eq* solved for U using the isolate command and the result simplified.

> eq2:=simplify(isolate(eq,U));

$$eq2 := U = -\frac{3\,c\left(\tanh\left(\dfrac{z\,\sqrt{c}}{2}\right)^2 - 1\right)}{\alpha}$$

Then, substituting $z = x - c\,t$ into the rhs of *eq2* yields the solitary wave ψ, which is converted into a more standard form with the sincos option.

```
>  psi:=subs(z=x-c*t,rhs(eq2));
```

$$\psi := -\frac{3\,c\left(\tanh\left(\frac{(x-c\,t)\sqrt{c}}{2}\right)^2 - 1\right)}{\alpha}$$

```
>  psi:=simplify(convert(psi,sincos));
```

$$\psi := \frac{3\,c}{\cosh\left(\frac{(x-c\,t)\sqrt{c}}{2}\right)^2 \alpha}$$

At $z \equiv x - ct = 0$, the solitary-wave solution has its maximum amplitude $3\,c/\alpha$. For a fixed value of α, the maximum amplitude is proportional to the velocity c, so by fixing this maximum amplitude, the velocity is also fixed. Clearly, a taller KdV solitary wave will have a larger velocity than a shorter solitary wave. On the other hand, the width of the pulse scales inversely with $\sqrt{c}$, so taller KdV solitary waves are thinner than shorter solitary waves.

To confirm that the solitary-wave solution is indeed localized and travels at constant velocity with unchanging shape, the analytic solution can be animated. Let's take, for example, $\alpha = 1$ and $c = \frac{1}{2}$,

```
>  psi:=eval(psi,{alpha=1,c=1/2});
```

$$\psi := \frac{3}{2}\,\frac{1}{\cosh\left(\frac{(x-t/2)\sqrt{2}}{4}\right)^2}$$

and animate the solitary-wave solution.

```
>   animate(plot,[psi,x=-20..70],t=0..100,numpoints=200,
    frames=50,axes=frame,thickness=2);
```

By running the animation, the reader will see that all the features that have been discussed are confirmed. Later, by looking at the collision of one solitary wave with another, we will confirm that the solitary wave is a soliton.

PROBLEMS:
Problem 7-10: Modified KdV equation
Derive a solitary-wave solution of the modified KdV equation
$$\frac{\partial \psi}{\partial t} + \alpha\,\psi^2\,\frac{\partial \psi}{\partial x} + \frac{\partial^3 \psi}{\partial x^3} = 0,$$
which appears in the theory of double layers in plasmas and as a model of ion acoustic solitons in a multicomponent plasma.

Problem 7-11: Boussinesq's equation
The Boussinesq water wave equation is
$$\frac{\partial^2 \psi}{\partial x^2} - \frac{\partial^2 \psi}{\partial t^2} + 6\frac{\partial^2}{\partial x^2}\left(\psi^2\right) + \frac{\partial^4 \psi}{\partial x^4} = 0.$$
Derive the analytic form of the bright solitary-wave solution and animate it.

7.2.2 Looking for a Kinky Solution

Masterpieces are not single and solitary births; they are the outcome of . . . thinking by the body of the people, so that the experience of the mass is behind the single voice.
Virginia Woolf, British novelist, *A Room of One's Own*, 1929

In our second example, we seek a kink solitary-wave solution of the SGE,

$$\frac{\partial^2 \psi}{\partial x^2} - \frac{\partial^2 \psi}{\partial t^2} = \sin \psi. \tag{7.3}$$

Recall that the moving kink has been used to model the motion of a magnetic domain wall separating two different spin regions in a ferromagnet.

To find a kink, the PDEtools package is loaded and the SGE entered.

```
> restart: with(plots): with(PDEtools):
> SGE:=diff(psi(x,t),x,x)-diff(psi(x,t),t,t)-sin(psi(x,t))=0;
```

$$SGE := \left(\frac{\partial^2}{\partial x^2} \psi(x, t) \right) - \left(\frac{\partial^2}{\partial t^2} \psi(x, t) \right) - \sin(\psi(x, t)) = 0$$

Then, *SGE* is reduced to a nonlinear ODE by assuming that $\psi(x,t) = U(z)$ with $x = z + c\tau$ and $t = \tau$, and applying the dchange command.

```
> tr:={x=z+c*tau,t=tau,psi(x,t)=U(z)}:
> ode1:=dchange(tr,SGE,[z,tau,U(z)]);
```

$$ode1 := \left(\frac{d^2}{dz^2} U(z) \right) - \left(\frac{d^2}{dz^2} U(z) \right) c^2 - \sin(U(z)) = 0$$

The second derivative terms are collected in *ode1*.

```
> ode1:=collect(ode1,diff(U(z),z,z));
```

$$ode1 := (1 - c^2) \left(\frac{d^2}{dz^2} U(z) \right) - \sin(U(z)) = 0$$

For a nontrivial solution, it is necessary that the velocity satisfy $c \neq 1$. Although c could be left as a general parameter, the analytic form is substantially simplified if a specific numerical value is chosen for c, e.g., $c = \frac{1}{4}$.

```
> c:=1/4:
```

Paralleling the procedure in the KdV recipe, let's set $V = dU/dz$, so that $d^2U/dz^2 = V (dV/dU)$, and substitute this result and $U(z) = U$ into *ode1*.

```
> ode2:=subs({diff(U(z),z,z)=V(U)*diff(V(U),U),U(z)=U},ode1);
```

$$ode2 := \frac{15}{16} V(U) \left(\frac{d}{dU} V(U) \right) - \sin(U) = 0$$

For the kink solitary wave, both U and $V \equiv dU/dz$ must go to zero as $z \to -\infty$. So *ode2* is solved for $V(U)$, subject to $V(0) = 0$.

```
> sol:=dsolve({ode2,V(0)=0},V(U));
```

$$sol := V(U) = \frac{4}{15} \sqrt{30 - 30\cos(U)}, \ V(U) = -\frac{4}{15} \sqrt{30 - 30\cos(U)}$$

The positive square root is chosen (the negative square root yields an antikink) and the trig identity $\cos U = 1 - 2\sin^2(U/2)$ substituted,

```
>  sol1:=subs(cos(U)=1-2*sin(U/2)^2,sol[1]);
```

$$sol1 := V(U) = \frac{4}{15}\sqrt{60}\sqrt{\sin\left(\frac{U}{2}\right)^2}$$

and the result simplified using the square root and symbolic options.

```
>  sol1:=simplify(sol1,sqrt,symbolic);
```

$$sol1 := V(U) = \frac{8}{15}\sqrt{15}\sin\left(\frac{U}{2}\right)$$

Since $V = dU/dz$, then $z \equiv x - ct$ is equal to the integral with respect to U of the reciprocal of the right-hand side of *sol1*.

```
>  z:=x-c*t=int(1/rhs(sol1),U);
```

$$z := x - \frac{t}{4} = \frac{1}{4}\sqrt{15}\ln\left(\csc\left(\frac{U}{2}\right) - \cot\left(\frac{U}{2}\right)\right)$$

Then, solving for U and recalling that $U(z) \equiv \psi(x - ct)$,

```
>  solwave:=psi(x,t)=solve(z,U);
```

$$solwave := \psi(x, t) =$$

$$2\arctan\left(\frac{2\,e^{\left(\frac{(4\,x-t)\,\sqrt{15}}{15}\right)}}{1 + \left(e^{\left(\frac{(4\,x-t)\,\sqrt{15}}{15}\right)}\right)^2}, -\frac{\left(e^{\left(\frac{(4\,x-t)\,\sqrt{15}}{15}\right)}\right)^2 - 1}{1 + \left(e^{\left(\frac{(4\,x-t)\,\sqrt{15}}{15}\right)}\right)^2}\right)$$

yields the analytic solitary-wave solution shown in *solwave*. The arctan function is expressed in terms of two arguments separated by a comma. The term to the left (right) of the comma is the numerator (denominator) of arctan.

The common denominators in the arguments can be removed and the numerator of the second argument simplified with the following assumption:

```
>  solwave:=simplify(solwave,assume=real);
```

$$solwave := \psi(x, t) = 2\arctan\left(2\,e^{\left(\frac{(4\,x-t)\,\sqrt{15}}{15}\right)}, -e^{\left(\frac{2\,(4\,x-t)\,\sqrt{15}}{15}\right)} + 1\right)$$

If you can't instantly see that this still complicated-appearing result is a kink solitary-wave solution of the SGE, you can be excused. However, that it is a solution can be confirmed by applying the following **pdetest** command.

```
>  pdetest(solwave,SGE);
```

$$0$$

To see that it is a kink solitary wave, let's animate the rhs of *solwave*.

```
>  animate(plot,[rhs(solwave),x=-10..50],t=0..200,frames=50,
     thickness=2,axes=framed);
```

On running the animation, we observe that a kink solitary wave travels to the right, maintaining its initial shape throughout. The kink varies in amplitude from 0 at $x = -\infty$ to 2π at $z = +\infty$.

PROBLEMS:

Problem 7-12: Antikink solitary waves

Show that the choice of the negative square root yields an antikink solution.

Problem 7-13: Relation between amplitude and velocity

Is there any relation between the maximum nonzero amplitude and the velocity? Is this the same sort of relationship as for the KdV solitary wave or is it different?

Problem 7-14: Burgers' equation

Burgers' nonlinear diffusion equation is of the form

$$\frac{\partial U}{\partial t} + U\,\frac{\partial U}{\partial x} = \sigma\,\frac{\partial^2 U}{\partial x^2},$$

where σ is the positive diffusion coefficient. Analytically derive an antikink solitary-wave solution to Burgers' equation and animate it. How do the width of the antikink region and the velocity depend on amplitude?

Problem 7-15: Sine–Gordon breather

The SGE permits a moving (velocity v) "breather"-mode solution, which is localized in space but oscillatory in time, of the form,

$$\psi = 4\arctan\left(\sqrt{\frac{m}{1-m}}\,\frac{\sin(\gamma\sqrt{1-m}\,(t-v\,x))}{\cosh(\gamma\sqrt{m}\,(x-v\,t))}\right),$$

with $\gamma = 1/\sqrt{1-v^2}$, $-1 < v < 1$, and $0 < m < 1$. The factor γ is the special Lorentz transformation of relativity with speed of light equal to one.

(a) Confirm that ψ is a solution of the SGE.

(b) Animate ψ for $m = \frac{1}{2}$ and (i) $v = 0$, (ii) $v = 0.5$, (iii) $v = -0.9$.

7.2.3   We Have Solitons!

No civilization... would ever have been possible without a framework of stability...
Hannah Arendt, German-born American political philosopher (1906–1975)

To establish whether a solitary wave is a soliton, one must check its ability to survive a collision with another solitary-wave solution of the same PDE. If it remains unchanged in shape after the collision, it's a soliton. Two main approaches are used to study the collision process: analytic and numerical. The analytic approach uses methods beyond the scope of this text (see, e.g., [EM00]) to establish *two-soliton* (more generally, multisoliton) solutions. In this recipe, we shall confirm and animate two-soliton solutions for the Korteweg–de Vries and sine–Gordon equations.

In 1971, Fred Tappert, of Bell Laboratories, found an exact two-soliton analytic solution for the KdV equation, this equation now being entered.

```
>  restart: with(plots):
>  KdVE:=diff(psi(x,t),t)+psi(x,t)*diff(psi(x,t),x)
        +diff(psi(x,t),x,x,x)=0;
```

$$KdVE := \left(\frac{\partial}{\partial t}\,\psi(x,\,t)\right) + \psi(x,\,t)\left(\frac{\partial}{\partial x}\,\psi(x,\,t)\right) + \left(\frac{\partial^3}{\partial x^3}\,\psi(x,\,t)\right) = 0$$

Tappert's two-soliton solution $\psi(x,t)$ is given.

```
>  psi(x,t):=72*(3+4*cosh(2*x-8*t)+cosh(4*x-64*t))
            /(3*cosh(x-28*t)+cosh(3*x-36*t))^2;
```

$$\psi(x,\,t) := \frac{72\,(3 + 4\cosh(2\,x - 8\,t) + \cosh(4\,x - 64\,t))}{(3\cosh(x - 28\,t) + \cosh(3\,x - 36\,t))^2}$$

The lhs of *KdVE* is extracted and simplified, $\psi(x,t)$ having been automatically substituted. A lengthy expression results, which is suppressed here in the text.

```
>  check1:=simplify(lhs(KdVE));
```

The combine command, with the trig option, is applied to the numerator of *check1*, the result being zero, confirming that $\psi(x,t)$ is a solution of *KdVE*.

```
>  check2:=combine(numer(check1),trig);
```

$$check2 := 0$$

The two-soliton solution is now animated.

```
>  animate(plot,[psi(x,t),x=-20..20],t=-1..1,frames=60,
    numpoints=250,axes=frame,thickness=2);
```

In the animation, one initially has a taller, narrower solitary wave to the left of a shorter, wider solitary wave. As the animation progresses, both pulses move to the right. Having a larger velocity, the taller pulse overtakes the shorter pulse and a collision occurs. During the collision, a *nonlinear superposition* takes place, the resultant amplitude being less than the linear sum of the two amplitudes. As time progresses, the taller, faster pulse passes through the shorter one and emerges unchanged in shape, as does the shorter pulse. The solitary waves are indeed solitons.

Next, we confirm that the sine–Gordon equation,

```
>  SGE:=diff(U(x,t),x,x)-diff(U(x,t),t,t)-sin(U(x,t))=0;
```

$$SGE := \left(\frac{\partial^2}{\partial x^2}\,U(x,\,t)\right) - \left(\frac{\partial^2}{\partial t^2}\,U(x,\,t)\right) - \sin(U(x,\,t)) = 0$$

is satisfied by the following two-soliton kink–kink solution. [Jac90]

```
>  U(x,t):=4*arctan(c*sinh(x/sqrt(1-c^2))
            /cosh(c*t/sqrt(1-c^2)));
```

$$U(x,\,t) := 4\arctan\left(\frac{c\sinh\left(\dfrac{x}{\sqrt{1 - c^2}}\right)}{\cosh\left(\dfrac{c\,t}{\sqrt{1 - c^2}}\right)}\right)$$

The lhs of *SGE* is simplified,

>   `check3:=simplify(lhs(SGE));`

and the numerator of *check3* expanded and further simplified.

>   `check4:=simplify(expand(numer(check3)));`

$$check4 := 0$$

The result is zero, so $U(x,t)$ satisfies the sine–Gordon equation. This two-soliton kink–kink solution is now animated, with $c = \frac{1}{4}$.

>   `animate(plot,[eval(U(x,t),c=1/4),x=-50..50],t=-100..100,`
>   `frames=50,thickness=2,axes=framed);`

On running the animation, you will observe that the two kinks travel in opposite directions, run into each other, and reverse directions after the collision, still maintaining their initial shapes.

PROBLEMS:

Problem 7-16: Kink–antikink collision

In the two-soliton kink–kink solution, replace the first c by $1/c$, x by ct, and ct by x. Animate the resulting solution and show that it represents a kink–antikink collision. Describe the observed behavior.

7.3 Simulating Soliton Collisions

Both authors of this text have spent an academic lifetime jousting with non-linearities in all mathematical shapes and sizes. In this text, we have tried to provide a glimpse of the excitement and complexity involved in the study of nonlinear dynamics, yet still present the bread-and-butter recipes necessary to solve linear ODE and PDE problems, the staple of most undergraduate science curricula. Whether the balance of linear and nonlinear recipes is right in our computer algebra menu, you will have to be the judge, but we could not resist presenting two numerical recipes that simulate soliton collisions. The first is for the Korteweg–de Vries equation, the second for the sine–Gordon equation.

7.3.1 To Be or Not to Be a Soliton

There is no means of proving it is preferable to be than not to be.
E. M. Cioran, French philosopher (1911–1995)

To prove that solitary-wave solutions are solitons, i.e., whether they survive collisions with each other unchanged in shape, is an important area of research in nonlinear dynamics. One approach is to numerically collide the solitary waves using a finite difference scheme to simulate the relevant nonlinear PDE. This was done by Norman Zabusky and Martin Kruskal [ZK65] for the KdV equation (taking $\alpha = 1$),

$$\frac{\partial \psi}{\partial t} + \psi \frac{\partial \psi}{\partial x} + \frac{\partial^3 \psi}{\partial x^3} = 0. \tag{7.4}$$

They used a CDA for each first derivative, approximated $\partial^3 \psi / \partial x^3$ by

$$\left(\frac{\partial^3 \psi}{\partial x^3}\right)_P = (\psi_{i+2,j} - 2\,\psi_{i+1,j} + 2\,\psi_{i-1,j} - \psi_{i-2,j})/(2\,h^3), \tag{7.5}$$

and averaged ψ in the nonlinear term equally over the three grid points $(i+1, j)$, (i, j), and $(i-1, j)$. Setting $r = k/h^3$, the Zabusky–Kruskal algorithm is

$$\psi_{i,j+1} = \psi_{i,j-1} - r\,h^2\,(\psi_{i+1,j} + \psi_{i,j} + \psi_{i-1,j})\,(\psi_{i+1,j} - \psi_{i-1,j})/3 \tag{7.6}$$

$$- r\,(\psi_{i+2,j} - 2\,\psi_{i+1,j} + 2\,\psi_{i-1,j} - \psi_{i-2,j}),$$

with $j = 1, 2, \ldots$. This scheme is numerically stable for $r < 0.3849$. [EM00]

As with the Klein–Gordon equation, the unknown value ψ on the time row $j + 1$ involves ψ values from the previous two time rows, j and $j - 1$. On the zero time row, the input will consist of separated solitary waves ordered so that taller pulses are to the left of shorter ones. As time progresses, the pulses move to the right, the taller, faster pulses overtaking the shorter, slower ones, and the pulses "collide." On the first time row, the values of ψ will be calculated from a similar time derivative condition to that used by Russell for the KGE. Let's now implement this numerical scheme with, believe it or not, not just two but three solitary waves. If the execution time is too long on your computer, remove the smallest pulse and shorten the number of time steps. For typing convenience, we shall use the symbol U instead of ψ in the recipe.

In the simulation, we take $M = 250$ spatial steps and $N = 450$ time steps.

```
>   restart: M:=250: N:=450: begin:=time():
```

An operator F is formed to produce a time-dependent KdV solitary-wave profile centered at time $t = 0$ at $x = X$ and having speed c.

```
>   F:=(X,c)->3*c *(sech((sqrt(c)/2)*((x-X)-c*t)))^2:
```

The spatial step size is taken to be $h = 1.0$ and the time step size $k = 0.25$. The ratio $r = k/h^3 = 0.25$ is less than 0.3849, so the numerical scheme will be stable.

```
>   h:=1.0: k:=0.25: r:=k/h^3;
                      r := 0.2500000000
```

The speeds c and pulse centers X of the three solitary waves are entered. The fastest ($c_1 = 0.95$) and therefore tallest and skinniest wave will be initially on the far left with center at $X_1 = 0.2\,M = 50$, the second-fastest ($c_2 = 0.5$) and tallest in the middle at $X_2 = 0.3\,M = 75$, and the slowest ($c_3 = 0.1$) and shortest at $X_4 = 0.4\,M = 100$.

```
>   c[1]:=0.95: c[2]:=0.5: c[3]:=0.1 :
>   X[1]:=0.2*M: X[2]:=0.3*M: X[3]:=0.4*M:
```

Using the operator F, we add the three time-dependent solitary waves in f, and
take the time derivative of f in g.

```
>   f:= add(F(X[i],c[i]),i=1..3):   g:=diff(f,t):
```

Setting the time to zero, we plot the input profile over the spatial range $x = 0$
to M, the resulting picture being shown in Figure 7.8.

```
>   t:=0: plot(f,x=0..M,thickness=2);
```

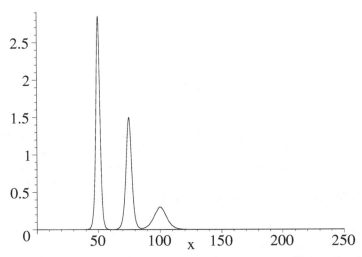

Figure 7.8: Input profile for the three-solitary-wave collision simulation.

Note that there is a slight overlap of the solitary-wave tails. One could place
them further apart initially, but this takes more computing time.

To evaluate f and g at the spatial mesh points, we use the **unapply** command
to turn them into operators in terms of the spatial coordinate x.

```
>   f2:=unapply(f,x):  g2:=unapply(g,x):
```

Using these two operators, we calculate the input amplitudes $U_{i,0} \equiv f(i,t = 0)$
and $U_{i,1} \equiv f(i,t = 0) + k\,g(i,t = 0)$ for $i = 0$ to M in the first and second
initial conditions, *ic1* and *ic2*.

```
>   ic1:=seq(U(i,0)=evalf(f2(i)),i=0..M):
```

```
>   ic2:=seq(U(i,1)=evalf(f2(i))+k*g2(i),i=0..M):
```

To avoid any possible unknown U values creeping into the double do loop that
will be used to iterate the numerical algorithm, we "initialize" all U values to
zero for $i = 0$ to M and $j = 2$ to N, i.e., for all remaining grid points. These
zeros will be overwritten as the loop is executed and new U values calculated.

```
>   init:=seq(seq(U(i,j)=0,i=0..M),j=2..N):
```

The two initial conditions and the initialization are assigned.

```
>   assign(ic1,ic2,init):
```

An operator G is introduced to calculate the rhs of the Zabusky–Kruskal algorithm (7.6) (with ψ replaced with U) for a specified i and j.

```
>   G:=(i,j)->U(i,j-1)-r*h^2*(U(i+1,j)+U(i,j)+U(i-1,j))
             *(U(i+1,j)-U(i-1,j))/3
             -r*(U(i+2,j)-2*U(i+1,j)+2*U(i-1,j)-U(i-2,j)):
```

The numerical algorithm is first iterated from $i = 2$ to $M - 2$ for a given j value and then from $j = 1$ to N. The spatial index i is started at 2 and ended at $M - 2$ to avoid unknown U values from the two edges of the grid entering into the double do loop calculation.

```
>   for j from 1 to N do;
>   for i from 2 to M-2 do
>   U(i,j+1):=G(i,j):
>   end do:
>   end do:
```

A graphing operator is formed to plot the entire profile on the jth time step.

```
>   gr:=j->plot([seq([i,U(i,j)],i=2..M-2)],thickness=2):
```

Using every second graph, the sequence of pictures is now animated.

```
>   plots[display]([seq(gr(2*j),j=0..N/2)],insequence=true);
```

You will have to execute the recipe, click on the resulting plot, and then on the start arrow to see the wonderful animation. Despite the initial overlap of the solitary-wave tails and the coarse spatial grid used (recall, $h = 1$), the solitary waves are remarkably stable, all three surviving the collision process apparently unchanged. After the collision the order of the waves is the reverse of the initial ordering, with the smallest pulse on the left and the largest on the right.

```
>   cpu:=time()-begin;
```

$$cpu := 29.072$$

The CPU time on a 3-GHz PC is about 29 seconds.

Although we have concentrated on soliton collisions here, the recipe may be easily modified to investigate the behavior of other input profiles. For most cases, an analytic solution will not exist and the numerical simulation route is the only feasible one to take.

PROBLEMS:

Problem 7-17: Third derivative
Using the Taylor expansion, derive the approximation (7.5) to $\partial^3 \psi / \partial^3 x$.

Problem 7-18: A different scheme
In the Zabusky–Kruskal finite difference scheme for the KdV equation, ψ in the nonlinear term $\psi (\partial \psi / \partial x)$ was approximated by the average of three ψ terms at the grid points $(i + 1, j)$, (i, j), and $(i - 1, j)$. Compare the results obtained in the text recipe with those you would obtain if $U \equiv \psi$ were approximated by $U_{i,j}$ alone. Discuss your result.

Problem 7-19: Amplification
Multiply the smallest of the three solitary waves in the text recipe by a factor of
3 and interpret the resulting behavior when the worksheet is executed. Explore
the effect of multiplying one or more pulses by numerical factors.

Problem 7-20: Radiative ripples
In the text recipe, change the sech^2 terms in the input pulses to sech^4 terms,
then execute the modified recipe, and discuss the results.

7.3.2   Are Diamonds a Kink's Best Friend?

I never hated a man enough to give him diamonds back.
Zsa Zsa Gabor, Movie actress (1919–)

Although the sine–Gordon equation,

$$\frac{\partial^2 \psi}{\partial x^2} - \frac{\partial^2 \psi}{\partial t^2} = \sin \psi, \tag{7.7}$$

can be numerically solved with an explicit scheme based on a rectangular mesh,
it lends itself more naturally to being tackled with a diamond-shaped mesh
chosen to follow the *characteristic directions* of the equation. To see how this
works, let's consider the general PDE

$$a \frac{\partial^2 U}{\partial x^2} + b \frac{\partial^2 U}{\partial x \partial y} + c \frac{\partial^2 U}{\partial y^2} + e = 0, \tag{7.8}$$

where a, b, c, and e are functions of U, $\partial U/\partial x$, $\partial U/\partial y$, but not of higher
derivatives. For the SGE, $U = \psi$, $y = t$, $a = 1$, $b = 0$, $c = -1$, and $e = -\sin \psi$.

Equation (7.8) can be quite generally solved by the *method of characteristics*.
Setting $p \equiv \partial U/\partial x$ and $q \equiv \partial U/\partial y$, equation (7.8) can be written in the form

$$a \frac{\partial p}{\partial x} + b \frac{\partial p}{\partial y} + c \frac{\partial q}{\partial y} + e = 0. \tag{7.9}$$

Since $p = p(x, y)$ and $q = q(x, y)$, then

$$\frac{dp}{dx} = \frac{\partial p}{\partial x} + \frac{\partial p}{\partial y} \frac{dy}{dx}, \quad \frac{dq}{dy} = \frac{\partial q}{\partial x} \frac{dx}{dy} + \frac{\partial q}{\partial y}. \tag{7.10}$$

Substituting $\partial p/\partial x$ and $\partial q/\partial y$ into (7.9), then multiplying through by dy/dx,
noting that $\partial q/\partial x = \partial p/\partial y$, and rearranging yields

$$\frac{\partial p}{\partial y} \left[a \left(\frac{dy}{dx} \right)^2 - b \left(\frac{dy}{dx} \right) + c \right] - \left[a \frac{dp}{dx} \frac{dy}{dx} + c \frac{dq}{dx} + e \frac{dy}{dx} \right] = 0. \tag{7.11}$$

At this stage, the resulting equation looks like a mathematical mess! However,
if we choose to work in the characteristic directions whose slopes are given by

$$a \left(\frac{dy}{dx} \right)^2 - b \left(\frac{dy}{dx} \right) + c = 0, \tag{7.12}$$

then equation (7.11) reduces to

$$a \left(\frac{dy}{dx} \right) dp + c \, dq + e \, dy = 0. \tag{7.13}$$

For the SGE, $a = 1$, $b = 0$, and $c = -1$, so that (7.12) yields $(dy/dx)^2 - 1 = 0$, or $dy/dx = \pm 1$. The two characteristic directions have slopes of $45°$ and $-45°$, respectively. Forming a diamond-shaped mesh with these slopes produces the grid illustrated in Figure 7.9. Given the new grid, how is U (or ψ) calculated?

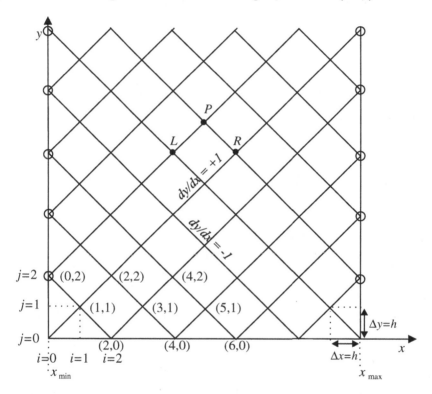

Figure 7.9: Characteristic directions and labels for solving the SGE.

Consider the mesh point P in Figure 7.9, where it is desired to calculate the unknown U_P from the known values U_L and U_R on the previous time step. The subscripts L and R denote advancing "from the left" along the characteristic direction $dy/dx = 1$ and "from the right" along the characteristic direction $dy/dx = -1$, respectively. Taking $dy/dx = \pm 1$ and replacing (7.13) with a finite difference approximation yields the following pair of equations,

$$(p_P - p_L) - (q_P - q_L) = -e_L (y_P - y_L),$$
$$\tag{7.14}$$
$$-(p_P - p_R) - (q_P - q_R) = -e_R (y_P - y_R),$$

which are easily solved for p_P and q_P,

$$p_P = \frac{1}{2}(p_R + p_L) + \frac{1}{2}(q_R - q_L) + \frac{1}{2}(e_R - e_L)\,\Delta y, \qquad (7.15)$$

$$q_P = \frac{1}{2}(p_R - p_L) + \frac{1}{2}(q_R + q_L) + \frac{1}{2}(e_R + e_L)\,\Delta y,$$

with $\Delta y = y_P - y_L = y_P - y_R = \Delta x \equiv h$.

To obtain U_P from the values at L and R, we note that $U = U(x, y)$, so that $dU = (\partial U/\partial x)\,dx + (\partial U/\partial y)\,dy = p\,dx + q\,dy$. Replacing this result with a finite difference approximation along the characteristic direction $dy/dx = 1$ yields

$$U_P = U_L + \frac{1}{2}(p_L + p_P)(x_P - x_L) + \frac{1}{2}(q_L + q_P)(y_P - y_L)$$

$$\text{or} \quad U_P = U_L + \frac{1}{2}\,h\,(p_L + p_P + q_L + q_P), \qquad (7.16)$$

where for improved accuracy, the "old" and "new" values of p and q have been averaged. Of course, we could also have calculated U_P along the other characteristic direction $dy/dx = -1$, viz.,

$$U_P = U_R + \frac{1}{2}(p_R + p_P)(x_P - x_R) + \frac{1}{2}(q_R + q_P)(y_P - y_R),$$

$$\text{or} \quad U_P = U_R + \frac{1}{2}\,h\,(-p_R - p_P + q_R + q_P). \qquad (7.17)$$

Since the two numerical values of U_P will differ slightly, an equally weighted average of the two results for U_P is usually taken.

This method of characteristics scheme is now applied to the collision of a sine–Gordon kink with an antikink solitary wave, the input profile taken to be,

$$U = 4\arctan\left(e^{(x - x_1 - c_1 t)/a}\right) + 4\arctan\left(e^{-(x - x_2 - c_2 t)/b}\right), \quad (7.18)$$

with $a = \sqrt{1 - c_1^2}$, $b = \sqrt{1 - c_2^2}$, $x_1 < 0$, $x_2 > 0$, $c_1 > 0$, and $c_2 < 0$, and the time $t = 0$. The first term in the input profile is the kink,[2] the second the antikink. Nonzero values of x_1 and x_2 are used to spatially separate the kink and antikink initially. The choice of signs puts the kink to the left of the antikink. Since the velocities are of opposite sign, the kink and antikink will move toward each other and a collision will ultimately take place.

So with this preamble behind us, let the final recipe of this chapter begin! Let's consider $M = 200$ spatial and $N = 100$ time steps.

```
>  restart: begin:=time(): M:=200: N:=100:
```

We take $x_1 = -5$, $x_2 = 5$, $c_1 = 0.8$, and $c_2 = -0.8$, and calculate a and b.

```
>  x[1]:=-5: x[2]:=5: c[1]:=0.8: c[2]:=-0.8:
>  a:=sqrt(1-c[1]^2); b:=sqrt(1-c[2]^2);
```

$$a := 0.6000000000 \quad b := 0.6000000000$$

---

[2]A more general form than derived earlier.

The spatial range is taken from $xmin = 3\,x_1 = -15$ to $xmax = 3\,x_2 = 15$, and the step size $h = (xmax - xmin)/M$ calculated.

```
>   xmin:=3*x[1]: xmax:=3*x[2]: h:=evalf((xmax-xmin)/M);
```

$$h := 0.1500000000$$

The kink–antikink profile at time t is entered.

```
>   U:=4*arctan(exp((x-x[1]-c[1]*t)/a))
      +4*arctan(exp(-(x-x[2]-c[2]*t)/b));
```

$$U := 4\arctan\left(e^{(1.666666667\,x + 8.333333335 - 1.333333334\,t)}\right)$$
$$+ 4\arctan\left(e^{(-1.666666667\,x + 8.333333335 - 1.333333334\,t)}\right)$$

The values of $p(x,0) \equiv (\partial U/\partial x)|_{t=0}$ and $q(x,0) \equiv (\partial U/\partial t)|_{t=0}$ are needed, so the relevant spatial and time derivatives are calculated and labeled P and Q.

```
>   P:=diff(U,x): Q:=diff(U,t): t:=0:
```

Setting $t = 0$, the following loop evaluates U, p, and q at each spatial grid point on the zeroth time step. Note that the numerical value of U at the ith spatial grid point is labeled $u_{i,0}$ and i is incremented in steps of 2 from 0 to M.

```
>   for i from 0 to M by 2 do #initial conditions
>   x:=xmin + i*h;
>   u[i,0]:=evalf(U); p[i,0]:=evalf(P); q[i,0]:=evalf(Q);
>   end do:
```

The input profile is now plotted and displayed in Figure 7.10.

```
>   plot([seq([xmin+2*i*h,u[2*i,0]],i=0..M/2)],view=
      [xmin..xmax,-1..15],tickmarks=[3,3],labels=["x","U"]);
```

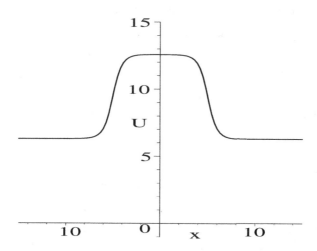

Figure 7.10: Input profile for the kink–antikink soliton collision simulation.

The kink is on the left, the antikink on the right, the two profiles being joined together in the middle. At the x boundaries, one has $U \approx 2\pi$ and $\partial U/\partial x \approx 0$. Since the kink is traveling to the right and the antikink to the left, a collision will take place before each reaches the opposite boundary. Provided that we do not let the kink and antikink come close to those boundaries, the above boundary conditions will prevail for all times in the run. This implies that $\partial U/\partial t \approx 0$ at the x boundaries. In the following do loop, the x-boundary grid points (circled points in Figure 7.9) are initialized, setting $U = 2\pi$ and $p = q = 0$.

```
>   for j from 2 to N by 2 do #boundary conditions
>   u[0,j]:=2*evalf(Pi): p[0,j]:=0: q[0,j]:=0;
>   u[M,j]:=2*evalf(Pi): p[M,j]:=0: q[M,j]:=0;
>   end do:
```

The following double do loop calculates the values of u at the other grid points.

```
>   for j from 0 to (N-1) do:
```

A conditional statement is inserted to start i at $i0 = 0$ for $j = 0, 2, 4, \ldots$ and $i0 = 1$ for $j = 1, 3, 5, \ldots$.

```
>   if j mod 2 = 0 then i0:=0: else i0:=1: end if;
```

The following do loop runs over the spatial grid points i, incrementing them from $i0$ to $M - 2$ in steps of 2.

```
>   for i from i0 to (M-2) by 2 do
```

Using equation (7.15), we calculate the values of p and q.

```
>   p[i+1,j+1]:=0.5*(p[i+2,j]+p[i,j]+q[i+2,j]-q[i,j]
                +(-sin(u[i+2,j])+sin(u[i,j]))*h);
>   q[i+1,j+1]:=0.5*(p[i+2,j]-p[i,j]+q[i+2,j]+q[i,j]
                +(-sin(u[i+2,j])-sin(u[i,j]))*h);
```

Then U is evaluated using equations (7.16) and (7.17), and the average taken.

```
>   uP1:=u[i,j]+0.5*h*(p[i,j]+p[i+1,j+1] +q[i,j]+q[i+1,j+1]);
>   uP2:=u[i+2,j]+0.5*h*(-p[i+2,j]-p[i+1,j+1]+q[i+2,j]
                +q[i+1,j+1]);
>   u[i+1,j+1]:=(uP1+uP2)/2;
>   end do: end do:
```

The results are plotted,

```
>   for j from 0 to N by 2 do
>   pl(j):=plot([seq([xmin+2*i*h,u[2*i,j]],i=0..M/2)],
                thickness=3,labels=["x","U"]):
>   end do:
```

and animated with the insequence=true option.

```
>   plots[display]([seq(pl(2*j),j=0..N/2)],insequence=true);
```

When the work sheet is executed, the kink–antikink hump flips upside down but the shape of the moving profile is identical to the input shape aside from a phase factor $e^{i\pi} = -1$, indicating that the kink and antikink solitary waves are indeed solitons. Finally, the CPU time is calculated.

> `cpu:=(time()-begin)*seconds;`

$$cpu := 3.024 \, seconds$$

and is about 3 seconds on a 3-GHz PC.

PROBLEMS:

Problem 7-21: Amplified kink–antikink input
In the text recipe, double the amplitudes of the input kink and antikink solitary-wave profiles. Remembering to also double the value 2π in the initialization statement to avoid causing an end-effect problem, run the file with the amplified input and interpret the outcome.

Problem 7-22: Kink–kink collision
Modify the recipe to simulate the collision of a kink solitary wave with another kink. Discuss the observed behavior.

Problem 7-23: Antikink–antikink collision
Modify the recipe to simulate the collision of an antikink solitary wave with another antikink. Discuss the observed behavior.

Problem 7-24: Rectangular mesh
Solve the problem of the text recipe using an explicit scheme based on a rectangular mesh.

Problem 7-25: Interacting laser beams
The interaction of two intense laser pulses of different frequencies as they pass through each other in opposite directions in a certain resonant absorbing fluid can be described [RE76] by the following normalized PDEs for the laser intensities U and V,

$$\frac{\partial U}{\partial x} + \frac{\partial U}{\partial y} = -g_1 \, U \, V - \alpha \, U, \quad \frac{\partial V}{\partial x} - \frac{\partial V}{\partial y} = -g_2 \, U \, V + \alpha \, V.$$

Here x is the normalized distance inside the fluid medium of length one unit, y the normalized time, $g_1 > 0$ and $g_2 > 0$ are the "gain" coefficients, and $\alpha \geq 0$ the absorption coefficient. The U pulse travels in the positive x direction, while the V pulse moves in the negative x direction.

(a) Find the characteristic directions along which the PDEs reduce to ODEs.

(b) Devise an explicit numerical scheme that integrates the ODEs along the characteristic directions assuming that there are no pulses initially inside the fluid $(U(x,0) = V(x,0) = 0$ for $0 < x < 1)$ and identical finite-duration U and V pulses are fed in at opposite ends $(U(0,y) = V(1,y) = f(y)$ for $0 \leq y \leq Y = \frac{1}{2}$ and zero for $y > Y$).

(c) Numerically solve the equations and animate the results, assuming that $f(y) = 1$, $g_1 = 0.4$, $g_2 = 20$, and (a) $\alpha = 0$, (b) $\alpha = 0.5$.

(d) Discuss the behavior of the two pulses as revealed in the animation.

(e) Repeat the calculation and animation for $f(y) = \sin(2\pi y)$, the parameter values and all boundary and initial conditions remaining the same. Compare the results with those obtained for the rectangular pulses.

Chapter 8

Nonlinear Diagnostic Tools

In the Appetizers, the reader was introduced to the concept of phase-plane analysis of nonlinear ODE models. This involved the creation of phase-plane portraits and the location and identification of the relevant stationary points of the ODE system. This graphical approach was extended in the last chapter to finding solitary wave solutions of physically important nonlinear PDEs.

Physicists and mathematicians have developed a wide variety of other graphical tools for exploring the frontiers of nonlinear dynamics and understanding what is observed. In this chapter, a few of the simpler diagnostic tools for nonlinear ODEs and nonlinear difference equations are presented for the reader who craves a final light but scrumptious intellectual dessert.

8.1   The Poincaré Section

An important approach to studying the forced motion of nonlinear oscillator systems is to create a *Poincaré section*. If the driving frequency is ω, one takes a "snapshot" of the phase plane after each period $T_0 = 2\pi/\omega$ of the driving force. After an initial transient time, the ODE system will settle down in steady state to either a periodic or a chaotic motion. For the periodic case, the system is said to display a *period-n* response if its period T equals $n T_0$, where $n = 1, 2, 3, \ldots$. In other words, the frequency response of a period-n solution is $\omega_n = \omega/n$.

If the system evolves to a period-1 solution, with $T = T_0$, the Poincaré section will consist of a single plot point in the phase plane that is reproduced at each multiple of the driving period. On the other hand, if the system evolves to a period-2 solution, with $T = 2 T_0$, the Poincaré section will have two points between which the system oscillates as multiples of T_0 elapse. And so on.

In contrast to the periodic situation, for chaotic motion a point is produced at a different location at each multiple of T_0, and the "sum" of the individual snapshots can produce strange, localized patterns ("strange attractors") of plot points with complex boundaries in the phase plane.

In the following recipe, the "period-doubling route to chaos" is explored once again for the forced Duffing oscillator, now from the Poincaré section viewpoint.

319

8.1.1   A Rattler Signals Chaos

Humor is emotional chaos remembered in tranquility.
James Thurber, American writer, humorist, and cartoonist (1894–1961)

As an illustrative example of how a Poincaré section is produced and how it changes character as a control parameter is varied, let's consider Duffing's ODE describing the force oscillations of a nonlinear spring system,

$$\ddot{x} + 2\,g\,\dot{x} + \alpha\,x + \beta\,x^3 = F\cos(\omega\,t). \tag{8.1}$$

Here x is the displacement, g the damping coefficient, α and β real parameters, F the force amplitude, and ω the driving frequency.

The solutions of Duffing's equation were examined in some detail in Section 1.2.1 as the control parameter F was varied, all other parameters being held fixed. In this recipe, we shall use exactly the same coefficient values and initial conditions as in the earlier treatment, including the same F values. This will allow us to compare the results of the Poincaré treatment with the conclusions reached previously on the response of the forced Duffing oscillator.

The number of F values is taken to be $N1 = 4$, and the maximum number of multiples of the driving period $T_0 = 2\pi/\omega$ considered is $N2 = 250$.

```
>  restart: with(plots): N1:=4: N2:=250:
```

The coefficient values are entered and the driving period calculated.

```
>  g:=0.25: alpha:=-1: beta:=1: omega:=1: T[0]:=2*Pi/omega;
```

$$T_0 := 2\pi$$

The force amplitudes are $F_1 = 0.325$, $F_2 = 0.35$, $F_3 = 0.356$, and $F_4 = 0.42$.

```
>  F[1]:=0.325: F[2]:=0.35: F[3]:=0.356: F[4]:=0.42:
```

Duffing's equation can be rewritten as a coupled set of first-order ODEs,

$$\dot{x} = y, \quad \dot{y} = -2\,g\,y - \alpha\,x - \beta\,x^3 + F_i\cos(\omega\,t),$$

which will be numerically solved with the initial conditions $x(0) = 0.09$, $y(0) = 0$.

```
>  ic:=x(0)=0.09,y(0)=0:
```

The coupled first-order differential equations are entered, an operator being used for the second one, the subscript i of the force amplitude F_i to be given.

```
>  de1:=diff(x(t),t)=y(t):
>  de2:=i->diff(y(t),t)=-2*g*y(t)-alpha*x(t)-beta*x(t)^3
         +F[i]*cos(omega*t):
```

An operator `sol` is introduced to numerically solve *de1* and *de2(i)*, subject to the initial conditions, for a given value of i. The option `maxfun=0` overrules any limit on the maximum number of function evaluations in Maple's numerical algorithm. The output is given as a `listprocedure`.

```
>  sol:=i->dsolve({de1,de2(i),ic},{x(t),y(t)},type=numeric,
         maxfun=0,output=listprocedure):
```

Operators X and Y are formed to evaluate $x(t)$ and $y(t)$

```
>  X:=i->eval(x(t),sol(i)): Y:=i->eval(y(t),sol(i)):
```

for the ith solution, the evaluations being carried out in the following do loop.

```
>   for i from 1 to N1 do xx[i]:=X(i); yy[i]:=Y(i); end do:
```

An operator `Gr` is created to graph a phase-plane point, represented by a size-16 blue cross, at time $t = n T_0$ for the ith amplitude.

```
>   Gr:=(i,n)->pointplot({[yy[i](n*T[0]),xx[i](n*T[0])]}
                  color=blue,symbol=CROSS,symbolsize=16):
```

Using this graphing operator in the following loop generates the Poincaré section for each of the $N1 = 4$ amplitude values. To eliminate the transient part of the solution, a certain number of initial points must be removed in each plot. This number will vary for each numerical run and must usually be determined by trial and error. Here, the first 24 points have been removed. Since $N2 = 250$, each plot still contains 225 points. A suitable viewing box is selected, which is the same for all four plots, and labels added and the minimum number of tick marks controlled.

```
>   for i from 1 to N1 do
>   display([seq(Gr(i,n),n=25..N2)],axes=boxed,view=
    [-0.4..0.8,-1.4..1.4],labels=["y","x"],tickmarks=[3,3]);
>   end do;
```

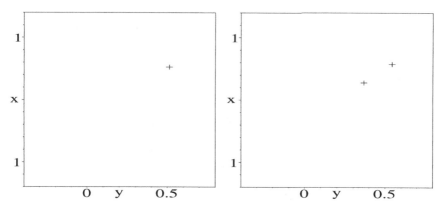

Figure 8.1: Period-1 (left figure) and period-2 (right) Poincaré sections.

The plot on the left of Figure 8.1 corresponds to $F_1 = 0.325$, the one on the right to $F_2 = 0.35$. Despite the fact that 225 points were plotted, we see only a single cross in the $F_1 = 0.325$ graph, indicating that the system has settled down to a period-1 solution. For $F_2 = 0.35$, the ODE system oscillates back and forth between the two crosses, the Poincaré section being characteristic of a period-2 solution. The observed periodicities agree with the results in Section 1.2.1.

For $F_3 = 0.356$, Figure 8.2 indicates a period-4 solution, again agreeing with our earlier conclusion about the periodicity for this forcing amplitude. So it is clear that the Poincaré section approach is a useful graphical tool for interpreting the periodicity of driven oscillator systems.

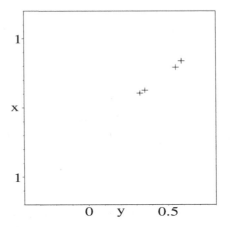

Figure 8.2: Period-4 Poincaré section.

Finally, for $F_4 = 0.42$, we had previously observed a localized chaotic trajectory, characteristic of a strange attractor. The corresponding "strange" Poincaré section is shown in Figure 8.3.

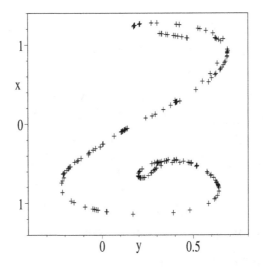

Figure 8.3: Chaotic "rattler."

Strange attractors are often given colorful or descriptive names. What name shall we give to the chaotic Poincaré section obtained above? This depends on the reader's experience and imagination. To coauthor Richard, it reminds him of an incident that occurred while he was hiking with his family in the Superstition Mountains of southern Arizona. While scrambling down a rocky scree slope, he had a feeling of impending chaos when his youngest daughter nearly stumbled into a coiled rattlesnake. To the author, this particular Poincaré section resembles that rattler poised to strike. Fortunately, the snake was just as frightened as the author's daughter and slithered away without striking. Some years later one could joke about the incident, but it was certainly not funny at the time.

PROBLEMS:

Problem 8-1: A different β value
Holding all other parameter values as in the text recipe, use the Poincaré section to determine the response of Duffings's ODE for each F value when $\beta = 2$.

Problem 8-2: Varying the frequency
For each of the four F_i, explore the response of the Duffing ODE as the frequency ω is varied, all other parameters being the same as in the text recipe.

Problem 8-3: Interchanging signs
Determine the response of the Duffing ODE for each F_i when all numerical values are the same as in the text recipe, but the signs of α and β are interchanged.

Problem 8-4: Periodicity?
Using the Poincaré section approach, determine the periodicity of the steady-state solution of the following forced oscillator equations:

(a) $\ddot{x} + 0.7\,\dot{x} + x^3 = 0.75\cos t$, with $x(0) = \dot{x}(0) = 0$;

(b) $\ddot{x} + 0.08\,\dot{x} + x^3 = 0.2\cos t$, with $x(0) = 0.25$, $\dot{x}(0) = 0$.

Problem 8-5: Varying force amplitude
Using the Poincaré section approach, determine the periodicity of the steady-state solution of the following ODE for $F = 0.357$ and $F = 0.35797$:

$$\ddot{x} + 0.5\,\dot{x} - x + x^3 = F\cos(t+1), \text{ with } x(0) = 0.09, \ \dot{x}(0) = 0.$$

8.1.2 Hamiltonian Chaos

Progress everywhere today does seem to come so very heavily disguised as Chaos.
Joyce Grenfell, British actor, writer, *Stately as a Galleon*, 1978

In the Hamiltonian formulation of classical mechanics the motion of a single particle of unit mass, with coordinates q_i and (generalized) momenta p_i, moving in a conservative potential $V(q_i)$ can be described by the *Hamiltonian*,

$$H = \sum_i^N \frac{1}{2} p_i^2 + V(q_i), \tag{8.2}$$

where the first term is the kinetic energy and N is the number of degrees of freedom. Hamilton's equations of motion then are given by

$$\dot{q}_i = \frac{\partial H}{\partial p_i}, \quad \dot{p}_i = -\frac{\partial H}{\partial q_i}. \tag{8.3}$$

For $N = 1$, the motion can be described by a trajectory in the q_1 vs. p_1 phase plane. For $N > 1$, the trajectory is in a $2N$-dimensional *phase space*.

Originally motivated to study the motion of a star inside a galaxy, Hénon and Heiles [HH64] introduced a conservative Hamiltonian describing the motion of a unit mass in the two-dimensional potential

$$V = \frac{1}{2} q_1^2 + \frac{1}{2} q_2^2 + q_1^2 \, q_2 - \frac{1}{3} q_2^3. \qquad (8.4)$$

The first two terms in V would generate a paraboloid of revolution characteristic of a two-dimensional harmonic oscillator. The force $\vec{F} = -\nabla V$ in this case is just the two-dimensional form of Hooke's law. The inclusion of the two cubic terms in V distort the shape of the potential away from a paraboloid, add nonlinear terms to Hooke's law, and Hamilton's equations generate nonlinear ODEs in the 4-dimensional phase space.

In this recipe, we shall use specialized commands found in the DEtools library package to generate these equations, numerically solve them to produce the trajectory for a specified energy and initial conditions, and produce a Poincaré section.

```
> restart: with(plots): with(DEtools):
```
Entering the Hénon–Heiles potential (8.4), a two-dimensional contour plot is generated, the contour lines given by $V = 0.04\,i$ with $i = 0$ to 9. To obtain smooth curves, the number of plotting points is taken to be 5000.

```
> V:=q1^2/2+q2^2/2+q1^2*q2-q2^3/3:
> contourplot(V,q1=-2..2,q2=-2..2,contours=[seq(0.04*i,
  i=0..9)],numpoints=5000,color=black);
```

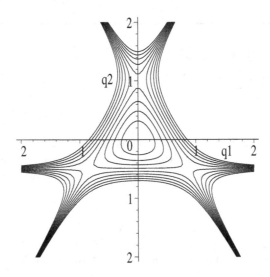

Figure 8.4: Contour plot of the Hénon–Heiles potential.

To aid in interpreting the contour plot, let's find the stationary points where the force vanishes. Differentiating V with respect to each coordinate and solving for the coordinate values that make the force components equal to zero,

```
>  sol:=solve({diff(V,q1),diff(V,q2)},{q1,q2});
```

$$sol := \{q2 = 0, \, q1 = 0\}, \, \{q2 = 1, \, q1 = 0\},$$
$$\{q2 = \frac{-1}{2}, \, q1 = \frac{1}{2} \, \text{RootOf}(-3 + \_Z^2, \, label = \_L2)\}$$

yields a stationary point at $q1 = q2 = 0$ and $q1 = 0$, $q2 = 1$, as well as others at $q2 = -\frac{1}{2}$ and $q1$ given by the RootOf placeholder. These latter values may be found by applying the allvalues command to the third entry in sol.

```
>  sol3:=allvalues(sol[3]);
```

$$sol3 := \{q1 = \frac{\sqrt{3}}{2}, \, q2 = \frac{-1}{2}\}, \, \{q2 = \frac{-1}{2}, \, q1 = -\frac{\sqrt{3}}{2}\}$$

There are two more fixed points at $q1 = \frac{\sqrt{3}}{2}$, $q2 = -\frac{1}{2}$ and at $q1 = -\frac{\sqrt{3}}{2}$, $q2 = -\frac{1}{2}$. The four stationary points are now extracted separately and labeled,

```
>  s1:=sol[1]; s2:=sol[2]; s3:=sol3[1]; s4:=sol3[2];
```

$$s1 := \{q2 = 0, \, q1 = 0\} \qquad s2 := \{q2 = 1, \, q1 = 0\}$$
$$s3 := \{q1 = \frac{\sqrt{3}}{2}, \, q2 = \frac{-1}{2}\} \qquad s4 := \{q2 = \frac{-1}{2}, \, q1 = -\frac{\sqrt{3}}{2}\}$$

and the potential energy at each stationary point determined.

```
>  U:=v->eval(V,v): V1:=U(s1); V2:=U(s2); V3:=U(s3); V4:=U(s4);
```

$$V1 := 0 \qquad V2 := \frac{1}{6} \qquad V3 := \frac{1}{6} \qquad V4 := \frac{1}{6}$$

The stationary point $s1$ at the origin is the minimum of what would be a parabolic potential well if the cubic terms were not present in V. Referring to the contour plot, the shape of the contour lines changes as one moves away from the origin, the contour lines near the other three stationary points being characteristic of saddle points.[1] If the particle has a total energy below the potential energy $\frac{1}{6}$ at the saddle point and starts inside the region bounded by the three saddle points, it will have a bounded orbit inside this region. If $E > \frac{1}{6}$, the particle could escape through one of the saddle points to infinity.

The Hamiltonian is entered, and the command hamilton_eqs used to generate Hamilton's equations and a list of the four dependent variables.

```
>  H:=p1^2/2+p2^2/2+V;
>  hamilton_eqs(H);
```

$$[\frac{d}{dt} p1(t) = -q1(t) - 2 \, q1(t) \, q2(t), \, \frac{d}{dt} p2(t) = -q2(t) - q1(t)^2 + q2(t)^2,$$
$$\frac{d}{dt} q1(t) = p1(t), \, \frac{d}{dt} q2(t) = p2(t)], \, [p1(t), \, p2(t), \, q1(t), \, q2(t)]$$

---

[1]You can confirm this by making a 3-dimensional contour plot using the plot3d command.

The result is a nonlinear system of four coupled ODEs, which must be solved numerically. Let's take as initial conditions $q1(t=0) = q10 = -0.1$, $q2(0) = q20 = -0.2$, $p2(0) = p20 = -0.05$, and a total energy $E0 = 1/16$. This will produce a bounded orbit inside the region bounded by the saddle points.

```
>   q10:=-0.1: q20:=-0.2: p20:=-0.05: E0:=1/16:
```

Note that it is not necessary to specify the initial value of $p1$, since it will be determined by energy conservation. In fact, $p1(0)$ can be obtained by entering the following generate_ic command and asking for one solution.

```
>   ic:=generate_ic(H,{t=0,p2=p20,q2=q20,q1=q10,energy=E0},1):
```

$$ic := \{[0., 0.2667708130, -0.05, -0.1, -0.2]\}$$

The second entry in the output tells us that $p1(0) \approx 0.267$. If desired, more initial conditions can be generated by stating, e.g., a range for $q2(0)$ and specifying the number of $p1(0)$ values wanted.

The following poincare command uses a fourth-order Runge–Kutta method with three iterations and a step size of 0.05 to produce a trajectory over the time interval $t = 0$ to 300 in the $q2$ vs. $p2$ vs. $q1$ phase space. The last number specifies that a 3-dimensional plot is to be produced.

```
>   poincare(H,t=0..300,ic,stepsize=.05,iterations=3,
    scene=[q2=-0.4..0.4,p2=-0.4..0.4,q1=-0.4..0.4],3);
```

$H = .62500000e - 1$,   *Initial conditions* :, $t = 0.$,
$p1 = 0.2667708130$, $p2 = -0.05$, $q1 = -0.1$, $q2 = -0.2$,
Maximum H deviation : $.1020000000e - 5$ %

Time consumed : *19 seconds*

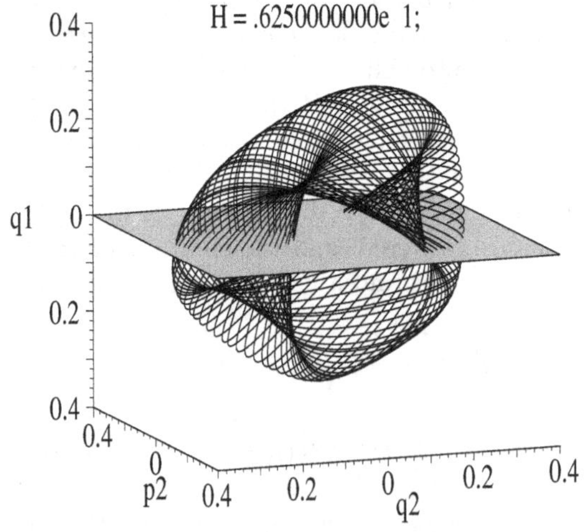

Figure 8.5: Quasiperiodic trajectory for $E = 1/16$.

The output informs us of the input H (energy) value, the initial conditions, the maximum percentage deviation from the input H value, and the computer time taken to produce a plot of the trajectory, which is shown in Figure 8.5. The trajectory executes quasiperiodic motion on the surface of a twisted torus, referred to as the *Kolmogorov–Moser–Arnold* (KAM) *torus*.

The Poincaré section for the $q1 = 0$ plane shown in Figure 8.5 can be obtained with the following `poincare` command. In addition to the same information as before, the number of points (96 here) in the $q2$-$p2$ plane where the trajectory crosses the plane is also given.

```
> poincare(H,t=0..300,ic,stepsize=.05,iterations=3,
  scene=[q2,p2]);
```

$H = .62500000e - 1,$ *Initial conditions* $:, t = 0.,$
$p1 = 0.2667708130,\ p2 = -0.05,\ q1 = -0.1, q2 = -0.2$
Number of points found crossing the $(q2, p2)$ *plane* :   96
Maximum H deviation : $.9800000000e - 6\ \%$

Time consumed :   *1 seconds*

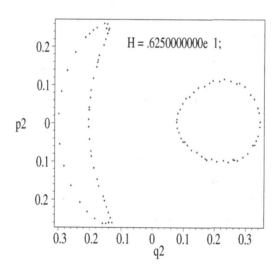

Figure 8.6: Poincaré section for $E = 1/16$.

The Poincaré section can be qualitatively understood. One can think of the KAM torus as a twisted "donut," with the trajectories confined to the donut's surface. The configuration of points in the Poincaré section resembles two distorted ellipses that one might expect from "slicing" the donut through its cross section with the $q1 = 0$ plane.

If the total energy is increased to $\frac{1}{8}$ and the viewing box increased in size, the "chaotic" trajectory shown in Figure 8.7 results.

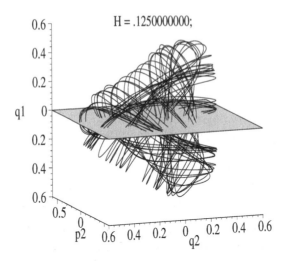

Figure 8.7: Chaotic trajectory for $E = \frac{1}{8}$.

A great deal of mathematics research has gone into understanding the on-set of chaos for the nonlinear Hénon–Heiles ODE system. The interested reader is referred to the nonlinear dynamics texts by Jackson [Jac90] and by Hilborn [Hil94].

PROBLEMS:
Problem 8-6: A different potential
Replace the potential in the text recipe with

$$V = \frac{1}{2} q_1^2 + \frac{1}{2} q_2^2 + q_1^4 \, q_2 - \frac{1}{4} q_2^3.$$

(a) Execute the modified recipe with $E = 1/16$ and initial conditions as in the text recipe. Discuss the resulting plots.

(b) Explore other initial conditions for the same total energy as in part (a). Discuss the results.

(c) Explore what happens as the energy is increased with the same initial conditions as in part (a). Discuss the results.

(d) Explore other potential energy functions and discuss the results.

Problem 8-7: Toda potential
The *Toda potential* [Jac90] is given by

$$V = \frac{1}{3} \left(e^{(q_2 + \sqrt{3} \, q_1)} + e^{(q_2 - \sqrt{3} \, q_1)} + e^{(-2 \, q_2)} \right) - 1.$$

(a) Create two- and three-dimensional contour plots of V, choosing suitable potential energy contours and viewing ranges.

(b) Locate and identify the nature of the stationary points.

(c) Generate the Hamiltonian equations.

(d) For $E = 1.0$, $p_2(0) = -0.05$, $q_2(0) = -0.2$, $q_1(0) = -0.2$, determine $p_1(0)$.

(e) Plot the system's trajectory in the q_1-q_2-p_2 space and discuss the result.

(f) Generate the Poincaré section in the q_2 vs. p_2 plane. Interpret the result.

(g) Explore the Toda Hamiltonian for other E values and discuss the results.

8.2 The Power Spectrum

Still another important diagnostic tool is the *power spectrum*, which, although perhaps better known for its use by engineers in digital signal processing [SK89], can be adapted to studying the frequency content of a solution $x(t)$ of a forced linear or nonlinear oscillator equation.

Suppose that in principle, a nonlinear ODE of physical interest has the time-dependent solution $x(t)$, valid for all t $(-\infty < t < \infty)$. To study the frequency spectrum of the solution, one can introduce the Fourier transform $X(f)$ of x and its inverse through the relations

$$X(f) = \int_{-\infty}^{\infty} x(t)\, e^{-2\pi I f t}\, dt, \quad x(t) = \int_{-\infty}^{\infty} X(f)\, e^{2\pi I f t}\, df, \qquad (8.5)$$

where f is the frequency in hertz (cycles/second), which is related to the frequency w in radians/second by the relation $w = 2\pi f$. With the help of (8.5), the scientifically important *Parseval's theorem* can be derived, having the form,

$$\int_{-\infty}^{\infty} |x(t)|^2\, dt = \int_{-\infty}^{\infty} |X(f)|^2\, df. \qquad (8.6)$$

Parseval's theorem has a simple physical interpretation in, for example, classical mechanics. In this case, $x(t)$ is the instantaneous displacement, and the left-hand side of (8.6) is proportional to the total energy. Thus, since the right-hand side must have the same dimensions, the quantity $|X(f)|^2$ represents the energy per unit frequency interval. Aside from a suitable normalization, which can be introduced, $S(f) \equiv |X(f)|^2$ is called the *power spectrum*. The power spectrum gives us information on the distribution of energy as a function of frequency. If, for example, all the energy is in a single frequency, the spectrum will consist of a single vertical "spike" at that frequency.

For a nonlinear forced oscillator ODE such as Duffing's equation, $x(t)$ cannot in general be determined analytically. For such situations, one must evaluate x numerically at discrete time steps. Instead of obtaining a continuous function $x(t)$, a sequence of x values is obtained for some finite time domain. For sufficiently small time steps, the sequence will approximate a continuous function. Although you might think that one could use the entire sequence to numerically evaluate the Fourier transform of $x(t)$ and then $S(f)$, this procedure would lead to an inordinately long computation time. In practice, one tries to obtain an accurate power spectrum using a limited number of x points. How this is done is now explained in some detail.

Assume that a sequence of N values of x, viz., $x_n \equiv x(t_n = n\,T_s)$ with $n = 0, 1, 2, \ldots, N - 1$ is recorded at evenly spaced time intervals T_s over some finite time range. As with the Poincaré section analysis, one starts recording at a sufficiently large time t_0, so as to ensure that all transients have died away. How large this time must be depends on the nature of the forced oscillator, the parameter values, initial conditions, etc., and is determined by trial and error.

For any choice of the sampling time interval T_s, there is a very special corresponding frequency, $f_{\text{Nyquist}} = 1/(2\,T_s)$, called the *Nyquist frequency*. Note that the sampling frequency $F_s = 1/T_s$ is twice the Nyquist frequency. Why is the Nyquist frequency important? The answer lies in the *sampling theorem*, due to Nyquist [Nyq28] and Shannon [Sha49], which states:

If a continuous signal $x(t)$, sampled at an interval T_s, is such that its Fourier transform $X(f)$ is equal to 0 for all frequencies $|f| > f_{\text{Nyquist}}$, then $x(t)$ is completely determined by the sampled values x_n.

In this case, the Nyquist frequency is clearly greater than the maximum frequency f_{max} in the signal's frequency spectrum. Thus, $F_s > 2\,f_{\text{max}}$.

In signal processing, engineers ensure that the sampling theorem prevails by using a low-pass analog filter on their signal to select $f_{\text{max}}(< f_{\text{Nyquist}})$, removing all higher frequencies by the process of attenuation. When $X(f)$ is zero outside the range $-f_{\text{Nyquist}}$ to f_{Nyquist}, they refer to $x(t)$ as being *bandwidth limited* to frequencies smaller in magnitude than f_{Nyquist}.

What happens if $x(t)$ is not bandwidth limited to this frequency range, i.e., its Fourier transform $X(f)$ does not vanish outside the range $-f_{\text{Nyquist}}$ to f_{Nyquist}? It turns out that the power outside this frequency range gets "folded back" into the range giving an inaccurate power spectrum. This phenomenon is called *aliasing*. To avoid aliasing, one should attempt to make $x(t)$ bandwidth limited by taking the sampling frequency $F_s > 2\,f_{\text{max}}$. Unfortunately, for forced oscillator problems it is usually not known a priori what the maximum frequency component is in the signal $x(t)$. On the other hand, suppose that the chosen sampling frequency or sampling interval $T_s = 1/F_s$ is such that $X(f)$ is not zero at the Nyquist frequency. Then increase F_s to check for possible aliasing. Increasing F_s pushes the Nyquist frequency up, allowing us to see whether there are indeed higher-frequency components present.

Keeping these important aspects in mind, let's continue with the formal derivation of the power spectrum from the N sampled values x_n. It follows from elementary mathematics that given these N values, we can generate the Fourier transform at only N frequencies. Assuming that $x(t)$ is bandwidth limited, we shall take these frequencies to be equally spaced between $-f_{\text{Nyquist}}$ and f_{Nyquist}, viz., the frequencies $f_k \equiv k/(N\,T_s)$ with $k = -N/2, \ldots, 0, \ldots, N/2$. The extreme k values generate $-f_{\text{Nyquist}}$ and f_{Nyquist}. It might seem on counting the

k values that we have $N + 1$ of them, but due to periodicity of the Fourier transform, the extreme k values are not independent, but in fact are equal.

We now approximate the continuous Fourier transform as follows,

$$X(f_k) = \int_{-\infty}^{\infty} x(t)\, e^{-2\pi I f_k t}\, dt \approx \sum_{n=0}^{N-1} x_n\, e^{-2\pi I f_k t_n}\, T_s = T_s\, X_k,$$

where X_k is the *discrete Fourier transform*,

$$X_k \equiv \sum_{n=0}^{N-1} x_n\, e^{-2\pi I k n/N}. \tag{8.7}$$

We can change the k range from $-N/2, \ldots, N/2$ to $0, \ldots, N-1$, thus making it the same as the n range, by noting that X_k is periodic in k with period N. With this standard convention for the range of k, $k = 0$ corresponds to zero frequency, $k = 1, 2, \ldots, N/2-1$ to positive frequencies, $k = N/2+1, \ldots, N-1$ to negative frequencies, and $k = N/2$ to both f_{Nyquist} and $-f_{\text{Nyquist}}$. Because of symmetry of the power spectrum about $k = N/2$, it suffices to plot only the positive frequency range, which is what we will do in our recipes.

In a similar manner, the inverse discrete Fourier transform can be derived:

$$x_n = \frac{1}{N} \sum_{k=0}^{N-1} X_k\, e^{2\pi I k n/N}. \tag{8.8}$$

From the discrete Fourier transform pair, Parseval's theorem then becomes

$$\sum_{n=0}^{N-1} |x_n|^2 = \frac{1}{N} \sum_{k=0}^{N-1} |X_k|^2 \equiv \sum_{k=0}^{N-1} S_N(k), \tag{8.9}$$

where $S_N(k) = |X_k|^2/N$ is the power spectrum.

In calculating the discrete Fourier transform X_k, we will make use of the *fast Fourier transform* (FFT), which is the default numerical algorithm [Bri74] in the FourierTransform command found in Maple's DiscreteTransforms library package. The FFT is based on the idea of splitting the data set in the discrete Fourier transform into even- and odd-labeled points and using the periodicity of the exponential function to eliminate redundant operations. A detailed discussion of this conversion may be found in standard numerical analysis texts, for example in Burden and Faires [BF89] and in *Numerical Recipes* [PFTV89]. Why use this routine? As suggested by the process of eliminating redundant calculations, the FFT is significantly faster than the straightforward evaluation of the discrete Fourier transform.

How much faster? A lot! If N is the number of data points, the discrete Fourier transform involves N^2 multiplications, while the FFT turns out to involve about $N \log_2(N)$ operations. If, for example, $N = 10000$ as in the following recipe, the "normal" discrete Fourier transform requires $(10^4)^2 = 10^8$ computations compared to about $10^4 \times \log_2(10^4) \approx 10^5$ for the FFT. In this case the FFT is about 1000 times faster.

8.2.1   Frank N. Stein's Heartbeat

The more powerful and original a mind, the more it will incline towards the religion of solitude.
Aldous Huxley, British writer (1894–1963)

Let's first illustrate how the power spectrum is calculated for a simple example. We are given Frank N. Stein's steady heartbeat described by

$$x = \sum_{i=1}^{2} A_i \sin(2\pi f_i t),$$

with fundamental frequency $f_1 = 1$ beat per second (60 beats per minute) and a small second harmonic $f_2 = 2$ beats per second. The suitably normalized amplitudes are $A_1 = 1$ and $A_2 = 0.4$. Pretending that we do not know the frequencies, we will extract them from the power spectrum. (We could be dealing with experimental data, not a known analytic form.)

To calculate the discrete Fourier transform, the DiscreteTransforms package must first be loaded. The number of sampling points is taken to be $N = 1000$. The numerical factor $d = 10$ will be used to determine the sampling time.

```
>   restart: with(DiscreteTransforms): N:=1000: d:=10:
```

The parameter values are entered and the time $T_2 = 1/f_2$ is calculated.

```
>   A[1]:=1: A[2]:=0.4: f[1]:=1: f[2]:=2.0: T[2]:=1/f[2];
```
$$T_2 := 0.5000000000$$

An operator is formed to calculate x at a specified time t.

```
>   x:=t->add(A[i]*sin(2*Pi*f[i]*t),i=1..2):
```

Then, $x(t)$ is plotted over the time interval $t = 0$ to $d\,T_2$.

```
>   plot(x(t),t=0..d*T[2],labels=["t","x"]);
```

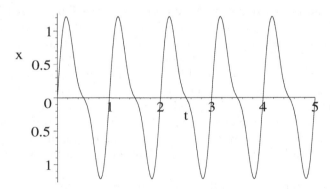

Figure 8.8: Frank N. Stein's heartbeat.

If one were given only the plot shown in Figure 8.8, it would not be obvious exactly what frequencies are contained in Frank's heartbeat. The power spectrum will now reveal what they are. The sampling time interval is taken to be $T_s = T_2/d$ and the sampling frequency $F_s = 1/T_s$ is calculated.

```
> T[s]:=T[2]/d; F[s]:=1/T[s];
```
$$T_s := 0.05000000000 \qquad F_s := 20.00000000$$

The continuous function $x(t)$ is sampled at times $t = n\,T_s$ with $n = 0$ to $N-1$. The sequence of sampled x values is then put into an array format.

```
> x:=Array([seq(x(n*T[s]),n=0..N-1)]):
```

The (discrete) Fourier transform of x is performed. Maple defines this transform such that the kth item in FT is $X_k/\sqrt{N}$ in our notation.

```
> FT:=FourierTransform(x):
```

The following operator will allow us to calculate $S(k)$ for a given k value.

```
> S:=k->abs(FT[k])^2:
```

A plotting point operator pt is introduced that gives the power spectrum $S(k)$ at the frequency $(k-1)\,F_s/N$.

```
> pt:=k->[F[s]*(k-1)/N,S(k)]:
```

Taking the power spectrum to be zero at zero frequency, we produce the points in the power spectrum for k ranging up to $N/2$.

```
> pts:=[[0,0],seq(pt(k),k=2..N/2)]:
```

The complete power spectrum is plotted, and shown in Figure 8.9. There are clearly two frequencies present at 1 and 2 beats per second, as expected.

```
> plot(pts,labels=["f","S"],view=[0..5,0..0.3]);
```

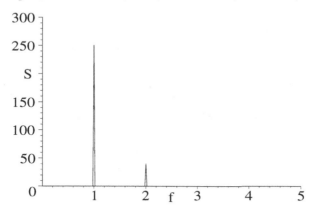

Figure 8.9: Frequencies in Frank N. Stein's heartbeat.

PROBLEMS:
Problem 8-8: Third Harmonic

Suppose that Frank N. Stein's heartbeat also contains the term $0.1 \sin(6\pi t)$. Plotting $\sqrt{S}$, show that the presence of the third harmonic is revealed.

8.2.2 The Rattler Returns

You cannot have power for good without having power for evil too. Even mother's milk nourishes murderers as well as heroes.
George Bernard Shaw, Anglo-Irish playwright, Cusins, in *Major Barbara*, act 3

To illustrate how the power spectrum is calculated for a nonlinear ODE, let's consider the Duffing oscillator of the previous section with exactly the same parameter values and initial conditions. We begin by loading the DiscreteTransforms package and taking the number of sample values of x to be $N = 10000$.

```
>   restart: with(DiscreteTransforms): N:=10000:
```

The Duffing oscillator ODE is entered,

```
>   ode:=diff(x(t),t,t)+ 2*g*diff(x(t),t) + alpha*x(t)
        + beta*x(t)^3=F*cos(omega*t);
```

$$ode := \left(\frac{d^2}{dt^2} x(t) \right) + 2\,g \left(\frac{d}{dt} x(t) \right) + \alpha\, x(t) + \beta\, x(t)^3 = F \cos(\omega\, t)$$

along with the parameter values (taking $F=0.325$ here) and initial conditions.

```
>   g:=0.25: alpha:=-1: beta:=1: omega:=1; F:=0.325; #change F
```

$$\omega := 1 \qquad F := 0.325$$

```
>   ic:=x(0)=0.09,D(x)(0)=0:
```

The force amplitude F will be adjusted to the other values, $F = 0.35, 0.356, 0.42$, used previously. The Duffing oscillator equation is numerically solved,

```
>   sol:=dsolve({ode,ic},{x(t)},numeric,maxfun=0,
        output=listprocedure):
```

and the solution used to evaluate $x(t)$ for arbitrary time t.

```
>   X:=eval(x(t),sol):
```

The driving frequency is $\omega=1$ rad/s, or $f=\omega/(2\,\pi)=1/(2\,\pi)$ Hz. The sampling frequency f_s is taken to be $4\,f$ and the sampling time $T_s=1/f_s$ is calculated.

```
>   f:=omega/(2*Pi): f[s]:=4*f; T[s]:=1/f[s]; #frequencies in Hz
```

$$f_s := \frac{2}{\pi} \qquad T_s := \frac{\pi}{2}$$

To eliminate the transient solution, the x values are recorded starting at $t_0 = 50\,\pi$ seconds and sampled every T_s seconds up to time $(N-1)\,T_s$. These values are entered with the Array command.

```
>   x:=Array([seq(X(50*Pi+T[s]*i),i=0..N-1)]):
```

The FourierTransform command is applied to the sampled x values. This command calculates the discrete Fourier transform (8.7), divided by $\sqrt{N}$, using the fast Fourier transform as the default algorithm.

```
>   FT:=FourierTransform(x):
```

The following operator F enables us to extract the ith term in FT and, since the result is generally complex, take the absolute value. Because of Maple's definition of the discrete Fourier transform, this is equivalent to taking the square root of the power spectrum for the ith frequency point. We take the square root here because it helps to accentuate smaller peaks in the spectrum.

```
>  F:=i->abs(FT[i]): #sqrt of S
```

An operator `pt` is introduced to form the ith plotting point for the power spectrum, the frequency being expressed in radians per second so a comparison can be easily made with the driving frequency $\omega = 1$ rad/s.

```
>  pt:=i->[2*Pi*f[s]*(i-1)/N,F(i)]:
```

With the aid of the `pt` operator, the plotting points are formed into a list of lists, with the zero frequency value of the power spectrum taken as zero. Since the power spectrum is symmetric about N/2, only the points up to this number will be plotted. In the present case, 5000 points will be included. The point, e.g., $i = 2501$ corresponds to an angular frequency of 1 rad/s, for which $S \approx 29$.

```
>  pts:=[[0,0],seq(pt(i),i=2..N/2)]: pts[2501];
```
$$[1, 28.82963799]$$

The points are plotted and labels are added,

```
>  plot(pts,tickmarks=[3,2],labels=["omega","S"]);
```

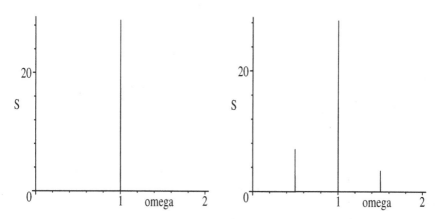

Figure 8.10: Power spectrum for period-1 (left) and period-2 (right).

the power spectrum for $F = 0.325$ being shown in the left plot of Figure 8.10. This spectrum shows a single sharp spike located at a frequency of 1 rad/s, i.e., exactly at the driving frequency. The period is $T = 2\pi/\omega = 2\pi$ seconds, so the response is period 1, in agreement with the conclusion reached earlier for the same Duffing oscillator on the basis of a Poincaré section analysis.

The plot on the right of Figure 8.10 results on executing the work sheet with $F = 0.35$. In addition to the tallest spike at the driving frequency, there is a second spike to the left, located at the frequency $\omega/2 = 0.5$ rad/s. This corresponds to a period $T = 2\pi/(\omega/2) = 4\pi$ seconds, i.e., twice as long as for $F = 0.325$. The spectrum is characteristic of a period-2 solution. Note that there is also a spike in the power spectrum to the right of the driving frequency, occurring at 1.5 rad/s. This is the third harmonic of the period-2 frequency. Harmonics are often present in power spectra, so remember that only those spikes lying to the left of the driving frequency spike should be used to determine the periodicity. It is because these harmonics are present that one must be sure to use a sufficiently high sampling frequency so that these harmonics do not get "folded" back (aliased) into the frequency range of interest.

On increasing F to 0.356, one sees three spikes to the left of the driving frequency in the left plot of Figure 8.11. The spike furthest to the left is located at $\omega/4 = 0.25$ rad/s, corresponding to a period of 8π seconds. The spectrum is characteristic of a period-4 response. Again, harmonics of these frequencies appear to the right of the driving frequency.

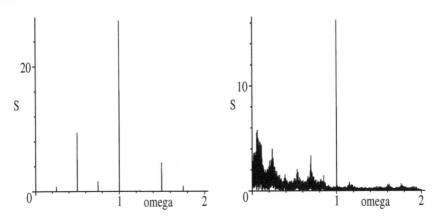

Figure 8.11: Power spectrum for period-4 (left) and chaotic response (right).

Finally, taking $F = 0.42$ produces the power spectrum shown on the right of Figure 8.11. Although a sharp spike is still seen at the driving frequency, the spectrum is very chaotic, which is not surprising, since it corresponds to the chaotic "rattler" Poincaré section seen earlier.

Recalling Richard's fright when his daughter nearly stumbled and rolled over a real rattlesnake, one might fancifully interpret this power spectrum as representing the rattling of his teeth or the shaking of his knees at the time of the incident.

PROBLEMS:

Problem 8-9: A different β value
Holding all other parameter values as in the text, use the power spectrum to determine the response of the forced Duffing ODE for each F value when $\beta = 2$.

Problem 8-10: Varying the frequency
For each of the four F values, use the power spectrum to explore the response of the Duffing oscillator as ω is varied, all other parameters being unchanged.

Problem 8-11: Interchanging signs
Use the power spectrum to determine the response of the forced Duffing oscillator for each F value when all numerical values are the same as in the text recipe, but the signs of α and β are interchanged.

Problem 8-12: Steady-state solution?
Using the power spectrum approach, determine the nature of the steady-state solution for the following forced oscillator equation:

$$\ddot{x} + 0.7\,\dot{x} + x^3 = 0.75\cos t, \quad x(0) = \dot{x}(0) = 0.$$

Problem 8-13: Solution?

Using the power spectrum approach, determine the nature of the steady-state solution for the following forced Duffing equation:

$$\ddot{x} + 0.08\,\dot{x} + x^3 = 0.2\cos t, \quad x(0) = 0.25, \ \dot{x}(0) = 0.$$

Problem 8-14: Solutions?

Using the power spectrum approach, determine the nature of the steady-state solution for the following oscillator equation for $F = 0.357$ and $F = 0.35797$:

$$\ddot{x} + 0.5\,\dot{x} - x + x^3 = 0.357\cos(t + 1), \quad x(0) = 0.09, \ \dot{x}(0) = 0.$$

Problem 8-15: Forced glycolytic oscillator
The equations describing forced oscillations of the glycolytic oscillator are

$$\dot{x} = -x + \alpha\,y + x^2\,y, \quad \dot{y} = \beta - \alpha\,y - x^2\,y + A + F\cos(\omega t).$$

Taking $\alpha = \beta = 0$, $A = 0.999$, $F = 0.42$, $x(0) = 2$, and $y(0) = 1$, determine the periodicity of the response using the Poincaré section approach for (a) $\omega = 2$ and (b) $\omega = 1.75$. Explore the frequency range in between and identify any interesting solutions.

8.3 The Bifurcation Diagram

Bifurcation diagrams can be generated for both nonlinear ODEs and nonlinear *difference equations* by plotting the system "response" versus a "control" parameter as the latter is varied. The word bifurcation is derived from the Latin word *furca* for fork. When period doubling occurs from period one to period two, the response curve resembles a two-pronged fork, the period-one portion being the "handle" and the period-two portion looking like two "prongs." In a typical period-doubling scenario, the two prongs then split into four and then into eight and so on as the control parmeter is further increased. Period doubling is not the only "route" to chaos, so bifurcation diagrams are useful in revealing the nature of the route.

8.3.1   Pitchforks and Other Bifurcations

Though you drive away nature with a pitchfork, she always returns.
Horace, Roman poet known for his odes (65–8 BC)

Consider the following difference equation, known as the *logistic map*,

$$x_{n+1} = a\,x_n\,(1 - x_n),\tag{8.10}$$

where $n = 0, 1, 2, \ldots, N$ and a is allowed to vary between 0 and 4. As a is increased over its range, a period-doubling sequence to chaos occurs that can be illustrated by plotting x_n (at large N to eliminate the transient) versus a.

The first part of the following recipe illustrates a period-2 solution of the logistic map for $a = 3.2$, $N = 119$, and initial value $x_0 = 0.1$.

```
> restart: a:=3.2: N:=119: x[0]:=0.1:
```

An operator **F** is formed to calculate the rhs of the logistic map for a given x.

```
> F:=x->a*x*(1-x):
```

The logistic map is iterated from $n = 0$ to N,

```
> for n from 0 to N do
> x[n+1]:=F(x[n]);
> end do:
```

and the sequence of points $[n, x_n]$ plotted, each point being represented by a size-12 black circle. The resulting picture is shown in Figure 8.12.

```
> plot([seq([n,x[n]],n=0..N)],style=point,symbol=circle,
  symbolsize=12,color=black,labels=["n","x"]);
```

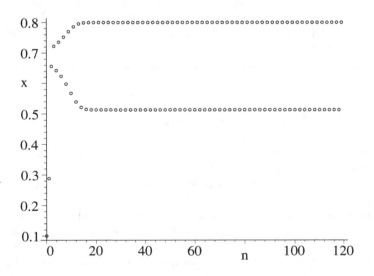

Figure 8.12: Period-2 solution for the logistic map.

After a short transient interval, the logistic system oscillates back and forth between the two branches shown in the figure. This is characteristic of a period-2 solution, there being only one branch in steady state for a period-1 solution. As a is increased, further bifurcations occur.

To create a bifurcation diagram for the logistic map, a particular initial value of x_0 is chosen, and the map iterated for a given a. A certain number of the initial points are thrown away to remove the transient part of the solution, and the subsequent steady-state points plotted. Then, one increments a by a small amount and repeats the process, and so on. The whole process is easily automated, as is now illustrated.

```
>  restart: with(plots):
```
As you may confirm by running the first part of the recipe, a period-1 solution prevails until $a = 3.0$, at which point period two begins. In the recipe, the range of a is taken from the starting value $Sa = 2.9$ up to the final a value $Fa = 4$. If this a range is divided into, say, $N = 200$ equal intervals,

```
>  Sa:=2.9: Fa:=4: N:=200: stepsize:=(Fa-Sa)/N; x:=0.2;
```
$$stepsize := 0.005500000000 \qquad x := 0.2$$
the *stepsize* is $\Delta a = 0.0055$. The initial value of x has been taken to be $x = 0.2$.

A total of 500 iterations will be considered and the first 100 points ignored in order to remove any transients. This leaves 400 points to be plotted for each a value. The quantity c is a counter to keep track of the plots, a graph being produced for each a value. The counter is initially set equal to zero.

```
> totalpts:=500: ignorepts:=100: pts:=totalpts-ignorepts; c:=0:
```
$$pts := 400$$
In the following double do loop, the first, or outer, loop increments a from the starting value Sa to the final a value Fa in incremental steps given by *stepsize*.

```
>  for a from Sa to Fa by stepsize do
```
The second, or inner, do loop iterates the logistic equation from 1 to *totalpts*.

```
>  for n from 1 to totalpts do
```
```
>  x:=a*x*(1-x);
```
The points for a given a value are formed into a list,

```
>  pts[n]:=[a,x];
```
```
>  end do;
```
and the inner do loop completed. Then, the counter is advanced by one,

```
>  c:=c+1;
```
and a list of lists is made using the final 400 (presumably) steady-state points.

```
>  points:=[seq(pts[k],k=ignorepts..totalpts)]:
```
A plot is made for each a value using a point style,

```
>  Gr[c]:=pointplot(points,symbol=point,color=black):
```
```
>  end do:
```

and the outer do loop ended. The $N = 200$ plots are superimposed with the
`display` command, yielding the bifurcation diagram shown in Figure 8.13.

```
>  display([seq(Gr[m],m=1..N)],view=[Sa..Fa,0..1],
   labels=["a","x"]);
```

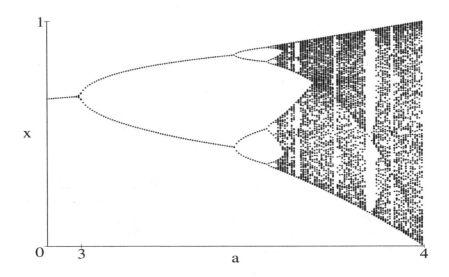

Figure 8.13: Bifurcation diagram for the logistic map for $a = 2.9$ to 4.

Starting at $a = 2.9$, the reader can observe period 1 occurring up to $a = 3.0$.
Then the steady-state response undergoes a so-called *pitchfork bifurcation* to
period 2, followed by clearly seen bifurcations to period four, period eight, and
a barely observable period-sixteen solution. At higher a values, the response is
generally chaotic, but narrow periodic windows also occur. The reader should
be able to see, for example, a period-3 solution for $a \approx 3.83$. In this case, the
bifurcation to period 3 is an example of a *tangent bifurcation*.

Since the periodic windows are often very narrow in terms of the range of a,
one should really increase the number N, but this leads to a longer computing
time. A better approach is to leave N unchanged and zoom in on a particular
range of a by changing the values of Sa and Fa.

Bifurcation diagrams are very useful diagnostic tools for studying the behavior of nonlinear maps as well as forced oscillator ODEs as one or more control
parameters are varied.

A complementary graphical approach is to calculate the Lyapunov exponent
as a function of a as illustrated in the next recipe.

PROBLEMS:

Problem 8-16: Finer structure

For the logistic map, produce the bifurcation diagram for the region $a = 3.54$ to $a = 3.6$, taking $x_0 = 0.2$, and dividing the a interval into 100 steps. Summarize the various periodic solutions that you observe.

Problem 8-17: More fine structure

Explore the periodic window in the vicinity of $a = 3.8$ and report on what periodicities you observe at each a value sampled.

Problem 8-18: Cubic map

Produce a bifurcation diagram for the *cubic map*

$$x_{n+1} = a\,x_n - x_n^3$$

over a suitable range of the parameter a. Determine the values of a in the diagram at which the periodicity changes.

Problem 8-19: Quartic map

With $x_0 = 0.2$, produce the bifurcation diagram for the *quartic map*

$$x_{n+1} = a\,x_n\,(1 - x_n^3)$$

over the range $a = 1.5$ to $a = 2.0$, taking as small an a step size as you can. Summarize the behavior of the map as a varies over the specified range.

Problem 8-20: The tent map

Produce the bifurcation diagram for the *tent map*

$$x_{n+1} = 2\,a\,x_n,\ 0 < x \le \tfrac{1}{2};\quad x_{n+1} = 2\,a\,(1 - x_n),\ \tfrac{1}{2} \le x < 1$$

with $0 < a < 1$. Take $x_0 = 0.2$ and $x_0 = 0.6$. Summarize the periodic solutions that you see in each case. Are there any differences in the bifurcation diagrams for the two inputs.

Problem 8-21: The sine map

Produce the bifurcation diagram for the *sine map*

$$x_{n+1} = a\sin(\pi\,x_n)$$

with $0 \le a \le 1$ and $0 \le x \le 1$. How does the bifurcation diagram qualitatively compare with that for the logistic map if only the range $a = 0.7$ to 1 is plotted?

Problem 8-22: Miscellaneous maps

Produce bifurcation diagrams for the following maps over suitable ranges of $a > 0$ and discuss the results:

(a) $x_{n+1} = x_n\,e^{a\,(1-x_n)}$ **(b)** $x_{n+1} = e^{-a\,x_n}$ **(c)** $x_{n+1} = a\,\cos(x_n)$
(d) $x_{n+1} = a + x_n^2$

8.4   The Lyapunov Exponent

Named after the Russian mathematician Aleksandr Mikhailovich Lyapunov
(1857–1918), the *Lyapunov exponent* λ is a measure of the very sensitive dependence on initial conditions that is characteristic of chaotic behavior in nonlinear
difference equations (maps) and ODEs. The discussion that follows is for maps,
a similar analysis applying to ODEs.

Consider a one-dimensional map

$$x_{n+1} = f(x_n), \tag{8.11}$$

with f some specified functional form, and let x_0 and y_0 be two initial values
very close to each other. In phase space, they would be represented by two very
close points. After n iterations, the values of x_n and y_n will be given by

$$x_n = f^{(n)}(x_0), \quad y_n = f^{(n)}(y_0), \tag{8.12}$$

where $f^{(n)}$ denotes the nth iteration of the map. Because the chaotic regime
is typically characterized by an extreme sensitivity to initial conditions, for a
chaotic situation nearby initial points will rapidly separate. On the other hand,
periodic solutions are insensitive to initial conditions, and nearby initial points
rapidly converge.

This suggests that for sufficiently large n, one might assume an approximately exponential dependence on n of the separation distance, viz.,

$$|x_n - y_n| = |x_0 - y_0|e^{\lambda n}, \tag{8.13}$$

with $\lambda > 0$ for the chaotic situation and $\lambda < 0$ for the periodic case. Taking n
large, λ can be extracted from (8.13):

$$\lambda = \lim_{n \to \infty} \frac{1}{n} \ln \left| \frac{x_n - y_n}{x_0 - y_0} \right|. \tag{8.14}$$

However, for trajectories confined to a bounded region such as the range $0 <
x < 1$ for the logistic map, such exponential separation for the chaotic case
cannot occur for very large n, unless the initial points x_0 and y_0 are very close.
Therefore, the limit $|x_0 - y_0| \to 0$ must also be taken.

Modifying (8.14) yields

$$\lambda = \lim_{n \to \infty} \frac{1}{n} \lim_{|x_0 - y_0| \to 0} \ln \left| \frac{x_n - y_n}{x_0 - y_0} \right| = \lim_{n \to \infty} \frac{1}{n} \lim_{|x_0 - y_0| \to 0} \ln \left| \frac{f^{(n)}(x_0) - f^{(n)}(y_0)}{x_0 - y_0} \right|,$$

or

$$\lambda = \lim_{n \to \infty} \frac{1}{n} \ln \left| \frac{df^{(n)}(x_0)}{dx_0} \right|. \tag{8.15}$$

Now, $f(x_0) = x_1$ and $f(x_1) = f^{(2)}(x_0) = x_2$, so that, for example,

$$\frac{df^{(2)}(x_0)}{dx_0} = \frac{df(x_1)}{dx_1} \frac{dx_1}{dx_0} = \frac{df(x_1)}{dx_1} \frac{df(x_0)}{dx_0}. \tag{8.16}$$

Generalizing (8.16) produces

$$\frac{df^{(n)}(x_0)}{dx_0} = \prod_{k=0}^{n-1} \frac{df(x_k)}{dx_k}, \tag{8.17}$$

and the Lyapunov exponent λ is given by

$$\lambda = \lim_{n \to \infty} \frac{1}{n} \sum_{k=0}^{n-1} \ln \left| \frac{df(x_k)}{dx_k} \right|. \tag{8.18}$$

For periodic solutions, which starting point x_0 is chosen doesn't matter, but for chaotic trajectories, the precise value of λ will depend on x_0, i.e., in general $\lambda = \lambda(x_0)$. One can, if desired, define an average λ, averaged over all starting points. Whether this is done or not, $\lambda > 0$ should correspond to chaos and $\lambda < 0$ to periodic behavior.

Let us now calculate the Lyapunov exponent as a function of a for the logistic map. This will allow the regions of periodicity and chaos observed in the Lyapunov exponent to be compared with those observed in the bifurcation diagram of the previous section.

8.4.1 Mr. Lyapunov Agrees

If you know a thing only qualitatively, you know it no more than vaguely. If you know it quantitatively—grasping some numerical measure that distinguishes it from an infinite number of other possibilities—you are beginning to know it deeply.
Carl Sagan, American astronomer (1934–1996)

This recipe directly programs equation (8.18), the number of iterations being taken to be 300.

```
>   restart: with(plots): numpts:=300:
```
The range of a is taken from the starting value $Sa = 2.8$ to the final a value $Fa = 4$, with the range divided into $N = 480$ equal steps.

```
>   Sa:=2.8: Fa:=4: N:=480: x:=0.2: stepsize:=(Fa-Sa)/N; c:=0:
```
$$stepsize := 0.002500000000$$

The *stepsize* is 0.0025 and the initial value of x is taken to be 0.2. The plots counter c has been "initialized" to zero. The outer loop in the following double loop increments a in units of *stepsize* from *Sa* to *Fa*.

```
>   for a from Sa to Fa by stepsize do
```
The sum in equation (8.18) is calculated for each value of a, and the sum is assigned the name *total*. To start off, *total* is set to zero.

```
>   total:=0;
```
The inner do loop runs over the total number of iterations, *numpts* = 300.

```
>   for j from 1 to numpts do
```
The logistic map is entered, and the absolute value of the derivative with respect to x formed and labeled d.

```
>   x:=a*x*(1-x);
>   d:=abs(a*(1-2*x));
```

If d is not equal to zero, then $f = \ln(d)$ is calculated. If $d = 0$, we would obtain $f = -\infty$. This latter situation is avoided by setting $f = 0$ in this case.

```
> if d<>0 then f:=ln(d) else f=0 end if;
```

To perform the sum in equation (8.18), the *total* is incremented by the value of f, and the inner loop ended.

```
> total:=total+f;
> end do:
```

The counter c is incremented by one,

```
> c:=c+1;
```

and a list of points formed with a as the horizontal coordinate and $\lambda = total/numpts$ as the vertical coordinate. This completes the evaluation of λ in equation (8.18) for a given a value.

```
> pts[c]:=[a,total/numpts]:
> end do:
```

The sequence of $N = 480$ points is plotted, the points being joined using a line style. Figure 8.14 shows the Lyapunov exponent λ for $a = 2.8$ to $a = 4$.

```
> plot([seq(pts[i],i=1..N)],style=line,view=[Sa..Fa,-1.5..1],
  tickmarks=[3,3]);
```

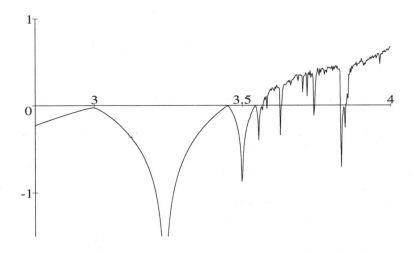

Figure 8.14: Lyapunov exponent (vertical axis) for $a = 2.8$ to $a = 4$.

If one compares the periodic windows where the Lyapunov exponent goes negative, there is good agreement with the bifurcation diagram, Figure 8.13, for the logistic map. Because the Lyapunov spikes are sometimes quite narrow, the agreement can be improved by zooming in on a particular a region by altering the values of Sa and Fa.

PROBLEMS:
Problem 8-23: Quartic map
With $x_0 = 0.2$, calculate the Lyapunov exponent for the quartic map

$$x_{n+1} = a\, x_n\, (1 - x_n^3)$$

over the range $a = 1.5$ to $a = 2.0$, keeping all other parameters as in the text recipe. Over what regions of a do periodic solutions occur?

Problem 8-24: The tent map
Plot the Lyapunov exponent λ versus a for the tent map

$$x_{n+1} = 2\, a\, x_n,\ 0 < x \le \tfrac{1}{2};\quad x_{n+1} = 2\, a\, (1 - x_n),\ \tfrac{1}{2} \le x < 1$$

with $0 < a < 1$. Take $x_0 = 0.2$. Analytically show that $\lambda = \ln(2\, a)$ and discuss your graph in terms of this result.

Problem 8-25: The sine map
Plot the Lyapunov exponent versus a for the sine map

$$x_{n+1} = a \sin(\pi\, x_n)$$

with $0 \le a \le 1$ and an x value of your choosing from the range $0 \le x \le 1$. Discuss your graph.

Problem 8-26: Another sine map
Plot the Lyapunov exponent for the following map over the range $a = 2.9$ to $a = 6$:

$$x_{n+1} = a \sin(x_n)\, (1 - \sin(x_n)).$$

Identify ranges of a where periodic windows occur.

8.5   Reconstructing an Attractor

An extremely important issue in many areas of modern science is how to distinguish between random noise and deterministic chaos. Typically, experimental data is acquired by sampling at regular time intervals, and the investigator would like to know whether there is some underlying chaotic attractor that can be described by a mathematical function that would therefore allow other behavior to be predicted or whether one is dealing with noise from which nothing can be foretold.

One approach is to assume that if there is some underlying chaotic attractor, perhaps it is possible to recover its geometric shape, and deduce the mathematical function that would produce the shape, from the sampled time series. Suppose that the sampling time interval is t_s and the data points are

$$x_0 = x(t=0),\ x_1 = x(t=t_s),\ x_2 = x(t=2\, t_s),\ \ldots,\ x_n = x(t=n\, t_s),\ x_{n+1},\ \ldots.$$

If a deterministic relation exists, then x_{n+1} will depend somehow on x_n, x_{n-1}, etc. The simplest assumption is to assume that a *one-dimensional map* exists so that x_{n+1} depends only on the previous value x_n. If this is the case, then a functional form f will exist such that $x_{n+1} = f(x_n)$.

Assuming that this is the case, one plots pairs of numbers (x_n, x_{n+1}) from the time series to form the two-dimensional space x_{n+1} versus x_n. If the points appear to lie on a definite geometrical line, this implies that there is an underlying attractor and an associated functional form f. If the geometrical shape has more structure to it, this could imply that there is an underlying "two-dimensional" map, namely,

$$x_{n+1} = f(x_n) + g(x_{n-1}), \quad \text{or} \quad x_{n+1} = f(x_n) + y_n, \ y_{n+1} = g(x_n).$$

If the dimensionality of the underlying map is higher than two, one must increase the dimensionality of the space accordingly. To "see" a three-dimensional map, for example, one must work in a three-dimensional space. The above procedure is then generalized by plotting triplets of numbers. For example, one might use (x_0, x_1, x_2), (x_1, x_2, x_3), etc.

8.5.1 Putting Humpty Dumpty Together Again

Humpty Dumpty sat on a wall, Humpty Dumpty had a great fall,
All the king's horses, And all the king's men,
Couldn't put Humpty Dumpty together again.
Lewis Carroll, *Alice's Adventures in Wonderland*, 1865

Suppose that we have been given the following lengthy data list x, where each entry $x(n)$ corresponds to a time $t = n \, t_s$, with $n = 1, 2, \ldots, N$.

```
> restart: with(plots):
> x:=[6.24, 9.15, 3.03, 8.24, 5.66, 9.58, 1.56, 5.14, 9.74, .980,
     3.45, 8.81, 4.09, 9.43, 2.10, 6.47, 8.91, 3.79, 9.18, 2.93, 8.07,
     6.07, 9.30, 2.53, 7.37, 7.56, 7.18, 7.89, 6.50, 8.88, 3.89, 9.27,
     2.65, 7.60, 7.12, 8.00, 6.24, 9.15, 3.03, 8.23, 8.23, 5.67, 9.57,
     1.60, 5.23, 9.73, 1.03, 3.59, 8.98, 3.58, 8.97, 3.61, 9.00, 3.51,
     3.86, 9.24, 2.74, 7.76, 6.77, 8.53, 4.89, 9.74, .969, 8.89, 3.41,
     8.77, 4.21, 9.51, 1.82, 5.81, 9.50, 1.87, 5.93, 9.41, 2.16, 6.61,
     8.74, 4.30, 9.56, 1.64, 5.33, 9.71, 1.11, 3.85, 9.23, 2.76, 7.80,
     6.70, 8.62, 4.65, 9.70, 1.13, 3.91, 9.28, 2.59, 7.48, 7.34, 7.61,
     7.10, 8.04, 6.16, 9.23, 2.78, 7.83, 6.62, 8.72, 4.34, 9.58, 1.56,
     5.13, 9.74, .977, 3.44, 8.80, 4.13, 9.45, 2.02, 6.29, 9.10, 3.18,
     8.46, 5.08, 9.75, .960, 3.39, 8.73, 4.31, 9.56, 1.62, 5.30, 9.72,
     1.08, 3.76, 9.15, 3.04, 8.25, 5.63, 9.60, 1.51, 5.00, 9.75, .951,
     3.36, 8.70, 4.43, 9.62, 1.42, 4.75, 9.73, 1.04, 3.63, 9.02, 3.45,
     8.81, 4.09, 9.43, 2.10, 6.47, 8.90, 3.81, 9.20, 2.88, 8.00, 6.23,
     9.16, 3.01, 8.21, 5.73, 9.54, 1.70, 5.49, 9.66, 1.30, 4.41, 9.61,
     1.45, 4.84, 9.74, .989, 3.48, 8.84, 3.98, 9.35, 2.38, 7.07, 8.07,
     6.07, 9.30, 2.52, 7.35, 7.59, 7.14, 7.97, 6.30, 9.09, 3.23, 8.53,
     4.88, 9.74, .972, 3.42]:
```

Although the data appears to be confined to the approximate range 0 to 10, it is not clear at first glance whether it represents deterministic chaos or is simply a "noisy" set of data. Can we confirm that it is the former and identify the probable identity of the underlying map? Calling this the "Humpty Dumpty map," can we then reconstruct his or her mathematical "appearance"? Or, at the risk of sounding rather melodramatic, can we put Humpty Dumpty back together again?

The number of operands command reveals that there are 200 entries in x.

```
>   N:=nops(x);
```

$$N := 200$$

First, let us plot x versus n and see whether that reveals any pattern that we might have missed. By default, the N points are joined by straight lines.

```
>   plot([seq([n,x[n]],n=1..N)],labels=["n","x[n]"]);
```

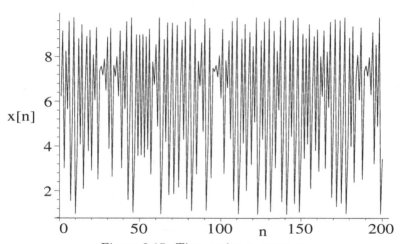

Figure 8.15: Time series x versus n.

Although there is a hint of some repetition, it certainly isn't conclusive that the series comes from a map producing deterministic chaos rather than just being noisy output. Following the procedure mentioned in the introduction, we will plot $x_{n+1}/10$ versus $x_n/10$, the factor of 10 being introduced to reduce both vertical and horizontal ranges to 0 to 1 in the graph. The points are represented by size-12 black circles. The graph is assigned a name, gr1, so it can first be displayed and then used to reveal Humpty Dumpty's true identity.

```
>   gr1:=pointplot([seq([x[n]/10,x[n+1]/10],n=1..N-1)],
        symbol=circle,symbolsize=12,color=black,
        labels=["x[n]","x[n+1]"]):
```

Entering gr1 with a command-line-ending semicolon,

```
>   gr1; #humpty dumpty?
```

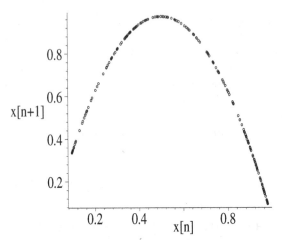

Figure 8.16: x_{n+1} versus x_n for time series.

produces Figure 8.16. The data points appear to lie on an inverted parabola, suggesting that Humpty Dumpty might be a one-dimensional logistic map. Adjusting the parameter a by trial and error to provide the best fit, the logistic curve is now plotted for $a = 3.9$ as a thick red line over the range $X = 0$ to 1.

```
>   a:=3.9: #adjust
>   gr2:=plot(a*X*(1-X),X=0..1,color=red,thickness=2):
```

The two graphs, **gr1** and **gr2**, are plotted in the same picture, the result being shown in Figure 8.17.

```
>   display({gr1,gr2}); #successful reconstruction
```

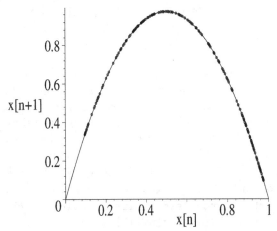

Figure 8.17: Best-fitting logistic curve (solid) and x_{n+1} vs. x_n points.

The points lie on the logistic curve, so we have been successful in identifying and reconstructing Humpty Dumpty. We can confirm from either of the previous two recipes that the logistic map produces deterministic chaos for $a = 3.9$.

PROBLEMS:
Problem 8-27: Reconstructing the quartic map
Taking $a = 1.97$ and $x_0 = 0.9$, iterate the quartic map

$$x_{n+1} = a\, x_n\, (1 - x_n^3)$$

to produce a time series with 200 points. Plot the time series with a line style. Attempt to reconstruct any possible underlying deterministic attractor by plotting x_{n+1} versus x_n using a point style. Show that these points lie on the curve $a\, X\, (1 - X^3)$ with $a = 1.97$. (This problem is a bit like a dog chasing its tail!)

8.5.2 Random Is Random

We humans have purpose on the brain. We find it hard to look at anything without wondering what it is "for," what the motive for it is, or the purpose behind it. When the obsession with purpose becomes pathological it is called paranoia–reading malevolent purpose into what is actually random bad luck. But this is just an exaggerated form of a nearly universal delusion. Show us almost any object or process, and it is hard for us to resist the "Why" question—the "What is it for?" question.
Richard Dawkins, British biologist, author, *River Out of Eden*, 1995

In this recipe, we shall use Maple's random-number generator to produce a time series and attempt to reconstruct any underlying deterministic attractor. The time series will have $N = 200$ points.

```
>   restart: N:=200:
```
The command `randomize()` sets the random-number seed to a different value based on the computer system clock.

```
>   randomize():
```
In the following do loop, N random decimal numbers lying between 0 and 1 are generated.

```
>   for n from 0 to N do
```
The command `rand()` produces a random positive twelve-digit number. A fractional number between 0 and 1 results on dividing by 10^{12}. The number is then put into decimal form by applying the floating-point evaluation command.

```
>   x[n]:=evalf(rand()/10^12);
>   end do:
```

The randomly generated time series is plotted in Figure 8.18.

```
>   plot([seq([n,x[n]],n=1..N)],labels=["n","x[n]"]);
```

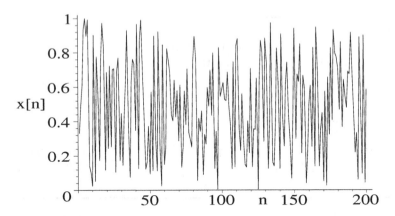

Figure 8.18: Time series for random data.

If we didn't know how the series was produced, we would probably still conclude from the figure that we are dealing with random noise. But one might have arrived at the same conclusion in the previous recipe from the time series plot. As in that recipe, let's plot x_{n+1} versus x_n using a point style, thus producing Figure 8.19.

```
>   plot([seq([x[n],x[n+1]],n=1..N-1)],style=point,symbol=
    circle,symbolsize=16,labels=["x[n]","x[n+1]"]);
```

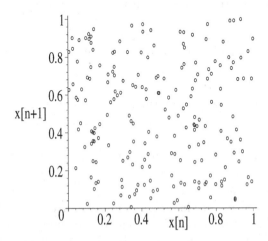

Figure 8.19: x_{n+1} versus x_n for time series.

There is no discernible pattern in the figure, unlike the situation in the previous recipe. By plotting other pairs, or even triplets, of numbers, one can conclude (not surprisingly) that there is no underlying chaotic attractor here.

PROBLEMS:

Problem 8-28: Rolling a die
Consulting Maple's Help, produce a random set of 500 numbers from the positive integers one to six inclusive. This might simulate the rolling of an honest die. Make a plot of the "time series" and x_{n+1} versus x_n to show the randomness.

Problem 8-29: Literature search
Apply the techniques of attractor reconstruction to some time series data (e.g., the Dow Jones index) extracted from newspapers, magazines, or whatever, and see whether a pattern emerges.

8.5.3 Butterfly Reconstruction

A people's literature is the great textbook for real knowledge of them. The writings of the day show the quality of the people as no historical reconstruction can.
Edith Hamilton, American classical scholar (1867–1963)

Our final recipe will illustrate how Lorenz's butterfly attractor can be reconstructed from time series data extracted from the governing system of three coupled nonlinear ODEs. The Lorenz system (which was discussed in Section 2.2.1) is entered, along with the initial condition $x(0) = 2$, $y(0) = 5$, $z(0) = 5$.

```
> restart: with(plots):
> sys:=diff(x(t),t)=sigma*(y(t)-x(t)),diff(y(t),t)=-x(t)*z(t)
       +r*x(t)-y(t),diff(z(t),t)=x(t)*y(t)-b*z(t);
```

$$sys := \frac{d}{dt}x(t) = \sigma\left(y(t) - x(t)\right),$$
$$\frac{d}{dt}y(t) = -x(t)\,z(t) + r\,x(t) - y(t),$$
$$\frac{d}{dt}z(t) = x(t)\,y(t) - b\,z(t)$$

```
> ic:=(x(0)=2,y(0)=5,z(0)=5):
```

The parameter values are taken to be $r = 28$, $b = 8/3$, and $\sigma = 10$.

```
> r:=28: b:=8/3: sigma:=10:
```

The total number of entries in the time series will be taken to be $N = 2000$. The total time is $T = 50$, so the step size $\Delta = T/N$ is 1/40.

```
> N:=2000: T:=50: Delta:=T/N;
```

$$\Delta := \frac{1}{40}$$

The system of ODEs is solved numerically, the solution being given as a list procedure, so that we can create a time series.

```
>  sol:=dsolve({sys,ic},{x(t),y(t),z(t)},numeric,maxfun=0,
          output=listprocedure):
```

The following line uses the numerical solution to evaluate $x(t)$ at an arbitrary time, which must be specified.

```
>  S:=eval(x(t),sol):
```

Then entering S(n*Delta) will yield $x(t)$ at $t = n\,\Delta$. Using this result, the x values are obtained at times $t = n\,\Delta$ for $n = 0$ to N.

```
>  for n from 0 to N do X[n]:=S(n*Delta); end do:
```

The time series is created and plotted, the default being to join the points with straight lines. The resulting picture is shown in Figure 8.20.

```
>  plot([seq([n*Delta,X[n]],n=0..N)],labels=["t","x"]);
```

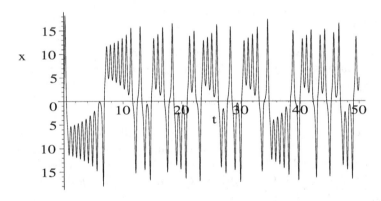

Figure 8.20: x time series for the Lorenz system.

To reconstruct the 3-dimensional Lorenz butterfly attractor, it is necessary to form triplets of numbers from the time series. By trial and error, we use the triplet combination involving n, $n + 3$, and $n + 6$ to create the plotting points.

```
>  points:=[seq([X[n],X[n+3],X[n+6]],n=0..(N-6))]:
```

Using the spacecurve command with shading=z to color the trajectory, the butterfly attractor is revealed.

```
>  spacecurve(points,style=line,shading=z,axes=framed,
    orientation=[-30,60],tickmarks=[3,3,3],
    labels=["X(n)","X(n+3)","X(n+6)"]);
```

A black-and-white version is shown in Figure 8.21. If one compares the picture, which can be rotated on the computer screen, with the butterfly picture obtained in Chapter 2, the reconstruction of the butterfly is quite good.

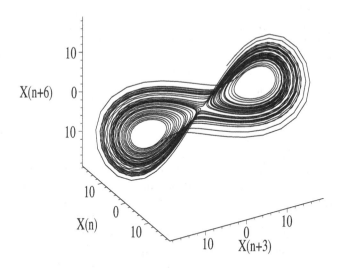

Figure 8.21: Butterfly attractor reconstructed from time series.

In the preceding recipes, the reconstruction attempt has been tested on known examples. So the question is, does the method work for real experimental time series data, particularly when the answer is not known? The answer is yes! The method has been applied to the Belousov–Zhabotinski chemical oscillator reaction (Section 1.2.2) [RSS83] [SWS82], to Taylor–Couette flow in hydrodynamics [AGS84], to ultrasonic cavitation in liquids [LH91], and to measles data for the cities of Baltimore and New York [SK86].

PROBLEMS:

Problem 8-30: Butterfly attractor

Try reconstructing the butterfly attractor from the $z(t)$ time series, and explain why you do not obtain two "wings."

Problem 8-31: Rössler attractor

The Rössler system is given by

$$\dot{x} = -(y + z), \qquad \dot{y} = x + a\,y, \qquad \dot{z} = b + z\,(x - c).$$

Using the time series for $x(t)$ with $a = 0.2$, $b = 0.2$, $c = 5.7$, $x(0) = y(0) = z(0) = 0.1$, construct the Rössler attractor. Hint: Your picture should resemble Figure 1.18.

Epilogue

Thus grew the tale of Wonderland.
Thus slowly, one by one,
Its quant events were hammered out—
And now the tale is done,
And home we steer, a merry crew,
Beneath the setting sun.
Lewis Carroll, *Alice's Adventures in Wonderland*, 1865

The storybook characters who formed the "merry crew of our "Wonderland,"
having earlier attended Mike and Vectoria's wedding ceremony in a sun-dappled
alpine meadow, are beginning to slowly drift away from the banquet table to
dance beneath the setting sun. Eavesdropping on their conversation, it appears
that everyone was pleased with the entire menu, the appetizers, the entrees,
and the desserts. The crew also chatted about the new friends that they have
made. They hope that you the reader, having shared some of their experiences
and recipes, will include yourself in this group.

Now your CAS chefs must reluctantly end this tale of Wonderland and
close this gourmet's guide to some of the advanced mathematical models of
science. Our guiding principle throughout both volumes of *Computer Algebra
Recipes* has been to introduce the reader to what we believe is an important
educational and scientific innovation, the use of a computer algebra system
to learn and explore science. This belief is supported by the fact that over
a million physicists, engineers, mathematicians, chemists, and other groups of
scientists, are already using one CAS or another to solve technological and
scientific problems of interest to them. If you have successfully applied the
computer algebra recipes in our menu to the text problems in both volumes,
you should be in good shape to join their ranks.

Richard and George, Your CAS chefs

Bibliography

[AGS84]   R. H. Abraham, J. P. Gollub, and H. L. Swinney. Testing nonlinear dynamics. *Physica D*, 11:252, 1984.

[AJMS81] R. M. Anderson, H. C. Jackson, R. M. May, and A. M. Smith. Population dynamics of fox rabies in Europe. *Nature*, 289:765, 1981.

[AS72]   M. Abramowitz and I. A. Stegun. *Handbook of Mathematical Functions with Formulas, Graphs, and Mathematical Tables*. National Bureau of Standards, Washington, DC, 1972.

[Bac69]   J. Backus. *The Acoustical Foundations of Music*. W. W. Norton, New York, 1969.

[Bel58]   B. P. Belousov. Oscillation reaction and its mechanism. In *Collection of Abstracts on Radiation Medicine*. Medgiz, Moscow, 1958.

[BEMS71] A. Barone, F. Esposito, C. J. Magee, and A. C. Scott. Theory and applications of the sine–gordon equation. *Riv. Nuovo Cimento*, 1:227, 1971.

[BEP71]   I. P. Batra, R. H. Enns, and D. Pohl. Stimulated thermal scattering of light. *Physica Status Solidi*, 48:11, 1971.

[BF89]   R. L. Burden and J. D. Faires. *Numerical Analysis*, 4th ed. PWS–KENT, Boston, MA, 1989.

[Boa83]   M. L. Boas. *Mathematical Methods in the Physical Sciences*, 2nd ed. John Wiley, New York, 1983.

[Bri74]   E. O. Brigham. *The Fast Fourier Transform*. Prentice-Hall Inc., New Jersey, 1974.

[Cha39]   S. Chandrasekhar. *An Introduction to the Study of Stellar Structure*. Dover Reprint, Chicago, 1939.

[Cho64]   W. F. Chow. *Principles of Tunnel Diode Circuits*. Wiley, New York, 1964.

[Dav62]   H. T. Davis. *Introduction to Nonlinear Differential and Integral Equations*. Dover, New York, 1962.

[EJMR81]   R. H. Enns, B. L. Jones, R. M. Miura, and S. S. Rangnekar. *Nonlinear Phenomena in Physics and Biology*. Plenum Press, New York, 1981.

[EK88]   L. Edelstein-Keshet. *Mathematical Models in Biology*. Birkhäuser, Boston, MA, 1988.

[EM00]   R. H. Enns and G. C. McGuire. *Nonlinear Physics with Maple for Scientists and Engineers*, 2nd ed. Birkhäuser, Boston, MA, 2000.

[EM01]   R. H. Enns and G. M. McGuire. *Computer Algebra Recipes: A Gourmet's Guide to the Mathematical Models of Science*. Springer-Verlag, New York, 2001.

[EM06]   R. H. Enns and G. M. McGuire. *Computer Algebra Recipes: An Introductory Guide to the Mathematical Models of Science*. Springer-Verlag, New York, 2006.

[Enn05]   R. H. Enns. *Computer Algebra Recipes for Mathematical Physics*. Birkhäuser, Boston, MA, 2005.

[ER79]   R. H. Enns and S. S. Rangnekar. The 3-wave interaction in nonlinear optics. *Phys. Status Solidi*, 94:9, 1979.

[Erl83]   H. Erlichson. Maximum projectile range with drag and lift, with particular application to golf. *Amer. J. Phys.*, 51:357, 1983.

[FC99]   G. R. Fowles and G. L Cassiday. *Analytic Mechanics*. Saunders College, Orlando, FL, 1999.

[FKN72]   R. J. Field, E. Kőrös, and R. M. Noyes. Oscillations in chemical systems, Part 2. Thorough analysis of temporal oscillations in the bromate–cerium–malonic acid system. *Journal of the American Chemical Society*, 94:8649, 1972.

[FN74]   R. J. Field and R. M. Noyes. Oscillations in chemical systems, IV. Limit cycle behavior in a model of a real chemical reaction. *Journal of Chemical Physics*, 60:1877, 1974.

[Gau69]   G. F. Gause. *The Struggle for Existence*. Hafner, New York, 1969.

[Gri95]   D. J. Griffiths. *Introduction to Quantum Mechanics*. Prentice Hall, Englewood Cliffs, N. J., 1995.

[Has90]   A. Hasegawa. *Optical Solitons in Fibers*. Springer-Verlag, New York, 1990.

[Hay64]   C. Hayashi. *Nonlinear Oscillations in Physical Systems*. McGraw Hill, New York, 1964.

[HH64]   M. Hénon and C. Heiles. The applicability of the third integral of motion: some numerical experiments. *Astrophys. J.*, 69:73, 1964.

[Hil94]   R. C. Hilborn. *Chaos and Nonlinear Dynamics*. Oxford University Press, Oxford, 1994.

[Jac90]   E. A. Jackson. *Perspectives of Nonlinear Dynamics*, Vol. 1 and 2. Cambridge University Press, Cambridge, 1990.

[Kau76]   D. J. Kaup. The three-wave interaction. *Studies Appl. Math.*, 55:9, 1976.

[KRB79]   D. J. Kaup, A. Rieman, and A. Bers. Space–time evolution of nonlinear three-wave interactions. *Rev. Modern Phys.*, 51:275, 1979.

[LH91]   W. Lauterborn and J. Holzfuss. Acoustic chaos. *International Journal of Bifurcation and Chaos*, 1:13, 1991.

[Lor63]   E. N. Lorenz. Deterministic nonperiodic flow. *J. Atmospheric Sci.*, 20:130, 1963.

[MA88]   J. J. McPhee and G. C. Andrews. Effect of sidespin and wind on projectile trajectory, with particular application to golf. *Amer. J. Phys.*, 56:933, 1988.

[Map05]   Maplesoft. *Maple 10 User Manual*. Waterloo Maple, Waterloo, Canada, 2005.

[May80]   R. M. May. Nonlinear phenomena in ecology and epidemiology. *Ann. New York Acad. Sci.*, 357:267, 1980.

[MGH+05]   M. B. Monagan, K. O. Geddes, K. M. Heal, G. Labahn, S. M. Vorkoetter, J. McCarron, and P. DeMarco. *Maple 10 Introductory (Advanced) Programming Guide*. Waterloo Maple, Waterloo, Canada, 2005.

[MH91]   W. M. MacDonald and S. Hanzely. The physics of the drive in golf. *Amer. J. Phys.*, 59:213, 1991.

[Mor48]   P. M. Morse. *Vibration and Sound*. McGraw-Hill, New York, 1948.

[Mur89]   J. D. Murray. *Mathematical Biology*. Springer-Verlag, New York, 1989.

[MW71]   J. Mathews and R. L. Walker. *Mathematical Methods of Physics*, 2nd ed. Addison-Wesley, New York, 1971.

[Nyq28]   H. Nyquist. Certain topics in telegraph transmission theory. *Trans. AIEE*, 47:617, 1928.

[Oha85]   H. C. Ohanian. *Physics*. W. W. Norton, New York, 1985.

[PFTV89]   W. H. Press, B. P. Flannery, S. A. Teukolsky, and W. T. Vetterling. *Numerical Recipes*. Cambridge University Press, Cambridge, 1989.

[PLA92]   M. Peastrel, R. Lynch, and A. Armenti. Terminal velocity of a shuttlecock in vertical fall. In Jr. Angelo Armenti, editor, *The Physics of Sports*. American Institute of Physics, New York, 1992.

[R76]   O. E. Rössler. An equation for continous chaos. *Phys. Lett. A*, 57:397, 1976.

[Rap60]   A. Rapoport. *Fights, Games and Debates*. University of Michigan Press, 1960.

[RE76]   S. S. Rangnekar and R. H. Enns. Numerical solution of the transient gain equations for stimulated backward scattering in absorbing fluids. *Canadian Journal of Physics*, 54:1564, 1976.

[RSS83]   J. C. Roux, R. H. Simoyi, and H. L. Swinney. Observation of a strange attractor. *Physica D*, 8:257, 1983.

[Rus44]   J. S. Russell. Report on waves. *British Assoc. Adv. Sci., 14th Meeting*, 1844.

[SCM73]   A. C. Scott, F. Y. F. Chu, and D. W. McLaughlin. The soliton: A new concept in applied science. *Proc. IEEE*, 61:1443, 1973.

[Sco87]   D. E. Scott. *An Introduction to Circuit Analysis*. McGraw-Hill, New York, 1987.

[Sel68]   E. E. Sel'kov. Self-oscillations in glycolysis. *European J. Biochem.*, 4:79, 1968.

[Sha49]   C. E. Shannon. Communication in the presence of noise. *Proc. IRE*, 37:10, 1949.

[SK86]   W. M. Schaffer and M. Kot. Differential systems in ecology and epidemiology. In *Chaos*. Princeton University Press, Princeton, N. J., 1986.

[SK89]   R. D. Strum and D. E. Kirk. *Discrete Systems and Digital Signal Processing*. Addison-Wesley, Reading, MA, 1989.

[Str88]   S. H. Strogatz. Love affairs and differential equations. *Math. Mag.*, 61:35, 1988.

[Str94]   S. H. Strogatz. *Nonlinear Dynamics and Chaos*. Addison-Wesley, Reading, MA, 1994.

[SWS82]   R. H. Simoyi, A. Wolf, and H. L. Swinney. One-dimensional dynamics in a multi-component chemical reaction. *Physical Review Letters*, 49:245, 1982.

[Tys76]   J. J. Tyson. The Belousov–Zhabotinskii reaction. In *Lecture Notes in Biomathematics, Vol. 10*. Springer-Verlag, New York, 1976.

[ZK65]   N. J. Zabusky and M. D. Kruskal. Interaction of "solitons" in a collisionless plasma and the recurrence of initial states. *Phys. Rev. Lett.*, 15:240, 1965.

Index